MYCOTOXINS AND FOOD SAFETY

ADVANCES IN EXPERIMENTAL MEDICINE AND BIOLOGY

Recent Volumes in this Series

Volume 495
PROGRESS IN BASIC AND CLINICAL IMMUNOLOGY
Edited by Andrzej Mackiewicz, Maciej Kurpisz, and Jan Żeromski

Volume 496
NONINVASIVE ASSESSMENT OF TRABECULAR BONE ARCHITECTURE AND THE COMPETENCE OF BONE
Edited by Sharmila Majumdar, Ph.D., and Brian K. Bay, Ph.D.

Volume 497
INTRACTABLE SEIZURES: Diagnosis, Treatment, and Prevention
Edited by W. McIntyre Burnham, Peter L. Carlen, and Paul A. Hwang

Volume 498
DIABETES AND CARDIOVASCULAR DISEASE: Etiology, Treatment, and Outcomes
Edited by Aubie Angel, Naranjan Dhalla, Grant Pierce, and Pawan Singal

Volume 499
FRONTIERS IN MODELING AND CONTROL OF BREATHING
Edited by Chi-Sang Poon and Homayoun Kazemi

Volume 500
BIOLOGICAL REACTIVE INTERMEDIATES VI: Chemical and Biological Mechanisms of Susceptibility to and Prevention of Environmental Diseases
Edited by Patrick M. Dansette, Robert Snyder, Marcel Delaforge, G. Gordon Gibson, Helmut Greim, David J. Jollow, Terrence J. Monks, and I. Glenn Sipes

Volume 501
BIOACTIVE COMPONENTS OF HUMAN MILK
Edited by David S. Newburg

Volume 502
HYPOXIA: From Genes to the Bedside
Edited by Robert C. Roach, Peter D. Wagner, and Peter H. Hackett

Volume 503
INTEGRATING POPULATION OUTCOMES, BIOLOGICAL MECHANISMS AND RESEARCH METHODS IN THE STUDY OF HUMAN MILK AND LACTATION
Edited by Margarett K. Davis, Lars A. Hanson, Charles E. Isaacs, and Anne L. Wright

Volume 504
MYCOTOXINS AND FOOD SAFETY
Edited by Jonathan W. DeVries, Mary W. Trucksess, and Lauren S. Jackson

---

MYCOTOXINS AND FOOD SAFETY

Edited by

Jonathan W. DeVries

Medallion Laboratories Division
General Mills, Inc.
Minneapolis, Minnesota

Mary W. Trucksess

Food and Drug Administration
Center for Food Safety and Applied Nutrition
Division of Natural Products
Washingon, D.C.

and

Lauren S. Jackson

Food and Drug Administration
National Center for Food Safety & Technology
Summit-Argo, Illinois

Kluwer Academic / Plenum Publishers
New York, Boston, Dordrecht, London, Moscow

Proceedings of the American Chemical Society Symposium, "Mycotoxins and Food Safety," held August 21–23, 2000 in Washington, D.C.

ISBN 0-306-46780-1

233 Spring Street, N.Y., NY 10013

http://www.wkap.nl/

10 9 8 7 6 5 4 3 2 1

A C.I.P. record for this book is available from the Library of Congress

Printed in the United States of America

PREFACE

The contents of this book are the proceedings of the American Chemical Society Symposium, "Mycotoxins and Food Safety," which was held August 21 to 23, 2000 at the 220th ACS National Meeting in Washington, DC. Researchers from diverse backgrounds in industry, academia, and government presented mycotoxin related topics ranging from occurrence and impact, analysis, reduction through processing and plant breeding, toxicology and safety assessments to regulatory perspectives. Speakers represented a range of international perspectives.

Mycotoxins, from the Greek *mukes* referring to fungi or slime molds and toxin from the Latin *toxicum* referencing a poison for arrows, have earned their reputation for being potentially deleterious to the health and well being of a consuming organism, whether it be animal or human. Unfortunately, mycotoxins are a ubiquitous factor in the natural life cycle of food producing plants. As such, control of the potential impact of mycotoxins on food safety relies heavily upon accurate analysis and surveys followed by commodity segregation and restricted use or decontamination through processing. Significant mycotoxin research activity began approximately four decades ago following the death of large numbers of turkeys ingesting peanut meal contaminated with aflatoxins. Research interest continues to be high as aflatoxins, deoxynivalenol, fumonisins, ochratoxins and patulin have been found to occur in crops around the world. Ongoing research that continues to elucidate the detrimental effects of mycotoxin exposure serves as a basis for risk management based regulatory decisions.

The purpose of this book is to provide the most comprehensive and current information on the topic of mycotoxins and assuring food safety. The first section of the book reviews the biology and ecology of the various mold species responsible for the contamination of foods and feeds by mycotoxins. It also provides a look at a unique approach to reduction of deoxynivalenol during crop production as well as a look at another potential route of mycotoxin exposure, namely from building environments.

Analytical methods for identifying and quantifying mycotoxins in foods and feeds are the subject of Part Two of the book. To adequately survey for the occurrence of mycotoxins, and to establish and enforce regulatory limits, effective analytical methods are necessary. Analytical procedures involve three basic steps, extraction, purification/separation, and identification/quantitation. Various aspects of the three basic steps are expanded upon and evaluated for effectiveness.

Handling and processing of commodities into food products can have a dramatic impact upon the potential for mycotoxin exposure in the food supply. Section Three of the book highlights research efforts aimed at understanding the effects of handling, storage, and processing on mycotoxin levels, and thereby providing potential means of reducing exposure risk.

Regulating mycotoxin levels for food safety makes up Part Four of the book. Ultimately, the results of scientific research must be evaluated with regard to the need for regulation of the mycotoxin in question as well as the appropriate level at which to regulate. Sound research and comprehensive understanding of the risks involved provide a solid basis for sound regulations. Public health implications resulting from different exposure scenarios must by thoroughly and thoughtfully considered to assure food safety, an adequate food supply, and public confidence in the regulatory process.

ACKNOWLEDGMENTS

The symposium organizers would like to thank the American Chemical Society Division of Agricultural and Food Chemistry for their technical and financial support of the Mycotoxins and Food Safety Symposium. We wish to express appreciation to the session chairs David Miller, Robert Eppley, Mary Trucksess, Sam Page, Lloyd Bullerman, Jane Robens, Jonathan DeVries, and Terry Troxell for their excellent jobs in organizing their sessions. We would also like to thank Dr. Ellen Hopmans for initiating the idea of a broad based food safety symposium related to mycotoxin occurrence, exposure, and control. We especially want to acknowledge the speakers and contributing authors, without whom the symposium and this book would not be possible. We also wish to express our gratitude to the following financial sponsors. Without their financial support, a symposium and book of this magnitude would not have been possible.

- Agilent Technologies
- AOAC Research Institute
- Frito-Lay
- General Mills
- Joint Institute-Food Safety and Nutrition
- Kellogg's
- Kraft Foods
- Medallion Laboratories
- National Center for Food Safety and Technology
- National Coffee Association-Scientific Advisory Group
- Neogen
- North American Millers Association
- R-Biopharm
- Romer Laboratories
- Strategic Diagnostics
- Thermoquest
- U.S. Department of Agriculture-Agriculture Research Service
- U.S. Food and Drug Administration
- U.S. Food and Drug Administration-Food Safety Initiative
- Vicam
- Waters Chromatography
- Woodson Tenant Laboratories

Jonathan W. DeVries
Mary W. Trucksess
Lauren S. Jackson

CONTENTS

RELEVANCE OF MYCOTOXINS IN THE FOOD SUPPLY AND IN THE BUILT ENVIRONMENT

ANALYTICAL ASPECTS OF MYCOTOXINS

PROCESSING EFFECTS ON MYCOTOXINS

TOXICITY, RISK ASSESSMENT AND REGULATORY ASPECTS OF MYCOTOXINS

RELEVANCE OF MYCOTOXINS IN THE FOOD SUPPLY AND IN THE BUILT ENVIRONMENT: INTRODUCTION

J. David Miller
Department of Chemistry
Carleton University
Ottawa, Ontario K1S 5B6

The unifying theme of this session was that an ecological perspective was taken towards fungi found on various cereals and groundnuts, and, perversely, on wet cellulosic building materials. Studies of the fungus *Stachybotrys chartarum*, hitherto only of interest from reports of human and animal toxicosis associated with straw in Europe. Very careful studies have shown that there are different chemotypes of *S. chartarum*, one of which produces very potent trichothecenes and the other not but both producing an array of other interesting metabolites. Both chemotypes co-exist in nature. This fungus grows well only on wet cellulose and is common in north temperate climates and a sister genus *Memnoniella echinata* is more prevalent on wet cellulose in subtropical areas.

The pattern is similar for the toxigenic fusaria. The species that cause head blight also have chemotypes that interestingly vary roughly according to continent: strains from North and South America are different than those from Asia. The occurrence of species that affect wheat and corn are strongly affected by temperature, followed by moisture (too much or too little) and other factors that stress the plants, particularly temperature and insects. Competition for nutrients and moisture and indeed space is a feature of the biology of these toxigenic fungi. The processes played out on the infection court of Fusarium head blight *(F. graminearum* or *F. culmorum* disease of wheat) are exploited in studies of the potential for biological control of this disease. Bacteria that can occupy the niche provided by the infection court can exclude *F. graminearum* by both interference competition and competitive exclusion.

The toxigenic penicillia are generally less well known than the fungi that produce deoxynivalenol and aflatoxin. However, the prospect of regulation of ochratoxin A at vanishingly small levels in various food products has brought these taxa into sharper focus. Some Aspergillus species also produce ochratoxin, notably those on coffee and grapes. The ecology of these fungi has been less studied than some others and deserves more attention.

Finally, aflatoxin remains a major problem for human health and agricultural economics despite more than 40 years of research. Although the broad features of the biology of the aflatoxigenic molds were resolved early, many gaps remain. Modern studies of the field biology of *A. flavus* in tropical areas i.e. outside the United States of America have shown that its ecology varies from dryland corn and groundnut production to the subtropical. New methods for tracking fungal spores and new studies of the metabolites of,

Mycotoxins and Food Safety, edited by Trucksess *et al.*
Kluwer Academic/Plenum Publishers, 2002

for example, *Aspergillus flavus* have also changed the way mycologically-oriented ecologists have understood the habits of aflatoxin-producing fungi.

These papers show many common elements of methods and scientific approach and similar and different ecologies.

BIOLOGY AND ECOLOGY OF MYCOTOXIGENIC ASPERGILLUS SPECIES AS RELATED TO ECONOMIC AND HEALTH CONCERNS

David M. Wilson,[1] Wellington Mubatanhema,[1] and Zeljko Jurjevic[2]

[1]Department of Plant Pathology
University of Georgia
Coastal Plain Experiment Station
Tifton, GA 31793
[2]Department of Phytopathology
University of Zagreb
Zagreb, Croatia

ABSTRACT

The fungal genus *Aspergillus* was established in 1729, and includes species that are adapted to a wide range of environmental conditions. Many aspergilli produce mycotoxins in foods that may be toxic, mutagenic or carcinogenic in animals. Most of the Aspergillus species are soil fungi or saprophytes but some are capable of causing decay in storage, disease in plants or invasive disease in humans and animals. Major agricultural commodities affected before or after harvest by fungal growth and mycotoxins include corn, peanuts, cottonseed, rice, tree nuts, cereal grains, and fruits. Animal products (meat, milk and eggs) can become contaminated because of diet. *Aspergillus flavus*, *A. parasiticus*, *A. ochraceus*, *A. niger*, *A. fumigatus* and other aspergilli produce mycotoxins of concern. These include the aflatoxins and ochratoxins, as well as cyclopiazonic acid, patulin, sterigmatocystin, gliotoxin, citrinin and other potentially toxic metabolites.

INTRODUCTION

There is a saying in the real estate industry that prices are governed by location, location, location. All other considerations including marketability and possible use of the property depend on the location and local needs. Mycotoxin contamination of foods and feeds and the associated problems also depend primarily on the production, storage and marketing location. The Aspergillus species that produce mycotoxins are more common in the warmer, subtropical and tropical areas than in the temperate areas of the world. However, stringent government regulations for mycotoxins and risk analyses are more common in temperate areas than in the warmer areas of the world. Therefore, one must be naive to think that the health and marketplace risk factors considered to be important for the control and management of unavoidable mycotoxin contamination are always science-based. Mycotoxin risk analysis, regulations and management practices to assure health of consumers and markets should be science-based as far as possible, but with the marketing and world trade issues, often deliberations on these factors are as much political as science based. The critical elements needed for mycotoxin management are interdisciplinary in nature.

The toxic secondary metabolites are produced by fungi so there is a critical need to have accurate fungal taxonomy. Fungal metabolites are often produced in low or trace concentrations so there is also a critical requirement for excellence in analytical assays. Mycotoxins occur in many crops and stored seed as well as in processed and formulated products. Therefore, many areas of crop production, mycology, plant pathology, entomology, engineering and chemistry as well as storage and food sciences have to be adequately addressed in relation to the biology and ecology of the mycotoxigenic fungi.

There are many mycotoxins produced by Aspergillus species. The mycotoxigenic aspergilli are usually considered to be saprophytes that can grow in products stored with high moisture content. Colonization of stored products by the aspergilli is primarily a function of location and moisture content complicated by insect and rodent activity. Some Aspergillus species produce mycotoxins in meats such as country ham, sausage, cheese or other fermented products; these processes are beyond the expertise of the present authors and will not be addressed. There are only a few mycotoxins that are regularly monitored in the marketplace and we will briefly consider the global and agronomic factors related to aflatoxin, sterigmatocystin, cyclopiazonic acid and ochratoxin A contamination of foods and feeds.

The genus *Aspergillus* was first described and named in 1729 by Pier Antonio Micheli in his monograph "Nova Planetarium Genera" (Micheli, 1729). Micheli named the genus after the holy water sprinkler (aspergillum). The aspergilli are most often soil fungi or saprophytes but several are important because they produce mycotoxins. The aspergilli can also cause decay and deterioration in stored products, and may cause disease in plants, insects, poultry, humans and other mammals. In addition to being disease agents some aspergilli are important aeroallergens. Some of the Aspergillus species and teleomorphs that potentially produce mycotoxins include *A. candidus, A. clavatus, A. flavus, A. fumigatus, A. niger, A. nomius, A. ochraceus, A. parasiticus, A. restrictus, A. tamarii, A. terreus, A. versicolor*, *Emericella nidulans*, and *Eurotium amstelodami* (Frisvad and Samson, 1991). A partial list of the mycotoxins includes the aflatoxins, citrinin, citreoviridin, cyclopiazonic acid, gliotoxin, ochratoxin, penicillic acid, sterigmatocystin, xanthomegnin and other potentially important metabolites (Frisvad and Samson, 1991).

COMMODITIES AND TISSUES CONTAMINATED BY MYCOTOXINS

Aflatoxin Contamination

The Food and Drug Administration in the United States, The Institute of Public Health in Japan and many other agencies around the world regularly test products in the marketplace for aflatoxins. The following products in the marketplace have been reported to be contaminated with aflatoxins or aflatoxin metabolites: almonds, Brazil nut, filberts, pistachio nut, cashew nut, walnut, pecan, figs, melon seed, pumpkin seed, sesame seed, sunflower seed, lotus seed, coix seed, marzipan, red pepper, white pepper, nutmeg, paprika, mixed spice, rice, corn (maize), corn products, mixed cereals, peanuts, peanut products, crude sugar, soybean meal, beans, buckwheat, chilies, chili powder, cocoa products, cotton, copra, sorghum, millet, snack foods, eggs, milk, cheese, yogurt and meats (Wilson and Abramson, 1992). Human tissues and fluids that have been shown to contain aflatoxin after exposure include urine, human milk, blood, liver, kidney and various other tissues. Aflatoxin and cyclopiazonic acid have been seen to occur together in corn and peanuts (Urano et al., 1992). These lists are not complete and are in no particular order. The products are not ranked as to the relative risk to consuming animals or people. The large number of products that may contain aflatoxins illustrates the severity of the problem and the low probability that easy and effective solutions will ever be possible. For aflatoxins there is no doubt that some crops are contaminated because they have been invaded and colonized by *A. flavus* and associated species before harvest. Some crops and products are contaminated in storage and processing so some manufactured products may contain aflatoxins from both preharvest and postharvest accumulation. Most often the manufacturing process will not detoxify the toxins and the products will be marketed and consumed. Humans are also exposed to the aflatoxins and their metabolites by consuming animal products contaminated because the animal ate feed containing aflatoxin.

Ochratoxin Contamination

Literature on contamination of crops and products with ochratoxin A and associated metabolites is not as easily located or as directly associated with a particular fungus as is the case with the aflatoxins. We have relied on the review by Wilson and Abramson (1992) and internet computer searches complemented with current literature to generate the list. Ochratoxin is also likely to occur with other mycotoxins such as citrinin and penicillic acid or patulin. The following products and tissues have been reported to be contaminated with ochratoxin or its metabolites: barley, wheat, coffee, table wine, grape juice, beer, meats, cocoa, baby foods, feed grains, corn, oats, cereals, rye, crackers, raisins, pecans, figs, pig blood, pig kidneys, other animal tissues and human blood. The principal crops affected tend to be those grown in temperate regions. Most likely the penicillia rather than the aspergilli are the major source of ochratoxin A in temperate zones. The aspergilli may be very important in the warmer zones of the world, but this distribution is not well documented, perhaps because no one has looked in these regions, or maybe the aspergilli that produce ochratoxin A are not particularly aggressive in preharvest and storage situations in these climates. There needs to be much more information generated on how and why ochratoxin A contaminates products, other than coffee, that are produced in the warmer and tropical climates.

ANALYTICAL CONSIDERATIONS

There are really two analytical method approaches needed, one for mycotoxin regulation and another for mycotoxin research. These approaches are actually quite different in their nature. The analytical needs for research often require a wide analytical range and those for regulatory purposes necessarily use a narrow concentration range, sometimes near the analytical detection limit. Consequently, methods used for mycotoxin regulation require a high degree of accuracy at extremely low concentrations. There are currently several official AOAC International methods for aflatoxins designed for the 20 ppb (nanogram/gram, ng/g) United States of America standard and a few methods for the European Union 2 ppb standard (Cunniff, 1995). None of these official methods take into account the tremendous sampling problems and mycotoxin distribution variations that are associated with aflatoxin measurement in large lots. However, there are elaborate sampling plans in many regulations (Coker, 1998). It is quite likely that the European Union regulations require aflatoxin concentrations in products that are far below the farmers' capability to manage mycotoxins when crops are grown between 35°N and 35°S. The European Union regulation for aflatoxin requires the shipper to pay for transporting the product back to the point of origin if the aflatoxin content is too high. The United States regulations take some of the natural variation into account and the United States allows products to be marketed on the basis of use, and in this way minimizes human exposure.

For research purposes, rapid and inexpensive analytical and screening methods for aflatoxins that are accurate between 0 and 40,000 ppb are critically needed. Similar research needs exist for methods for ochratoxin A, cyclopiazonic acid and sterigmatocystin assays. Analytical costs and ease of use are also important factors in the development and implementation of regulatory methods as well as applied research methods. There is a critical need for methods designed for research as well as regulatory applications if we are indeed serious about mycotoxin management worldwide.

FUNGAL CONSIDERATIONS

Aflatoxin and Cyclopiazonic Acid Producing Fungi

The taxonomy of Aspergillus species is not easy but it is relatively straightforward when compared to the taxonomy of Penicillium species. The book by Raper and Fennell (1965) remains the major guide to taxonomy of the Aspergillus genus. Klich and Pitt (1988) have incorporated subsequent taxonomic revisions and new species into a laboratory guide to the common Aspergillus species. The Aspergillus species that produce the aflatoxins are all in the subgenus Circumdati, section Flavi. The species in this section include: *A. flavus, A. oryzae, A. parasiticus, A. nomius, A. sojae* and *A. tamarii*. The only fungi that have been unequivocally demonstrated to produce the aflatoxins are *A. flavus, A. parasiticus*, and *A. nomius*. The other three species in the Flavi section do not produce aflatoxins. It is interesting that Kurtzman et al. (1986) used DNA relatedness to suggest that *A. flavus, A. parasiticus, A. sojae* and *A. oryzae* should be a single genus. Kurtzman et al. (1986) proposed that *A. flavus* be used with the varieties being flavus, parasiticus, sojae and oryzae. Klich and Pitt (1988) did not think this suggested change in nomenclature was warranted.

Miss Dorothy Fennell prepared an unpublished outline in 1976 for distinguishing *A. flavus* and *A. parasiticus*. She stated in this manuscript that "Cultural and morphological differences between strains identified unequivocally as *A. flavus* by our criteria cannot be correlated with aflatoxin production. Strains that appear identical may produce much or no B. It is almost safe to say they produce no G. In general *A. flavus* produces aflatoxin B_1, and B_2 when the isolate is aflatoxigenic, many isolates of *A. flavus* produce no aflatoxins in culture." Some aflatoxigenic as well as some non-aflatoxigenic isolates of *A. flavus* also produce cyclopiazonic acid which is independent of the ability to produce aflatoxin (Horn and Dorner, 1999). Wicklow and Shotwell (1983) reported that the sclerotium of some isolates of *A. flavus* may contain aflatoxin B_1 and G_1 and Cotty and Cardwell (1999) reported that S strain isolates of *A. flavus* from West Africa produced aflatoxin B_1 and G_1 while no North American S strain isolates produced aflatoxin G_1. Almost all aflatoxigenic *A. parasiticus* and *A. nomius* isolates produce the B and G aflatoxins. For these reasons we do not think that the use of aflatoxin production patterns is a stable character that is useful for nomenclature in the section Flavi.

Calvert et al. (1978) demonstrated that the B to G ratios in preharvest corn were affected by mixed inoculum of *A. flavus* and *A. parasiticus* and Wilson and King (1995) showed similar results in mixed cultures. Wilson and King (1995) reported that in mixed cultures of *A. flavus* and *A. parasiticus* the production of G_1 and G_2 by *A. parasiticus* was depressed when there was at least 25% of *A. flavus* in the original conidial mixture. Perhaps *A. flavus* suppresses the production of aflatoxin by *A. parasiticus*. Cotty and Bayman (1993) demonstrated that an atoxic *A. flavus* competed successfully with a toxic isoplate when they were grown in mixed culture. The mechanism of action is unknown at this time.

A. flavus and *A. parasiticus* have both been reported to grow and produce toxins in preharvest and stored products. Many people do not identify the *A. flavus* group (equivalent to the Aspergillus subgenus Circumdati section Flavi) to the species level, so much of the literature cannot be evaluated as to the importance of the individual species in relation to observed real world contamination. The FDA data referred to in the chapter by Wilson and Abramson (1992) illustrate the variable nature of the B to G ratios that naturally occur in various products.

Cotty (1997) has grouped *A. flavus* into S and L strains based on sclerotial size. In southern United States cotton producing areas, Cotty (1997) found that *A. flavus* and *A. tamarii* incidence differed between geographic areas and that *A. flavus* incidence was correlated with high minimum temperatures and inversely correlated with latitude. Odum et al. (1997) looked at the spatial and temporal patterns of the S and L strains in Yuma County, Arizona. The highest incidence of S strain isolates was during cotton boll formation and the highest *A. flavus* propagule density in the soil was in the summer. Horn and Dorner (1999) looked at soil populations of Aspergillus species of the section Flavi along a transect from eastern New Mexico through Georgia to eastern Virginia. There was a large population variation in the 166 cultivated fields with high densities of the section Flavi centered from east-central Texas to south-central Georgia. *A. flavus* was the dominant species along most of the transect. The S strain was mostly in the cotton growing regions of east-central Texas and Louisiana. *A. parasiticus* incidence was highest from south-central Alabama to eastern Virginia. *A. nomius* was detected only in Louisiana and Mississippi. Georgia and Alabama had the highest Flavi section soil populations and western Texas had the lowest populations. Horn and Dorner (1999) suggested that the population differences were related to drought and soil temperatures. It would be interesting if there were similar population changes in transects from cultivated soils taken in the United States from the Canadian border to south Florida. Most likely soil populations and the risks of aflatoxin contamination are influenced by weather patterns, soil moisture and soil temperature, with hot dry soils favoring the Aspergillus section Flavi.

Cyclopiazonic Acid Producing Fungi

Cyclopiazonic acid is produced by several Aspergillus and Penicillium species in addition to being produced by *A. flavus*. The fungi that have been reported to produce cyclopiazonic acid include *P. verrucosum, P. camembertii, P. patulum, P. puberulum, A. versicolor, A. oryzae* and *A. tamarii* (Wilson and Abramson, 1992). There is little information on the importance of cyclopiazonic acid in crops, so further investigation on factors affecting cyclopiazonic acid incidence is warranted.

Sterigmatocystin Producing Fungi

Little is known about sterigmatocystin and the fungi that produce the toxin. Studies in Japan and Canada by Sugimoto et al. (1976) and Abramson et al. (1983) determined the relationship between *A. versicolor* and production of sterigmatocystin in postharvest storage. The other aspergilli that produce sterigmatocystin include *A. rugulosus, A. nidulans, A. flavus, A. parasiticus, A. chevalieri, A. ruber, A. amstelodami, A. aurantobrunneus, A. quadrilineatus, A. sydowii* and *A. ustus*. Sterigmatocystin can also be produced by penicillia including *P. griseofulvum, P. commune* and *P. camembertii* (Wilson and Abramson, 1992). Little is known about the potential for sterigmatocystin production by these fungi in nature. Non-aflatoxigenic *A. flavus* and *A. parasiticus* could possibly have the aflatoxin biosynthesis pathway blocked in such a way as to be potentially efficient producers of sterigmatocystin or sterigmatocystin derivatives.

Ochratoxin Producing Fungi

A. ochraceus is likely the most important Aspergillus species involved with ochratoxin A contamination in storage. Possibly this fungus occurs before harvest or during drying and curing which could result in ochratoxin A contamination at harvest in some situations. Ochratoxin A found in cereals in storage is primarily produced by growth of *Penicillium verrucosum*, and perhaps the associated Aspergillus species, *A. ochraceus* and *A. ostianus* (Frisvad, 1995). However, Frisvad (1995) questions whether *A. ostianus* actually produces ochratoxin A *sensu stricto*. Naturally occurring ochratoxin A has not been linked to Aspergillus species in grain stored in warm climates but has been found in stored coffee beans invaded by *A. ochraceus* (Frisvad, 1995).

A partial list of the Aspergillus and Penicillium species that have been reported to produce ochratoxin A includes: *A. alliaceus, A. carbonarius, A. melleus, A. niger var. niger, A. sulphureus, A. sclerotiorum, A. niger var. awamori, A. ostianus, A. petrakii, P. verrucosum, P. palitans, P. commune, P. purpurascens, P. variable* and *P. chrysogenum*. (Wilson and Abramson, 1992; Hocking, 2001; Abarca et al., 1994; Heenan et al., 1998; Horie, 1995). Most of the Aspergillus species are closely related to *A. ochraceus* and are in the subgenus Circumdati, section Circumdati, with *A. carbonarius, A. niger var. niger* and *A. niger var. awamori* being in the subgenus Circumdati, section Nigri. Apparently the biosynthetic pathway for ochratoxin production is common in several Aspergillus and Penicillium species.

Most likely there are only a few species responsible for ochratoxin A contamination of foods. The Penicillium species that occur and produce ochratoxin A generally occur in temperate zones whereas the Aspergillus species are most likely more important in warmer zones of the world. It is not clear which of the Aspergillus species in addition to *A. ochraceus* may be important. In addition, it is not certain if ochratoxin A contamination is only a storage problem or perhaps a dual field and storage problem. The report of *A. ochraceus* infested coffee berry borer acting as a possible vector of *A. ochraceus* (Vega and Mercadier, 1998) suggests that the possibility of preharvest ochratoxin A contamination of coffee needs to be investigated in the same ways as preharvest aflatoxin was investigated. It is not apparent exactly which of the ochratoxigenic aspergilli produce ochratoxin A in warm subtropical and tropical climates, but *P. verrucosum* and *A. ochraceus* are both likely to be important in temperate climates.

ECOLOGICAL AND PLANT PATHOLOGICAL CONSIDERATIONS

Preharvest, Harvest and Drying Factors

The aflatoxins, cyclopiazonic acid and possibly ochratoxin A can be produced in the field, during the growing season or in the curing and drying processes prior to storage. Little is known about ochratoxin A contamination in the field. Vega and Mercadier (1998) isolated *A. ochraceus* from adult coffee berry borers isolated from coffee beans in Uganda and Benin. The coffee berry borer is a major insect pest of coffee. Preharvest and postharvest insect vectors of *A. ochraceus* and insect damage may be important in increasing the risk of ochratoxin A contamination. Bucheli et al. (2000) stated that sun drying of coffee cherries consistently led to ochratoxin A formation in the pulp and parchment (husks) of the cherries in Thailand. Dried beans contained about 1% of the ochratoxin A found in husks. Maturity of the coffee cherries at harvest affected the susceptibility of the crop to ochratoxin A contamination. Many more studies are needed to understand the factors that affect ochratoxin A contamination in coffee and all other susceptible crops.

The aflatoxins, cyclopiazonic acid and sterigmatocystin can all be produced by *A. flavus* while *A. parasiticus* is not known to produce cyclopiazonic acid. Wilson (1995), Wilson and Abramson (1992), Wilson and Payne (1994), Payne (1998), Dowd (1998), Widstrom (1992) and others have reviewed preharvest mycotoxin contamination in crops and we will only add a few observations to the generally accepted views on contamination. Aflatoxin and most likely cyclopiazonic acid contamination before harvest is common in many crops with an increased probability of occurrence of these mycotoxins in the warmer, sub-tropical and tropical climates compared to the temperate climates of the world.

Climate in the production area is the most important factor influencing the preharvest contamination of corn, peanut, cottonseed and tree nuts with aflatoxin. Dry weather and droughts are very important prerequisites for contamination of corn, cotton and peanut. Thus in Iowa, aflatoxins are occasionally seen in corn before harvested during the summers with late season drought. In Georgia, weather related aflatoxin contamination is chronic, but is most severe in dry years. Irrigation of corn in North Carolina is effective in reducing aflatoxin contamination in corn while in Georgia, irrigation is not always effective in reducing contamination. Insect damage is related to increased aflatoxin content in many crops. However, temperature and moisture effects override the insect effects. The insect effects depend on the time of planting and maturity of corn in Georgia. Late planted corn often has less aflatoxin contamination but more insect damage than early planted corn. Other stress factors such as plant density, fertility and plant disease interact with the environment to increase or decrease the risk of aflatoxin contamination. Widstrom (1992) described the infection and colonization period of corn by *A. flavus* between silking at about 60 days after planting until physiological maturity at about 120 days. There was increasing damage to the ear by insects from silking to harvest at 135 days after planting. There was a continuous post-infection aflatoxin accumulation from silking to harvest. The optimal temperature for corn production is lower than the optimal temperature for *A. flavus* growth and insect damage. Consequently, there are positive correlations between the average temperature, aflatoxin contamination and insect damage. This is especially evident when temperatures become too high for optimum corn growth. Widstrom (1992) used data from planting date studies in Georgia to illustrate the effects of temperature on aflatoxin contamination at harvest. Ears that are developing and maturing when the temperature is 28 - 32°C are much more susceptible to aflatoxin contamination than ears that develop at lower temperatures when the corn is planted later. The insect damage effects were not as important as the temperature effects in the planting date study. However, planting corn late is not an effective control measure because late planting of corn results in a reduced yield and it is not profitable to grow corn under these conditions. In corn as well as in peanuts and cottonseed, cyclopiazonic acid contamination most likely parallels the incidence of aflatoxin contamination. We do not have any absolutely effective management strategies for aflatoxin control other than growing crops in low risk areas.

Management of aflatoxin contamination of preharvest peanuts was reviewed by Wilson (1995). The most important factors favoring and controlling contamination include location, soil moisture, soil temperature and insect activity. Mites, lesser corn stalk borers in the United States and probably termites in Africa are efficient vectors of the *A. flavus* group. Damage by these insects creates a favorable habitat for growth of the fungi and aflatoxin production. Aflatoxin content of peanuts is most affected by the soil moisture and temperature but aflatoxin contents may also be increased by insect damage. Late season irrigation increases soil moisture and decreases soil temperature and is an effective way to lower aflatoxin content in mature kernels. Management factors other than irrigation, such as plant nutrition, fungicide applications, insecticide applications and harvesting techniques can all have effects on the final quality and aflatoxin content. Aflatoxin contamination of peanut is quite dependent on the climate and the geographic location. The damage and the rate of drying of peanuts after digging until storage can also have major effects on quality including aflatoxin contamination. The probability of successful preharvest control in peanut is highest in irrigated fields because soil moisture and temperature can be affected by irrigation. Sorting the peanuts to remove damage is also a fairly effective way to remove aflatoxin contaminated kernels. The United States marketing system is designed to facilitate aflatoxin management and the system is somewhat effective in years when there is not a severe hot and dry growing season. In years with severe drought, the buying-point marketing system is not very effective and aflatoxin removal methods have to be carefully implemented by peanut shellers and manufacturers of peanut products.

Aflatoxin contamination in cottonseed is primarily related to the environment (Marsh et al., 1973; Gardner et al., 1974). It is mostly associated with desert environments in the United States and elsewhere where the temperature and water stresses on the growing plants are outside the optimum for the plant to develop properly. Scientists at the USDA laboratory in New Orleans have made many advances in understanding the relationships of the cotton plant to the insect and environmental stresses, but the environment overrides all practical control measures. Therefore, detoxification and alternative use research studies are sorely needed to insure that the contaminated cottonseed has a market outlet.

Designing management strategies for tree nuts, especially the Brazil nut, is an even greater challenge than the dilemma faced by cotton producers in Arizona. The Brazil nut tree is a major overstory component of the South American Amazon basin forest region and the Brazil nut is an extractable product of Amazonia. The very large trees grow in the forest, not in plantations, so the production of nuts cannot be managed in any practical manner. The nuts fall to the ground enclosed in cannonball-like structures about the size of a grapefruit during the rainy season. The seeds are arranged in the woody coconut-like fruit similar to the way orange slices are arranged in oranges. The nuts are gathered by hand after the coconut is opened using a machete and then they are carried in bags to collection places in the forest for transport to processing facilities after the rainy season. The nuts are very resistant to decay and insects, but *A. flavus* group species frequently grow on the stored product which results in aflatoxin contamination. There is a tremendous need to develop postharvest strategies for sorting and other aflatoxin control measures in this tropical region. Marketing of the Brazil nut is a good example of the conflict we often see between governmental regulations and global needs for the protection of the environment (Williams and Wilson, 1999). The earth's climate benefits from the Amazon forest while local consumers, as well as those in other countries, need to be able to purchase products that are safe. Excluding the Brazil nut from the markets in either the developed or developing countries may have a minimal effect on the aflatoxin risk in the developed and developing countries, but cutting all the Brazil nut trees for timber would have a negative global weather consequence. The biological factors leading to aflatoxin contamination of corn, peanuts, cottonseed and Brazil nut are similar in some respects and very different in others. Designing ways to insure that aflatoxin contamination is not a human health risk is complex. Multi-disciplinary approaches are needed to accomplish the goal of providing safe wholesome food worldwide, not just in the developed countries.

Storage Factors

Fungal deterioration of foodstuffs in storage is a complex issue depending on the habitat and the storage conditions. The growth of fungi in stored grains and oilseeds has been well described and reviewed many times and the important factors need not be repeated here. Recent reviews include those by Ominski et al. (1994), Frisvad, (1995), Lacey and Magan, (1991), Abramson, (1991) and the books edited by Chelkowski, (1991), Sauer, (1992), and Jayas et al. (1995).

The literature on storage is extensive and the basic biological, moisture and structure requirements are readily available elsewhere. There are a few considerations relating to storage that are worth mentioning in relation to the aspergilli. First, the concept of field versus storage fungi as elaborated by Christensen (1974) was based on work done in a temperate climate in Minnesota. The incidence of many of the so-called storage fungi, particularly the aspergilli, that occur in the field increases as crops are produced from the higher to lower latitudes, so location is an important factor. The aspergilli that are present on crops as they enter storage are more common in the warmer and humid climates so the fungal succession in storage is dependent on location and temperature. Therefore, *A. flavus* is a storage fungus in Minnesota and Kansas and a field and storage fungus in warmer climates like Georgia. The climatic factors related to colonization and warm temperatures can result in a rapid accumulation of aflatoxins in the field as well as in storage in the warm humid, subtropic and tropical climates. There is a great need to do field and storage studies from high to low latitudes to define the location and environmental effects on mycotoxin contamination.

Lacey and Magan (1991) critically examined the relationships between the fungi, water content and temperature in cereal grains. Beti et al. (1995) found that maize weevils facilitate the growth of *A. flavus* in stored corn by acting as vectors for inoculum and by increasing the moisture content as well as increasing the surface area susceptible to fungal colonization. Aflatoxin accumulation can be quite profound in corn with weevil activity even in stored corn with initially low populations of *A. flavus*. We have seen that same situation in Georgia, but it is interesting that once the stored corn accumulates high aflatoxin concentrations the active weevil population rapidly decreases. In many parts of the world maize is stored locally by smallholder producers and consumers. Udoh et al. (2000) studied the storage structures in five agroecological zones of Nigeria and related the storage practices to aflatoxin content. Corn stored in polyethylene bags was particularly prone to aflatoxin contamination because of moisture retention as was the corn with insect infestations. Farmers in Nigeria were not eager to accept new methods for storage and there were many storage situations that could be improved to help maintain quality. The major problems observed usually related to the maize moisture content, insect activity, ear damage and the lack of sorting the maize for quality before storage. There is a critical need for research on effective and acceptable ways to improve smallholder produced maize and other products when they are stored in developing countries in the warmer regions of the world.

There is not enough information about the conditions affecting accumulation of cyclopiazonic acid, sterigmatocystin and ochratoxin A by aspergilli in storage situations worldwide to draw meaningful conclusions that have not already been addressed by authors like Abramson (1991) and Mababe and Tsurta (1991)

BIOLOGICAL CONTROL STRATEGIES

Biological control of mycotoxigenic fungi by competition, replacement or exclusion of toxin producing fungi has a great appeal. However, is this a practical approach that farmers can safely and profitably implement? The use of competitive microorganisms and mycoparasites to control the growth of fungi in field and storage situations has not often proved practical.

Wilson et al. (1986) used several *A. flavus* and *A. parasiticus* mutants to study infection and aflatoxin accumulation in preharvest corn. One white *A. flavus* mutant did not produce any of the aflatoxins in culture and was somewhat effective in reducing aflatoxin contamination in wound inoculated ears when compared to the untreated check. However, the aflatoxin content rapidly increased over all other treatments in ears inoculated with the non-toxic mutant 20 days post-silk and then infested at 55 days post-silk with maize weevils exposed to a wild type *A. flavus*. Corn plots have been regularly inoculated for many years in Georgia and the aflatoxin content at harvest can easily be affected by the toxin producing ability of the *A. flavus* or *A. parasiticus* used as inoculum. Thus, inoculation with low aflatoxin producers may lower aflatoxin content in certain situations.

Unfortunately nature does not always cooperate and the environmental effects on the corn plant and associated microflora cannot be predicted for all situations. Most often in years that favor preharvest aflatoxin accumulation the *A. flavus* that is the most succesful in colonizing the ears is the one that has the major effect. If the primary invader is a weak aflatoxin producer, then little aflatoxin may be found. However, if the principal invader is a proficient producer then high concentrations of aflatoxin may be seen. Biological control is not particularly needed in years with little risk of preharvest contamination but biological control needs to be extremely effective in high risk years. We have never been able to obtain consistent results in our inoculation trials using aflatoxin producing isolates in preharvest corn and peanut trials. Therefore, it may be unlikely that field applied non-toxic isolates will consistently reduce contamination, particularly in the years where there is a very high risk of contamination.

Will et al. (1994) evaluated field inoculation techniques for screening peanut genotypes for resistance to aflatoxin contamination. The primary goal was to decrease the number of plots that did not accumulate aflatoxin in the peanuts. Application of a mixture of aflatoxigenic *A. flavus* and *A. parasiticus* inoculum in Georgia did not significantly affect the aflatoxin content of the harvested peanuts but it usually decreased the number of plots that had little aflatoxin in the harvested peanuts. The use of the *A. flavus/A. parasiticus* infested corn matrix at planting did not affect soil populations of the *A. flavus* group at harvest but when the matrix was applied about 60 days after planting, the soil populations of the A. flavus group were increased.

Dorner et al. (1992) used a non-aflatoxigenic *A. parasiticus* isolate to inoculate soil in peanut plots in 1987 for studies carried out in 1987, 1988 and 1989. They reported that the biocompetitive agent was not found in untreated soil but that soil populations of this fungus at harvest were dependent on the initial inoculum level. The fungus persisted in the soil from 1987 to 1988 and the soil populations of the introduced fungus increased from the time of planting until harvest. The 1987 peanuts contained from 4 to 222 ppb of aflatoxin in the edible categories from the treated plots with various drought stress periods. Dorner et al. (1992) reported that aflatoxin concentrations in the edible peanut categories were reduced from 96 to 1 ppb in 1998 and from 241 ppb to 17-40 ppb in 1989 using the biocompetitive approach, but their experiments were not designed for statistical analysis so they could not analyze the data using statistics. Unfortunately, there was also from 577-11,783 ppb of aflatoxin in the peanuts in the damaged category in 1987, 3,908 ppb in the damaged category in 1988 and from 6,700 to 28,500 ppb in the damaged category in 1989. The *A. parasiticus* isolate they used produced O-methylsterigmatocystin and they were able to measure this compound in the edible and damaged categories. The edible categories contained 172 ppb O-methylsterigmatocystin in 1988 while the damage category contained over 13,000 ppb of O-methylsteigmatocystin.

Based on the work published by Dorner et al. (1992), it appears that unless more effective non-toxigenic isolates have been identified this approach will not result in a Georgia peanut crop that will meet the United States 20 ppb guideline as well as the 2 ppb European guideline in drought years. If this is the case then there will be little economic incentive for peanut growers to use the technique. There are other non-aflatoxigenic *A. flavus* and *A. parasiticus* isolates that could be used in peanut or other crops for biological control. All potentially useful *A. flavus* isolates need to be tested for their ability to produce cyclopiazonic acid, sterigmatocystin and its derivatives as well the other toxic intermediates in the aflatoxin biosynthetic pathway. Toxicity and disease tests in insects, mammals and birds as well an assessment of potential human disease will also need to be done. The allergic properties of the fungus to humans should be an important issue in the risk considerations.

Cotty (1990) has spent many years developing biological control strategies for aflatoxin contamination reduction in cottonseed in Arizona. He has been able to show a reduction of aflatoxin B_1 contamination from 66,240 ppb to 650 ppb using his best non-aflatoxigenic *A. flavus* isolate, strain 36, to inoculate bolls 30 minutes after inoculation with a toxin producing strain. He reported that strain 36 could be used to drastically reduce aflatoxin contamination in cottonseed when it was used to artificially inoculate bolls. Cotty and Bayman (1993) reported that co-inoculation with the non-toxic strain 36 and the toxic strain 13 resulted in a reduction of the aflatoxin content by 82-100% in three tests. The numeric comparisons in ppb of aflatoxin B_1 from the tests are in test one - 76,000 ppb reduced to 13,680 ppb, in test two - 176,000 ppb reduced to not detectable and in test three - 444,000 ppb reduced to 4,400 ppb. Apparently Cotty's biological control approach has been somewhat more effective in the field and he has been able to obtain an experimental use permit from the EPA for use in Arizona.

A. flavus usually produces B_1 and B_2. It is possible that the TLC method used by Cotty (1990) used did not separate the compounds completely because thin layer plates or development solvents do not always separate the aflatoxins. It is also possible that Cotty preferred to report only B_1 results. However, it is very unclear why there was a change in the way the results were reported from ppb as in Cotty (1989) to ppm in later publications by Cotty (1990) and Cotty and Bayman (1993). If the results are not read carefully the reduction from 76 to 13.7 micrograms/gram of cottonseed (ppm) in test one can easily be misinterpreted by thinking this is a reduction from 76 to 13.7 nanograms/gram of cottonseed (ppb). Aflatoxin concentrations of 76 and 13.7 ppm are both acutely toxic to animals while 76 and 13.7 ppb are below the United States FDA Action Levels for cottonseed to be used for specific animal feeds. The threshold for aflatoxin concentrations in cottonseed to be used for immature and dairy animals is 20 ppb while there is a maximum of 400 ppb allowed for cottonseed to be used for several other animal feeds.

The major pitfall in the widespread application of *A. flavus* in the environment may be unrelated to aflatoxin production. The potential human, animal, bird and insect disease risks from *A. flavus* have to be carefully examined. Furthermore, the possibility of increasing the incidence of human allergy problems related to *A. flavus* in the areas where the *A. flavus* is applied has to be determined. Certainly this is a worthwhile research approach. But we must be careful that in our research efforts to alleviate one marketing problem associated with aflatoxins or other mycotoxin contamination that we do not create other severe long term health or environmental problems. There are also serious food adulteration questions that will have to be addressed. When a farmer intentionally applies a known human allergen that is also a potential human pathogen, such as a non-aflatoxigenic *A. flavus* to his crop, does he adulterate his product as defined by the federal pure food laws and make it unmarketable in the United States? Certainly if a farmer is judged to have adulterated his crop he will lose the market and make no profit.

There are currently available commercial detoxification methods for aflatoxin contamination of cottonseed that can be used. There should also be research on alternative uses of cottonseed that will not be as affected by the excessive aflatoxin content associated with particular desert locations.

MARKETING, REGULATION AND RISK MANAGEMENT FACTORS

The farmer looks at mycotoxin management in his crops and products in a much different way than the buyers, regulatory agencies, processors and consumers of the products. From a human health standpoint there has to be a minimal risk to the consumer of foods and feeds. However, for mycotoxins like the aflatoxins and ochratoxin there is no way to eliminate them from all products and we are forced into developing management strategies.

The strategies have to be tailored for the mycotoxin of concern, the product, the marketplace and the geographical location of the consuming public. If the developed countries do not maintain markets and uses for many of the products then we just shift the health and economic risks to the developing countries. Therefore, there is a need for suitable sorting, detoxification and utilization method development to lower the risk worldwide.

First and foremost, there is a need for worldwide harmonization of mycotoxin regulations. Currently there is one set of guidelines for the United States, another for the European Union, and many other guidelines in developed and developing countries. The current situation creates chaos in the international marketplace. Even in the United States the FDA guidelines for shelled raw peanuts in the market are higher than the peanut industry requires in purchase contracts. Given the fact that the sampling error in peanut lots always contributes the greatest variation in the analysis, one must ask is there really a difference in the risk to human health in raw shelled peanut lots containing 25 ppb, 20 ppb, 15 ppb, 10 ppb, 5 ppb or even 2 ppb total aflatoxins? It is doubtful that we can reliably measure aflatoxin concentrations in peanut lots with great enough accuracy to answer this question (Whitaker and Park, 1994). The second question that is also logical is why the European Union regulations are 2 ppb in finished products? Is this regulation based on a realistic risk analysis or is this a way to manage the marketplace and protect locally produced products? One difference between the two approaches may be that the United States produces products that are at risk for aflatoxin contamination and the European Union is at a very low risk for aflatoxin contamination in the products produced in the member countries. The different regulations make sense from the standpoint of the risk related to domestically produced products. However, both the United States and European Union regulations may exclude many products such as Brazil nuts, pistachio nuts, peanuts, corn, figs and other commodities that may be very important to the welfare of the developing countries' economies as well as to worldwide community health and global ecology. We do not currently have easy solutions that will eliminate the risk of mycotoxins in our food. We need to have logical and fair approaches that are effective in managing the actual health risks rather than perceived risks. At the sixth ICRISAT regional groundnut meeting for western and central Africa held in Mali in 1998, Mr. P. Dimanche from CIRAD-CA in France talked about recent trends in European aflatoxin regulations and said "The regulations for tolerance levels have been laid down in response to concerns for human health. These regulations have continued to become increasingly stringent, as the tolerance level is directly linked to the development of increasingly refined methods of analysis." (Dimanche, 2000). The danger may be that the regulatory community may be willing to set the maximum limits in international commerce based on current analytical capabilities, rather than on realistic health factors.

There can be no argument that there is no defined "no effect level" for aflatoxin but there are multiple ideas about what the realistic allowable risk should be. It seems that a large part of the regulatory approach often depends on latitude and the probability of contamination of locally produced products. It is not possible to remove *A. flavus* or other fungi from the environment so we must develop mechanisms to minimize aflatoxin exposure while understanding that mankind cannot totally eliminate the health risks. There are similar risk issues about ochratoxin and there will be continuing risk considerations about other mycotoxins. Many of the developed countries have been slow in establishing guidelines on locally produced products when a particular mycotoxin occurs in the marketplace. The European Union has published proposed tolerable daily intakes for ochratoxin A while also considering local and import regulations for commodities such as coffee. This approach seems logical and makes much more sense in the long run than basing mycotoxin regulation on analytical capabilities.

CONCLUSIONS

Mycotoxin contamination of foods by the aspergilli is a very complicated issue. The biological factors allowing the growth of the Aspergillus species that produce mycotoxins are dependent on the fungus, the product and the location. Aspergillus species tend to be more prevalent in the warm, sub-tropical and tropical climates than in temperate climates. We cannot eliminate mycotoxin contamination from our foodstuffs, but we should strive to minimize their adverse effects by using the best practices possible. Mycotoxin contamination is a global issue and perhaps the developed countries tend to overemphasize the risks while the developing countries may tend to underestimate the risks. Therefore we need to work together to improve food security and food safety worldwide in order to foster international cooperation and develop or maintain markets and protect human health.

Mankind cannot manage all aspects of the food supply and there is a point where we have to admit some risks are presently unavoidable. The August 2000 issue of the National Geographic Magazine article "Fungi" included a sentence that seems to be very appropriate: "Intentional or not humans and fungi are partners from cradle to grave." (Murawski, 2000). The challenge posed is simply how can we devise ways to utilize minimally contaminated products in the marketplace even if it means that effective resorting, processing and detoxification treatments may be necessary?

ACKNOWLEDGMENTS

This work was supported by the University of Georgia Agricultural Experiment Stations, The USAID Peanut CRSP program and the Fulbright fellowship program. The opinions of the authors are their own and do not necessarily reflect the opinions of the University of Georgia or the University of Zagreb.

REFERENCES

Abarca, M.L., Bragulat, M.R., Castella, G., and Cabanes, F.J., 1994, Ochratoxin production by strains of *Aspergillus niger var. niger*, *Appl. Env. Microbiol.*, 60:2650.

Abramson, D., 1991, Development of molds, mycotoxins and odors in moist cereals, during storage in: *Cereal Grain Mycotoxins, Fungi and Quality in Drying and Storage*, Chelkowski, J., ed. Elsevier, Amsterdam.

Abramson, D., Sinha, R.N. and Mills, J.T., 1983, Mycotoxin and odor formation in barley stored at 16 and 20% moisture in Manitoba. *Cereal Chem.* 60:350.

Beti, J.A., Phillips, T.W. and Smalley, E.B., 1995, Effects of maize weevils (Coleoptora:Curculionidae) on production of aflatoxin B_1 by *Aspergillus flavus* in stored corn, *Stored-Product Entomol.*, 88:1776.

Bucheli, P., Kanchanomai, C., Meyer, L., and Pittet, A., 2000, Development of ochratoxin A during robusta (*Coffee canephora*) coffee cherry drying, *J. Agric. Food Chem.*, 48:1358.

Calvert, O.H., Lillehoj, E.B., Kwolek, W.F., and Zuber, M.S., 1978, Aflatoxin B_1 and G_1 production in developing *Zea mays* kernels from mixed inocula of *Aspergillus flavus* and *A. parasiticus*. *Phytopathology* 68:501.

Chelkowski, J., 1991, *Cereal Grain Mycotoxins, Fungi and Quality in Drying and Storage*, Elsevier, Amsterdam.

Christensen, C.M., 1974, *Storage of Cereal Grains and Their Products*, 2nd Edition, American Association of Cereal-species Chemists, St. Paul, MN.

Coker, R.D. 1998, Design of sampling plans for determination of mycotoxins, in: *Mycotoxins in Agriculture and Food Safety*, K.K. Sinha and D. Bhatnagar, eds., Marcel Dekker, Inc., New York.

Cotty, P.J., 1989, Virulence and cultural characteristics of two *Aspergillus flavus* strains pathogenic in cotton, *Phytopath.* 79:808.

Cotty, P.J., 1990, Effect of atoxigenic strains of *Aspergillus flavus* on aflatoxin contamination of developing cottonseed, *Pl. Dis.*, 74:233.

Cotty, P.J., 1997, Aflatoxin-producing potential of communities of Aspergillus section Flavi from cotton producing areas in the United States, *Mycol. Res.*, 6:698.

Cotty, P.J., and Bayman, P., 1993, Competitive exclusion of atoxigenic strain of *Aspergillus flavus* by an atoxigenic strain, *Phytopath.* 83:1283.

Cotty, P.J., and Cardwell, K.F., 1999, Divergence of West African and North American communities of Aspergillus section Flavi, *Appl. Env. Microbiol*, 65:2264.

Cunnif, P., 1995, Official Methods of Analysis of AOAC International 16th Edition, AOAC International, Arlington, VA.

Dimanche, P., 2000, Recent developments and trends in European regulations croncerning groundnut aflatoxin contamination: Consequences for groundnut production in Africa in: Summary Proceedings of the Sixth ICRISAT Regional Groundnut Meeting for Western and Central Africa, Waliyar, F. and Ntare, B.R., eds, ICRISAT, Patancheru, India.

Dorner, J. W., Cole, R.J., and Blankenship, P.D., 1992, Use of a biocompetitive agent to control preharvest aflatoxins in drought stressed peanuts, *J. Food Protection*, 55:888.

Dowd, P.F., 1998, Involvement of arthropods in the establishment of mycotoxigenic fungi under field conditions in: *Mycotoxins in Agriculture and Food Safety*, Sinha, K.K., and Bhatnagar, D., eds., Marcel Dekker, Inc., New York.

Frisvad, J.C., 1995, Mycotoxins and mycotoxigenic fungi in storage in: *Stored-Grain Ecosystems*, Jayas, D.S., White, N.D.G, and Muir, W.E., eds., Marcel Dekker, Inc.,New York.

Frisvad, J.C., and Samson, R.A., 1991, Mycotoxins produced by species of Penicillium and Aspergillus occurring in cereals, in: *Cereal Grain Mycotoxins, Fungi and Quality in Drying and Storage*, J. Chelkowski, ed., Elsevier, Amsterdan.

Gardner, D.E., McMeans, J.L. , Brown, C.M., Bilbrey, R.M., and Parker, L.L., 1974, Geographic localization and lint fluorescence in *Aspergillus flavus* infected cottonseed. *Phytopathology* 64:452-455.

Heenan, C.N., Shaw, K.J. and Pitt, J.I., 1998, Ochratoxin A production by *Aspergillus carbonarius* and *A. niger* and detection using coconut cream agar. *J. Food Mycol.* 1:67.

Hocking, A.D., 2001, Toxigenic Aspergillus species, in: *Food Microbiology Fundamentals and Frontiers*, Doyle, M.P. Beuchat, L.R., and Montville, T.J., eds. ASM press, Baltimore.

Horie, Y., 1995, Production of Ochratoxin A of *Aspergillus carbonarius* in Aspergillus section Nigri, *Nippon Kingakukai Kaiho*, 36:73.

Horn, B.W., and Dorner, J.W., 1998, Soil populations of Aspergillus species from section Flavi along a transect through peanut-growing regions of the United States, *Mycologia*, 90:767.

Horn, B.W., and Dorner, J.W., 1999, Regional differences in production of aflatoxin B_1 and cyclopiazonic acid by soil isolates of *Aspergillus flavus* along a transect within the United States, *Appl. Env. Microbiol*, 65:1444.

Jayas, D.S., White, N.D.G., and Muir, W.E., 1995, Stored-Grain Ecosystems, Marcel Dekker, Inc., New York.

Klich, M.A., and Pitt, J.I., 1998, A Laboratory Guide to the Common Aspergillus Species and their Teleomorphs, CSIRO Division of Food Processing, North Ryde, Australia.

Kurtzman, C.P., Smiley, M.J., Robnett, C.J. and Wicklow, D.T., 1986, DNA relatedness among wild and domesticated species in the *Aspergillus flavus* group, *Mycologia*, 78:955.

Lacey, J., and Magon, N., 1991, Fungi in cereal grains: Their occurrence and water and temperature relationships, in: *Cereal Grain Mycotoxins, Fungi and Quality in Drying and Storage*. Chelkowski, J. ed., Elsevier, Amsterdan.

Manabe, J., and Tsuruta, O., 1991, Mycoflora and mycotoxins in stored rice grain in: *Cereal Grain Mycotoxins, Fungi and Quality in Drying and Storage*, ed; Chelkowski, J., Elsevier, Amsterdan.

Marsh, P.B., Simpson, M.E. and Filsinger, E.C., 1973, *Aspergillus flavus* boll rot in the U.S. cotton belt in relation to high aqueous-extract p of fiber, and indicator of exposure to humid conditions. *Pl. Dis. Reporter* 57:664-668.

Micheli, P.A., 1729, Nova Plantarum Genera Juxta Tournefortii Disposita. Florence.

Murawski, D.A., 2000, Fungi, *Nat. Geographic*, 198:58.

Odum, T.V., Bigelow, D.M., Nelson, M.R., Howell, D.R., and Cotty, P.J., 1977, Spatial and temporal patterns of *Aspergillus flavus* strain composition and propagule density in Yuma County, Arizona soils, *Pl. Dis.* 81:911.

Ominski, K.H., Marquardt, R.R., Sinha, R.N., and Abramson, D., 1994, Ecological aspects of growth and mycotoxin production by storage fungi in: *Mycotoxins in Grain, Compounds Other Than Aflatoxin*, Miller, J.D., and Trenholm, H.L., eds., Eagan Press, St. Paul, MN.

Payne, G.A., 1998, Process of contamination by aflatoxin-producing fungi and their impact on crops in: Mycotoxins in Agriculture and Food Safety, Sinha, K.K. and Bhatnagar, D., eds., Marcel Dekker, Inc., New York.

Pons, W.A., and Goldblatt, L.A., 1965, The determination of aflatoxins in cottonseed products, *J. Am. Oil Chem. Soc.*, 42:471.

Raper, K.B., and Fennell, D.I., 1965, The Genus Aspergillus, Williams and Wilkins, Baltimore.

Sauer, D.B., 1992, Storage of Cereal Grains and their Products, 4th Edition, American Association of Cereal Chemists, St. Paul, MN.

Sugimoto, T., Minamisawa, M., Takano, K., Furukawa, Y., and Tsuruta, O., 1976, Natural occurrence of ochratoxin and sterigmatocystin in moldy rice, *Proc. Jpn. Assoc. Mycotoxicol* 3/4:3.
Udoh, J.M., Cardwell, K.F., and Ikotun, T., 2000, Storage structures and aflatoxin content of maize in five agroecological zones of Nigeria, *J. Stored Prod. Res.*, 36:187.
Urano, T., Trucksess, M.W., Beaver, R.W., Wilson, D.M., Dorner, J.W., and Dowell, F.E., 1992, Co-occurrence of cyclopiazonic acid and aflatoxins in corn and peanuts, *J. AOAC Int.* 75:838.
Vega, F.E., and Mercadier, G., 1998, *Florida Entomologist*, 81:543.
Whitaker, T.B., and Park, D.L., 1994, Problems associated with accurately measuring aflatoxins in food and feeds: Errors associated with sampling, Sample preparation and analysis, in: *The Toxicology of Aflatoxins*, Eaton, D.L. and Groopman, J. D., eds., Academic Press, San Diego, CA.
Wicklow, D.T., and Shotwell, O.L., 1983, Intrafungal distribution of aflatoxins among conidia and sclerotia of *Aspergillus flavus* and *Aspergillus parasiticus*. *Can. J. Microbiol.* 29:1.
Widstrom, N.W., 1992, Aflatoxin in developing maize: Interactions among involved biota and pertinent econiche factors, in: *Handbook of Applied Mycology*, Volume 5: Mycotoxins in Ecological systems, Bhatnagar, D., Lillehoj, E.B., and Arora, D.K., eds., Marcel Dekker, Inc. New York.
Will, M.E., Holbrook, C.C., and Wilson, D.M., 1994, Evaluation of field inoculation techiques for screening pea nut genotype: for reaction to preharvest *A. flavus* group infection and aflatoxin contamination, *Peanut Sci.*, 21:122.
Williams, J.T. and Wilson D.M., 1999, Report on the Brazil nut aflatoxin problem is Bolivia, Upublished 24 page report to Chemonics and US AID.Wilson, D.M., 1995, Management of mycotoxins in peanuts, in: *Peanut Health Management*, Melouk, H.A., and Shokes, F.M., eds., APS Press, St. Paul, MN.
Wilson, D.M., and Abramson, D., 1992, Mycotoxins, in: *Storage of Cereal Grains and their Products*, Sauer,D.B., ed., American Association of Cereal Chemists, St. Paul, MN.
Wilson, D.M., and King, J.K., 1995, Production of aflatoxins B_1,B_2,G_1 and G_2 in pure and mixed cultures of *Aspergillus parasiticus* and *Aspergillus flavus*. *Food Additives and Contaminants* 12:521.
Wilson, D.M., McMillian, W.W., and Widstrom, N.W., 1986, Use of *Aspergillus flavus* and *A. parasiticus* color mutants to study aflatoxins contamination in: *Biodeterioration* VI, Llewellan, G., O=Rear, C., and Barry, S., eds., International Biodeterioration Society, London.
Wilson, D.M., and Payne, G.A., 1994. Factors affecting *Aspergillus flavus* group infection and aflatoxin contamination of crops in: *The Toxicology of Aflatoxins*, Eaton, D.L., and Groopman, J.D., eds., Academic Press, Inc. San Diego, CA.

ASPECTS OF THE ECOLOGY OF FUSARIUM TOXINS IN CEREALS

J. David Miller

Department of Chemistry
Carleton University
Ottawa, Ontario K1S 5B6

ABSTRACT

Species of the genus *Fusarium* account for three of the five agriculturally important mycotoxins which are deoxynivalenol, aflatoxin, fumonisin, zearalenone and ochratoxin. The toxigenic fusaria have been complicated to study because morphologically-similar strains represent different biologies: saprophytes, pathyotypes and endophytes. This might explain the difficulties with systems of taxonomy for Fusarium species and increasing reliance on molecular techniques to characterize taxa. Another remarkable feature of the toxigenic fusaria is that each species produces compounds that cross several species as well as families of compounds that are species specific. In addition, reproductively-isolated strains (from different continents) of important species such as *F. graminearum* produce different compounds, and even produce the same compounds by different biosynthetic pathways.

INTRODUCTION

Although there are hundreds of fungal metabolites that are toxic in experimental systems, there are only five for which there is material exposure to humans and domestic animals: deoxynivalenol, aflatoxin, fumonisin, zearalenone and ochratoxin. Coincidently, five species of fungi are responsible for vast majority of the production of these toxins: *Fusarium graminearum* (Schwabe) Petch, *F. culmorum* (WG Smith) Sacc., (deoxynivalenol, zearalenone), *Aspergillus flavus* Link:Fr. (aflatoxin), *Fusarium verticillioides* (Saccardo) Nirenberg (fumonisin), and *Penicillium verrucosum* Dierckx (ochratoxin) (Miller, 1995). Three of the five important taxa are Fusarium species. A brief consideration of the ecology of these species involves some discussion of the impact of temperature, rainfall and other biota on their distribution on cereals, the impact of the host plant on toxin in the edible portion and the influence of these toxins on associated biota including those consuming the crop. This brief review will attempt to cover these three lines of thought as they pertain to the situation in North America.

FUSARIUM HEAD BLIGHT AND GIBERELLA EAR ROT

Fusarium head blight affects small grains, mainly wheat, but also barley and triticale. As noted, this disease is mainly caused by *F. graminearum* and *F. culmorum*. These are closely related species that produce deoxynivalenol or nivalenol and zearalenone depending on geographic origin of the isolate (Miller et al., 1991). This crop disease is associated with temperate grain-growing regions. Five or six species are consistently isolated from small grains affected by fusarium head blight. The most pathogenic species, *F. graminearum* and *F. culmorum* are the most common. *F. graminearum* was common in wheat from North America. *F. culmorum* is the dominant species in cooler wheat growing areas such as The Netherlands. *F. avenaceum* is also common in wheat from all regions studied. *F. crookwellense* Burgess, Nelson & Toussoun is rare in wheat from Canada and the U.S., but was the primary cause of head blight in irrigated land in the Orange Free State, South Africa. *Fusarium poae* (Peck) Wollenweber, *F. equiseti* (Corda) Saccardo and *F. sporotrichioides* Sherbakoff are also isolated from wheat kernels in low to moderate frequencies more often under cooler conditions. The distribution of head blight species is affected by pathogenicity where: *F. graminearum* > *culmorum* >> *crookwellense* > *avenaceum*. The regional and annual variation of the pathogenic species is most affected by temperature (from coldest to warmest): *F. culmorum* > *crookwellense* > *avenaceum* > *graminearum* (Miller, 1994).

Fusarium graminearum has a very narrow optimum temperature range of 26-28°C. Growth at 24 and 30°C is approximately 30% of that of the optimum (Miller, 2001). Monitoring of the growth of *F. graminearum* in experimentally-infected ears showed the growth rate of the fungus was sensitive to temperature. A period where the average "growing degree days >5" was approximately ten, virtually halted growth. An average value of about 15 resulted in rapid growth (Miller et al., 1983). Although increased rainfall promotes fusarium head blight incidence is most affected by moisture at anthesis as long as the temperature remains in the correct range (Miller, 1994). An example of this phenomenon can be seen in data on the recovery of *Fusarium* from surface-sterilized wheat kernels in Ottawa (Table 1). When temperatures were too warm and precipitation was average, the relative percentage of *F. graminearum* was low. When temperatures were closer to normal, the prevalence of this species increased markedly with relatively more *F. culmorum* seen during the coolest of the three years. The data also make the point that it is rainfall at anthesis rather than the total value. These species also occur on corn along with *F. subglutinans* (Miller, 1994). The disease associated with *F. graminearum*, Gibberella or pink ear rot is affected by moisture at silk emergence and prevalence is increased with wet weather later the season.

Table 1. Recovery of some *Fusarium* species from surface sterilized wheat kernels

Year	Percent of 30 average		Total	Percent of Total Fusarium			
	Corn Heat Units	Rainfall	*Fusarium*	*graminearum*	*culmorum*	*equiseti*	*sporotrichiodes*
1991	+19	0	67	3	1	15	24
1992	-3	+18	26	10	4	6	9
1993	+5	-11	31	10	1	7	29

Adapted from Miller et al. (1998)

In culture or in the field, the metabolite families that dominate depend on O_2, pH, osmotic tension and sometimes temperature. The best studied fermentation conditions are for the metabolite families for *F. graminearum* and *F. culmorum* (Greenhalgh et al., 1985; Hidy et al., 1977; Miller and Blackwell, 1986), *F. sporotrichioides* (Greenhalgh et al., 1986), *F. verticillioides* (Miller et al., 1994) and *F. avenaceum* (Corda) Fr. (enniatins) (Blais et al., 1992).

Only the terminal few cells of a mycelium are biologically active. It is known that mycotoxins are produced by the penultimate cells of each mycelium as it grows. The accumulation of mycotoxin occurs as the result of millions of cells in the kernel and is a function of total fungal biomass rather than of time. In stirred jar fermentations, this same process takes place except that billions cells are grown more or less in a synchronous fashion. Deoxynivalenol is produced under conditions of low oxygen tension (Miller and Blackwell, 1986). In growing plants, oxygen is tightly managed (because it poisons photosynthesis). Deoxynivalenol is not found in growing crops (Miller et al., 1983; Miller and Young, 1985). In contrast, zearalenone requires oxygen saturation for optimal production (Hidy et al., 1977) a condition seen after field crops senesce (Miller et al., 1983).

Among the very first measurements of deoxynivalenol under field conditions in 1983 showed that the toxin concentration declined (Scott et al., 1984), and thus the symptoms present in wheat did not strictly relate to the deoxynivalenol present. Cultivars with equivalent symptom expression or fungal biomass had up to 2-10 fold differences in the amount of deoxynivalenol present. This led to the conclusion that there were factors that either prevented the synthesis of deoxynivalenol, degraded it, or both (Miller et al., 1984). Groups all over the world ultimately reported similar findings (see Mesterhazy et al., 1999). Snijders and colleagues showed that deoxynivalenol appeared in wheat kernels from susceptible cultivars in advance of fungal mycelia (Snijders and Perkowski, 1990) (Table 2).

Table 2. Appearance of deoxynivalenol (DON) in wheat kernels before detectable fungal biomass as determined by ergosterol (ERG) concentration.

***F. cumorum*-**	**Susceptible**		**Resistant**	
infected spikes	**ERG**	**DON**	**ERG**	**DON**
0%	-	76 ppm	16 ppm	-
7.5	5 ppm	22	15	-
17.5	6	62	11	29 ppm
27.5	10	61	15	21
57.5	20	164	27	77

Adapted from Snijders and Kreching (1994)

Trichothecenes were recognized to be phytotoxic during the course of their initial discovery (Brian et al., 1961). The phytotoxicity of deoxynivalenol was explored when it was realized that there was a strong difference between wheat cultivars resistant to Fusarium head blight with respect to the phytotoxic properties of this toxin. Coleoptile tissue of cultivars that were resistant to Fusarium head blight were 10 times more resistant to deoxynivalenol (and some other metabolites including 3 acetyl deoxynivalenol and dihydroxycalonectrin) than disease susceptible cultivars (Wang and Miller, 1988). This was shown to be due to the presence of a modified peptidyl transferase (at protein synthesis) (Miller & Ewen 1997) and to unknown functional changes in the membranes of more resistant types (Cossette and Miller, 1995; Miller and Ewen, 1997; Snijders and Kreching, 1992). In addition, compounds were isolated that inhibit toxin synthesis (Miller et al.,

1997). Again, early studies had suggested that wheat appeared to be able to metabolize deoxynivalenol in the field (Miller and Young 1985; Scott et al., 1984). This was shown later to be the case *in vitro* in a head blight-resistant cultivar (Miller and Arnison, 1986). Strains that produce high concentrations of deoxynivalenol were more virulent (Snijders, 1994; Mesterhazy et al., 1999). This implied that one component of resistance to Fusarium head blight related to reducing the phytotoxic impact of deoxynivalenol. This has been examined from the other perspective i.e. by showing that strains of *F. graminearum* with the trichodiene synthetase removed have reduced virulence (Desjardins et al., 1996). Most crucially, deoxynivalenol results in massive electrolyte loss in plant cells upon exposure. *Fusarium graminearum* is a necrotrophic pathogen. Such pathogens invade by killing the host cells in advance. This necrotrophic pathogen produces a compound that results in cell lysis, and thus the release of sugars.

Deoxynivalenol is more toxic and important than meets the eye. "Red mold poisoning" was reported in rural Japan throughout the 1950's which lead to investigations on the cause. Eventually, deoxynivalenol was discovered by Japanese researches from grain that had made humans ill (Morooka et al., 1972). The same chemical was subsequently re-reported as "vomitoxin" by Vesonder and colleagues studying problems in swine fed *F. graminearum* -contaminated corn in 1973 (Vesonder et al. ,1973).

The array of metabolites that are produced by *F. graminearum* and related species has important ecological significance. Swine are the most sensitive domestic animal species to the effects of trichothecenes (Prelusky et al., 1994). Experiments feeding pure deoxynivalenol to swine suggested that the diets containing < 2 mg/kg deoxynivalenol have little impact on growth. However, many experiments using naturally-contaminated grains often demonstrated that such grain, was more toxic than indicated from the deoxynivalenol content (Prelusky, 1997 and citations therein). Several of the minor toxins from *F. graminearum* were present in such grain, sometimes at concentrations similar to deoxynivalenol (Foster et al., 1986; Langseth unpublished data). These co-occurring toxin including culmorin, dihydroxycalonectrin and sambucincol, were shown to increase the toxicity of deoxynivalenol when fed in combination to insects (Figure 1) (Dowd et al., 1989). This might be also true in swine (Rotter et al., 1992). In addition, culmorin, found originally from *F. culmorum* is antifungal and has been found from a number of unrelated marine fungi from bio-assay directed fractionations looking for antifungal compounds (Pedersen and Miller, 2000). In infected corn ears, fungi other than *F. graminearum* (and *F. verticillioides)* are uncommon (Miller et al., 1983), perhaps because of material concentrations of culmorin.

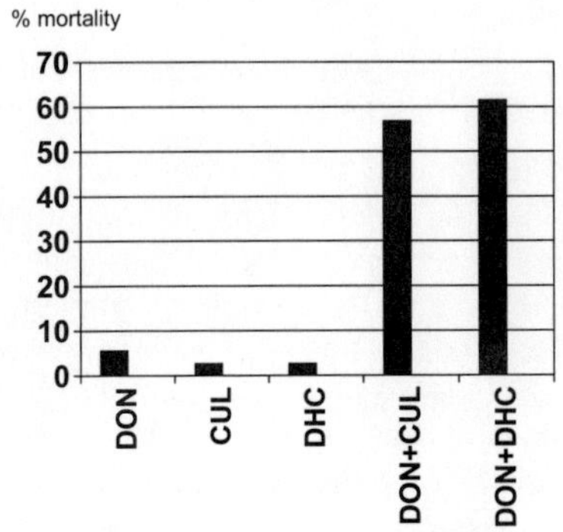

Figure 1. Affect of *F. graminearum* metabolites on *H. zea;* DON= deoxynivalenol, CUL = culmorin, DHC= dihydroxycalonectin (from Dowd et al., 1989).

Zearalenone is an estrogen analogue and causes hyperestrocism in female pigs at low levels; the dietary no effect level is < 1 mg/kg. Cows and sheep are also sensitive to the estrogenic effects of this toxin with depressed ovulation and lower lambing percentages. The no effect dietary levels are not clearly known. (Prelusky et al., 1994). Non-human primates are also sensitive to the estrogenic effects of zearalenone (Kuiper-Goodman, 1994).

FUSARIUM EAR ROT

Unlike the situation for *F. graminearum* and allied species, where both small grains and corn are affected, *F. verticillioides* is associated with corn only. The weight of evidence is that *F. verticillioides* is endemic in corn kernels. The biology of fumonisin production suggests that fumonisin is produced in material quantities only senescent corn tissue. This is consistent with the view that *F. verticillioides* is a mutualistic endophyte. Whereas the optimal range temperature for growth of *G. zea* is rather narrow, the growth of *F. verticillioides* appears to extend well above 30°C (Miller, 2001). After temperature, there are four factors that contribute to the occurrence of *F. verticillioides* ear rot of corn: drought stress, insect damage, other fungal diseases and corn genotype.

Studies of the occurrence of fumonisin from natural occurrence and experimental infections demonstrate the importance of drought rather than temperature stress on the occurrence of fumonisin. The occurrence of fumonisin in Ontario was limited to drought-stressed fields. Comparing highest and lowest fumonisin counties, temperature was similar at 104 and 107% of the 30-year average, respectively. Rainfall in the high group was only 49% of normal and in the low group, 95% of normal values (Miller et al., 1995). In experimental inoculations of corn in Poland, the year of highest FB_1 accumulation, the temperature was 117% of the 30-year mean; the year of lowest fumonisin accumulation had a temperature that was 102% of the mean. Rainfall in the highest year was 6% of normal and in the lowest year was 65% of normal (Pascale et al., 1997). In the U.S., fumonisin concentrations were inversely proportional to June rainfall (Shelby et al., 1994), again suggesting the important role of drought stress.

Drought stress results in greater insect herbivory on corn, hence it is not possible to totally separate these variables. It appears that kernel damage *per se* promotes the disease and fumonisin accumulation (Schaafsma et al., 1993; Drepper and Renfro, 1994). However, there is a strong relationship between insect damage and Fusarium ear rot. A field survey in Austria demonstrated that the incidence of the European corn borer increased *F. verticillioides* disease and fumonisin concentrations (Lew et al., 1991). Corn genotypes containing the anti-insectan Bt protein had lowered fumonisin (Munkvold et al., 1997; 1999).

Corn infected by other ear-damaging pathogens such as *F. graminearum* may be predisposed to *F. verticillioides* "infection" and fumonisin accumulation. Ears wound inoculated by *F. graminearum* and *F. subglutinans* produced severe disease. Despite the fact that *F. graminearum* and *F. subglutinans* do not produce fumonisin, these ears contained 42 and 3 μg/g FB_1, respectively (Schaafsma et al., 1993).

Fumonisins are potent phytotoxins that cause electrolyte loss and interfere with the formation of complex phytosphingolipids. Fumonisin exposure did not reduce maize seed germination, but reduced radicle elongation when exposed to solutions above 10^{-4} M and seed amylase production was inhibited (Doehlert et al., 1994). Fumonisin decreased shoot length, shoot dry mass and root length in the 10 μM range (Lamprecht et al., 1994). Fumonisin incorporated into plant tissue culture media reduced the growth of maize callus also at 10^{-6} M (Van Asch et al., 1992). Fumonisin B_1 has been shown to disrupt sphingolipid metabolism in maize seedlings (Riley et al., 1996). In crosses of high- and low- fumonisin-producing strains of *F. verticillioides*, only progeny that produced high concentrations of fumonisin *in vitro* caused significant stem rot (Nelson et al., 1993).

Even from regions where fumonisin accumulations in corn are less versus more likely, strains of *F. verticillioides* isolated from corn have the potential to produce this family of toxins (Miller 2001). Strains of *F. verticillioides* produce fumonisins and fusarins and those of *F. proliferatum* also result in the accumulation of moniliformin (Miller et al., 1995). There are also many minor derivatives of fumonisins that have been discovered (Musser et al., this volume). It is assumed that the fumonisins other than fumonisin B_1 and B_2 are much less potent and not material to the toxicity seen under field conditions where fumonisin B_1 is always dominant. Fusarins are heat unstable and are not present in food, but it cannot be ruled out that their toxicitity (Li et al., 1992) contributes to problems in animal health. In chickens, co-exposure to moniliformin and fumonisin was slightly additive (Javed et al., 1993). Unlike the situation for toxins from *F. graminearum* and related species, the potential for interactions between *F. verticillioides* toxins to interact has not been properly explored.

FUSARIUM SPOROTRICHIODES TOXINS

T-2 toxin was isolated from "strain T-2" of *F. sporotrichioides* misidentified as *F. tricinctum* isolated from corn associated with cow mortalities (Ueno, 1983). This compound has been the subject of considerable toxicological study because it is easy to isolate and purify. During World War II, there were large-scale poisonings of the rural population in the former Soviet Union caused by the consumption of grains left in the field over winter. Today, T-2 toxin is not common in grains (IARC, 1993) because, as noted above, it is not a cereal pathogen and occurs only in abused grain. T-2 and related compounds are somewhat common in grain dusts (De Mers, 1994). Early colonization of the lemmae by *F. poae* and *F. sporotrichioides* (Table 1) may promote subsequent colonization by *F. graminearum* (Sturz and Johnston, 1983), accounting for the T-2 and related compounds in grain dusts even when there are none in the kernel.

SUMMARY

In North America, Fusarium head blight, Gibberella ear rot and Fusarium ear rot are substantial problems. In the case of the former two diseases, the primary mycotoxin is also a potent phytotoxin, and, symptoms of disease are not simply related to the amount of toxin present. Additionally, the material presence of compounds such as culmorin, dihydroxycalonectrin and sambucinol is important in understanding the toxicity of naturally-contaminated grains. An incomplete understanding exists of the factors that affect the occurrence of fumonisin. It is excluded from environments where night-time temperatures are greater than 25°C, conditions where F. graminearum can prevail. Where temperatures are in excess of this, and where drought, insect, disease and other stressors occur, this endophyte grows in kernel tissue and produces fumonisin.

As a post-script, occupation exposures to dusts from Fusarium head blight may be more dangerous than we thought. From the beginnings of the *Fusarium* mycotoxin program at the Central Experimental Farm, Dr. Locksley Trenholm recognized the need to be careful handing grains and feeds contaminated with mycotoxins. By 1983, there were written guidelines on this issue. This perhaps was in part due to the knowledge that epidemiological studies of feed workers handling grains contaminated with aflatoxin had elevated risks for liver cancer, later shown to be unambiguously true in 1991 (IARC, 1993). Further, the US Army had supported considerable research on the impact of inhalation exposure to pure trichothecenes. Such exposures were found to be 20-50 more potent than iv or ip exposures (Creasia et al., 1989). In 1993, a formal risk assessment was done including information on exposures in grain elevators (De Mers, 1994). Based on the best information available, the judgement was that the 1993-94 exposures did not pose a material risk. Since then, Finnish workers reported the concentrations of dusts, deoxynivalenol and spores associated with on-farm handing of grains. Deoxynivalenol contents were similar to those found in grain from the 1993 western crop (Table 3) (De Mers, 1994).

Table 3. Airborne dusts during farm operations

Farm operation	*Spore concentration in air*	*DON concentration in air*
Grain drying	10^7 spores/m^3	1 mg/m^3
Milling	10^6 spores/m^3	2 mg/m^3
Cattle feeding	10^6 spores/m^3	1.4 mg/m^3

from Lappalainen et al. (1996)

Our studies from the handling of the 1993 crop found 0.5 to 5.8 ppm DON + 1 ppm T2 + HT2 toxin in such airborne dusts. In Finland, grain handlers were exposed to 2-6 times the values associated with farming activities, at least in Finland (De Mers, 1994).

Also since our analysis, Norwegian epidemiologists have been studying occupational disease in cereal farmers. In Norway, grains are contaminated by *Fusarium* toxins similar to those found in Canadian (and U.S.) grains. Perinatal health in woman farmers was at greater risk after harvest and after a poor crop. Occupational exposure to mycotoxins in grain was associated with miscarriage at an early stage of pregnancy (odds ratio 1.67-2.85) (Kristensen et al., 1997). More recent data from this group suggest that there are also non-respiratory health impacts on male farmers and that both effects increase during weather conditions associated with Fusarium head blight (Kristensen et al., 2000). Extensive exposure data are in the process of collection and further refinements of possible association are anticipated in the future. This means that the theoretical risk of non-respiratory (i.e. toxins associated) diseases from handling *Fusarium*-contaminated grain that we thought existed in 1994 are low, but may be occurring after a long time (many years) at current dust exposures.

REFERENCES

Blais, L.A., ApSimon , J.W., Blackwell, B.A., Greenhalgh, R., and Miller, J.D., 1992, Isolation and characterization of enniatins from *Fusarium avenaceum* DAOM 196490. *Can. J. Chem.* 70:1281-1287.

Brian, P.W., Dawkins, A.W., Grove, J.F., Hemming, H.G., Lowe D, Norris, G.L.F., 1961, Phytotoxic compounds produced by *Fusarium equiseti. J. Exp. Botany* 12:1-12.

Cossette, F., Miller J.D., 1995, Phytotoxic effect of deoxynivalenol and gibberella ear rot resistance of corn. *Natural Toxins* 3:383-388.

Creasia, D.A., and Lambert, R.J., 1989, Acute respiratory tract toxicity of the trichothecene mycotoxin T-2 toxin. In *Trichothecene Mycotoxins: Pathophysiologic Effects*, Beasley, V.R., ed., vol. 1. CRC Press, Boca Raton, FL, pp. 161-170.

De Mers, F., 1994, Preliminary assessment of the risk of exposure to trichothecenes produced by *Fusarium graminearum* in grain dust. Occupational Safety and Health Branch, HRD, Ottawa.

Desjardins, A.E., R.H. Proctor, G. Bai, S.P. McCormick, G. Shaner, G. Buechley and T.M. Hohn, 1996, Reduced virulence of trichothecene non-producing mutants of *Gibberella zea* in wheat field tests. *Molecular Plant-Microbe Interactions* 9:775-781.

Desjardins, A.E., Plattner , R.D., 1998, Distribution of fumonisins in maize ears infected with strains of *Fusarium moniliforme* that differ in fumonisin production. *Plant Disease* 82:953-958.

Doehlert, D.C., Knutson, C.A., and Vesonder, R.F., 1994, Phytotoxic effects of fumonisin B_1 on maize seedling growth. *Mycopathologia* 127:117-121.

Dowd, P.F., Miller, J.D. and Greenhalgh, R., 1989, Toxicity and interactions of some *Fusarium graminearum* metabolites to caterpillars. *Mycologia* 81:646-650.

Drepper, W.J. and Renfro, B.L., 1990, Comparison of methods for inoculation of ears and stalks of maize with *Fusarium moniliforme*. *Plant Disease* 74:952-956.

Foster, B.C., Neish, G.A., Lauren, D.L., Trenholm, H.L., Prelusky, D.B. and Hamilton, R.M.G., 1986b, Fungal and mycotoxin content of slashed corn. *Microbiol. Aliment. Nutr.* 4:199-203.

Greenhalgh, R, Levandier, D., Adams, W., Miller, J.D., Blackwell, B.A., McAlees, A.J. and Taylor, A., 1985, Production and characterization of deoxynivalenol and other secondary metabolites of Fusarium culmorum (CMI 14764, HLX 1503). *J. Agric. Food Chem.* 34: 98-102.

Greenhalgh, R., Blackwell, B.A., Paré, J.R.J., Miller, J.D., Levandier, D., Meier, R.M., Taylor, A., and ApSimon, J.W., 1986, Isolation and characterization by mass spectrometry and NMR spectroscopy of secondary metabolites of some *Fusarium* species. Proceedings of the 6th IUPAC international symposium on mycotoxins and phycotoxins. Steyn, P.S. (ed). Elsevier Press, Amsterdam, pp. 137-152.

Hidy, P.H., Baldwin, R.S., Greasham, R.L., Kieth, C.L., McMullen, J.R., 1977, Zearalenone and derivatives: production and biological activities. *Adv. Appl. Microbiol.* 22:54-82.

IARC, 1993, Monograph 56: Some naturally-occurring substances: some food items and constituents, heterocyclic amines and mycotoxins. International Agency for Research on Cancer, Lyon, France.

Javed, T., Bennett, G.A., Richard, J.L., Dombrink-Kurtzman, M.A.,Cote, L.M., Buck, W.B., 1993, Mortality in broiler chicks on feed amended with *Fusarium proliferatum* culture material or with purified fumonisin B_1 and moniliformin. *Mycopathologia* 123:171-184.

Kuiper-Goodman, T., 1994, Prevention of human mycotoxicoses through risk assessment and management. In *Mycotoxins in Grain: Compounds other than Aflatoxin*, Miller, J.D. and Trenholm, H.L., eds,. Eagan Press, St. Paul, MN. pp. 439-470.

Kristensen, P., Irgens, L.M., Andersen, A., Bye, A.S., and Sundheim, L., 1997, Gestational age, birth weight and perinatal death among births to Norwegian farmers. *American J. Epidemiology* 146:329-338.

Kristensen, P., Andersen, A., and Irgens, L.M., 2000, Hormone dependent cancer and adverse reproductive outcomes in farmers' families- effects of climatic conditions favouring fungal growth in grain. *Scan. J. Work Environ. Health* 26:331-337.

Lamprecht, S.C., Marasas, W.F.O., Alberts, J.F., Cawood, W.E., Gelderblom, W.C.A., Shephard, G.S., Thiel, P.G., and Calitz, F.J., 1994, Phytotoxicity of fumonisins and TA-toxin to corn and tomato. *Phytopathology* 84:383-391.

Lappalainen, S., Nikulin, M., Berg, S., Parikki, P., Hintikka, E.-L., and Pasanen, A,-L, 1996, *Fusarium* toxins and fungi associated with handling of grain on eight Finnish Farms. *Atmospheric Environment* 30:3059-3065.

Lew, H., Adler, A., and Edinger, W., 1991, Moniliformin and the European corn borer (*Ostrinia nubilalis*). *Mycotoxin Research* 7:71-76.

Li, S.J., Ronai, Y.Z., Han, N.J., Fan, W.G., Ma, J.L., and Bjeldanes, L.E., 1992, Fusarin C induced esophageal cancer and forestomach carcinoma in mice and rats. *Chin. J. Oncol.* 14:27-29.

Mesterhazy, A., Bartok, T., Mirocha, C.G., Komoroczy, R., 1999, Nature of wheat resistance to Fusarium head blight and the role of deoxynivalenol for breeding. *Plant Breeding* 118:97-110.

Miller, J.D., 1994, Epidemiology of *Fusarium graminearum* diseases of wheat and corn. In *Mycotoxins in Grain: Compounds other than Aflatoxin*, Miller, J.D., and Trenholm, H.L., eds, Eagan Press, St. Paul, MN, pp. 19-36.

Miller, J.D., 1995, Fungi and mycotoxins in grain: implications for stored product research. *J. Stored Product Research* 31:1-6.

Miller, J.D., 2000, Factors that affect the accumulation of fumonisin. *Environmental Health Perspectives* 109:321-324.

Miller, J.D., Young, J.C., Trenholm, H.L., 1983, Fusarium toxins in field corn. I. Parameters associated with fungal growth and production of deoxynivaneol and other mycotoxins. *Can. J. Bot.* 61:3080-3087.

Miller, J.D., Young, J.C., Sampson, D.R., 1984, Deoxynivalenol and Fusarium head blight resistance in spring cereals. *Phytopathol. Zeitschrift.* 113:359-367.

Miller, J.D., and Young, J.C., 1985, Deoxynivalenol in an experimental *Fusarium graminearum* infection of wheat. *Can. J. Plant Pathology* 8:147-150.

Miller, J.D., and Blackwell, B.A., 1986, Biosynthesis of 3-acetyl- deoxynivalenol and other metabolites by *Fusarium culmorum* HLX 1503 in a stirred jar fermentor. *Can. J. Bot.* 64: 1-5.

Miller, J.D., and Arnison, P.G., 1986, Degradation of deoxynivalenol by suspension cultures of the Fusarium head blight resistant cultivar Frontana. *Can. J. Plant Pathol.* 8:147-150.

Miller, J.D., Greenhalgh, R., Wang, Y.Z. and Lu, M., 1991, Mycotoxin chemotypes of three *Fusarium* species. *Mycologia* 83:121-130.

Miller, J.D., Savard, M.E., and Rapior, S., 1994, Production and purification of fumonisins from a stirred jar fermentor. *Natural Toxins* 2: 354-359.

Miller, J.D., Savard ,M.E., Schaafsma, A.W., Seifert, K.A., and Reid, L.M., 1995, Mycotoxin production by *Fusarium moniliforme* and *Fusarium proliferatum* from Ontario and occurrence of fumonisin in the 1993 corn crop. *Can. J. Plant Pathol.* 17:233-239.

Miller, J.D., Fielder, D.A., Dowd, P.F., Norton, R.A., and Collins, F.W., 1997, Isolation of 4-acetyl-benzoxazolin-2-one(4-ABOA) and diferuloylputrescine from an extract of gibberella ear rot-resistant corn that blocks mycotoxin biosynthesis, and the insect toxicity of 4-ABOA and related compounds. *Biochemical Systematics and Ecology* 24:647-658.

Miller, J.D., and Ewen, M.A., 1997, Toxic effects of deoxynivalenol on ribosomes and tissues of the spring wheat cultivars Frontana and Casavant. *Natural Toxins* 5:234-237.

Miller, J.D., Culley, J., Fraser, K., Hubbard, S., Meloche, F., Ouellet, T, Seaman, L., Seifert, K.A., Turkington, K., Voldeng, H., 1998, Effect of tillage practice on fusarium head blight of wheat. *Can. J. Plant Pathol.* 20:95-103.

Morooka, N., Uratsuji, N., Yoshizawa, T., and Yamamoto, H., 1972, Studies on the toxic substances in barley infected with *Fusarium* species. *J. Food Hygiene Soc. Japan* 13:368-375. [in Japanese].

Munkvold, G.P., Hellmich, R.L., Showers, W.B., 1997, Reduced Fusarium ear rot and symptomless infections of maize genetically engineered for European Corn Borer Resistance. *Phytopathology* 87:1071-077.

Munkvold, G.P., Hellmich, R.L., and Rice, L.P., 1999, Comparison of fumonisin concentrations in kernels of transgenic BT maize hybrids and nontransgenic hybrids. *Plant Disease* 83:130-138.

Nelson, P.E., Desjardins, A.E. and Plattner, R.D., 1993, Fumonisins, mycotoxins produced by Fusarium species: biology, chemistry and significance. *Ann. Rev. Phytopathol.* 31:233-252.

Pascale, M., Visconti, A., Pronczuk, M., Wisniewska, H., and Chelkowski, J., 1997, Accumulation of fumonisins in maize hybrids inoculated under field conditions with *Fusarium moniliforme* Sheldon. *J. Science Food Agriculture* 74:1-6.

Pedersen, P.B., and Miller, J.D., 2001, The fungal metabolite culmorin and related compounds. *Natural Toxins* 109(S2):321-324.

Prelusky, D.B., 1997, Effect of intraperitoneal infusium of deoxynivalenol on feed consumption and weight gain in the pig. *Natural Toxins* 5:121-125.

Prelusky, D.B., Rotter, B.A., and Rotter, R.G., 1994, Toxicology of mycotoxins. In: *Mycotoxins in Grain: Compounds other than Aflatoxin*., Miller, J.D., and Trenholm, H.L., eds, Eagan Press, St. Paul, MN, pp. 359-404.

Riley, R.T., Wang, E., Schroeder, J.J., Smith, E.R., Plattner, R.D., Abbas, H., Yoo, H.-S., and Merrill, A.H., Jr., 1996, Evidence for disruption of sphingolipid metabolism as a contributing factor in the toxicity and carcinogenicity of fumonisins. *Natural Toxins* 4:3-15.

Rotter, R.G., Thompson, B.K., Trenholm, H.L., Prelusky, D.B., Hartin, K.E. and Miller, J.D., 1992, A preliminary examination of potential interactions between deoxynivalenol and other selected Fusarium metabolites in growing pigs. *Can. J. Animal Sci.* 72:107-116.

Schaafsma, A.W., Miller, J.D., Savard, M.E., and Ewing, R., 1993, Ear rot development and mycotoxin production in corn in relation to inoculation method and corn hybrid for three species of *Fusarium*. *Canadian J. Plant Pathology* 15:185-192.

Shelby, R.A., White, D.G., and Burke, E.M., 1994, Differential fumonisin production in maize hybrids. *Plant Disease* 78:582-584.

Scott, P.M., Nelson, K., Khanhere, S.R., Karpinski, K.F., Hayward, S., Neish, G.A., and Teich, A., 1984, Decline in deoxynivalenol (vomitoxin) concentrations in 1983 Ontario winter wheat before harvest. *Appl. Environ. Microbiol.* 48:884-886.

Snijders, C.H.A., 1994, Breeding for resistance to Fusarium diseases in wheat and maize. In *Mycotoxins in Grain: Compounds other than Aflatoxin*., Miller, J.D., and Trenholm, H.L., eds, Eagan Press, St. Paul, MN, pp. 37-58.

Snijders, C.H.A., and Perkowski, J., 1990, Effects of head blight caused by *Fusarium culmorum* on toxin content and weight of wheat kernels. *Phytopathology* 80:566-570.

Snijders, C.H.A., and Krechting, C.F., 1992, Inhibition of deoxynivalenol translocation and fungal colonization in Fusarium head blight resistant wheat. *Can. J. Botany* 70:1570-1576.

Sturz, A.V., and Johnston, H.W., 1983, Early colonization of the ears of wheat and barley by *Fusarium poae*. *Can. J. Plant Pathology* 5:107-110.

Ueno, Y., 1983, *Trichothecenes,* Elsevier, Amsterdam.

Van Asch, M.A.J., Rijkenberg, F.H.J., Coutinho, T.A., 1992, Phytotoxicity of fumonisin B_1, moniliformin and T-2 toxin to corn callus cultures. *Phytopathology* 82:1330-1332.

Vesonder, R., Ceigler, A., and Jensen, A.H., 1973, Isolation of the emetic principle from *Fusarium graminearum*-infected corn. *Applied Microbiology* 26:1008-1010.

Wang, Y.-Z., and Miller, J.D., 1988, Effects of *Fusarium graminearum* metabolites on wheat tissue in relation to Fusarium head blight resistance. *J. Phytopathology.* 122: 118-125.

BIOLOGY AND ECOLOGY OF TOXIGENIC *PENICILLIUM* SPECIES

John I. Pitt

Food Science Australia
North Ryde, NSW 2113
Australia

ABSTRACT

Many *Penicillium* species produce mycotoxins. The importance of these toxic compounds varies widely, and is governed as much by the biology and ecology of the species concerned as by the inherent toxicity of the compounds themselves. For example, *P. citreonigrum* and *P. islandicum* make potent toxins, but as both species are rare in nature, the toxins are not important. Although *P. janthinellum* and *P. simplicissimum* are very widely distributed and make potent toxins, these species are rarely found outside soils so again, the toxins are of little practical importance. The very common *P. crustosum* makes a potent tremorgenic mycotoxin, fortunately, the toxin is only produced at very high water activities. On the other hand, *P. verrucosum*, unknown in the tropics, is widespread in cereals in cold climates. Consequently, ochratoxin A production by this species causes a major toxicosis. The biology and ecology of these and other *Penicillium* mycotoxins will be described in this paper.

INTRODUCTION

The history of mycotoxins and mycotoxicoses started with *Penicillium*. The first reliable account of toxin production by a microfungus in pure culture was that of Alsberg and Black (1913). They reported isolation of *Penicillium puberulum* from mouldy maize in Nebraska, an extract of which was toxic to animals when injected at 200-300 mg kg^{-1} body weight. They called the toxin penicillic acid. Their study aimed to resolve the question of whether common fungi or fungal products could have an injurious effect on animals, and was far ahead of its time. Such direct evidence that common fungi could be toxic was largely ignored.

Miyake et al. (1940) published a study on *Penicillium toxicarium*, a species they isolated and described from yellow rice. It produced a highly toxic metabolite, subsequently named citreoviridin. Perhaps because it was published in wartime, this study also failed to alert the world to the potential or actual danger of the toxicity of common fungi. However, it did provide an explanation for the toxicity of "yellow rice". In 1891, in Japan, Sakaki demonstrated that an ethanol extract from mouldy, unpolished yellow rice was fatal to dogs, rabbits and guinea pigs, with symptoms indicating paralysis of the central nervous system (Ueno and Ueno, 1972). In consequence, the sale of yellow rice was banned in Japan in 1910, and acute cardiac beriberi, the human disease caused by yellow rice, disappeared.

The discovery of penicillin in 1929 gave impetus to a search for other *Penicillium* metabolites with antibiotic properties, and ultimately to the recognition of citrinin, patulin and griseofulvin as "toxic antibiotics" or, later, mycotoxins.

By 2000, the literature on toxigenic Penicillia has become quite vast. No fewer than 85 *Penicillium* species were listed as toxigenic by Cole and Cox (1981), and more have been implicated since. This literature has accumulated in a rather random fashion, with emphasis in the majority of papers on chemistry or toxicology rather than on mycology. The impression gained from the literature is that toxin production by Penicillia lacks species specificity, i.e. most toxins are produced by a variety of species. This commonly accepted viewpoint is largely inaccurate. For example, citrinin production has been reported from at least 22 species (Leistner and Pitt, 1977; Cole and Cox, 1981; Frisvad, 1987), but the true number is less than five. Explanations include the fact that until recently the taxonomy of toxigenic *Penicillium* species has been uncertain, and that many literature reports on toxigenicity of particular species were based on misidentifications.

Recently, *Penicillium* taxonomy has profoundly changed (Pitt and Samson, 1990), and the names of *Penicillium* species have largely been stabilised (Pitt and Samson, 1993). Also more definitive lists of species-mycotoxin relations in *Penicillium* have been published (Frisvad and Filtenborg, 1989; Pitt and Leistner, 1991; Svendsen and Frisvad, 1994).

BIOLOGY AND ECOLOGY OF *PENICILLIUM* SPECIES AND MYCOTOXINS

Ecologically, *Penicillium* species are mostly saprophytes: the majority of species show optimal growth at moderate to low temperatures and are capable of growth at water activities (a_w) below 0.9. Many species are soil inhabitants; some, with their natural habitats in decaying vegetation, are important biodeterioative agents; and some specialise in colonising food commodities, especially grains and products derived from them. A few have become capable of invading plants or animals, but none is obligately parasitic. A summary of the overall biology and ecology of the genus is given in Table 1.

Penicillium species produce an exceptionally diverse range of chemical compounds, and a number of these compounds have been shown in laboratory experiments to be moderately to highly potent toxins. However, the importance of particular *Penicillium* species as producers of mycotoxins depends as much on the ecology of the species as on the potency of the toxin produced. An often overlooked criterion is that for a mycotoxin to be important, the species producing it must grow in foods or feeds, not just occur as contaminant spores. The ability of the particular fungus to invade plants before harvest (when temperatures are usually favourable and water activities near optimal), or to grow during drying, or to invade foods or feeds in storage, must be considered in any discussion of the significance of a particular toxin. So the topic of *Penicillium* mycotoxins is logically approached from an understanding of the biology and ecology of the species that make them.

The evolution and taxonomy of mycotoxin formation by *Penicillium* species is complex. The majority of species produce a single toxin in a well-defined one-to-one relationship. However, some species produce more than one toxin: for example, *P. canescens, P. expansum* and *P. crustosum* make two, *P. griseofulvum* and *P. viridicatum* make three, while *P. islandicum* and *P. simplicissimum* make four. Some species use a single biochemical pathway to produce two related toxins, e.g., cyclochlorotine and islanditoxin, viomellein and xanthomegnin, fumitremorgin B and verruculogen. Others, e.g. *P. crustosum* and *P. simplicissimum*, utilise two or more pathways to make a toxin. Some toxin molecules are very simple, e.g. citrinin and patulin, while others, especially the tremorgenic penitrems and janthitrems, are very complex, with years of work needed to elucidate their structures.

Table 1. Overall biology and ecology of *Penicillium* species and related teleomorphs

Genus or subgenus	Common substrates	Minimum water activities for growth	Temperature range for growth	Mycotoxin production
Eupenicillium	Soil, a few in foods	Few studies; most >0.85; a few xerophiles	Variable; most grow well at 37°C	Rare
Talaromyces	Soil	Not known; no species likely to be <0.85	Min. > 5°C Max. most >37°C	Very rare
Subgenus Aspergilloides	Soil; some common species foods	Some species xerophilic	Common species psychrotrophic; some grow >37°C	Uncommon
Subgenus Furcatum Section Furcatum	Soil, decaying vegetation and foods	Species often xerophilic	Usually psychrotrophic; few grow >37°C	Some species toxigenic
Subgenus Furcatum Section Divaricatum	Soil	Some species xerophilic	Usually psychrotrophic; few grow >37°C	Half of common species toxigenic
Subgenus Penicillium	Foods	Most species xerophilic	Psychrotrophic; growth at 37°C exceptional	Most species produce one or more mycotoxins
Subgenus Biverticillium	Soil, decaying vegetation and a few in foods	No species likely to be <0.85	Min. most > 5°C Max. most 37°C or a little higher	Uncommon

From another viewpoint, some toxins are made by more than a single species, e.g., penitrem A and patulin by four species, griseofulvin by at least five, and cyclopiazonic acid by seven or more. In the case of cyclopiazonic acid, all seven species listed here (Table 2) are classified in a single section of a single subgenus. In contrast, the species producing griseofulvin, viridicatumtoxin, and penitrem are each classified in two distinct subgenera. It is difficult to comprehend the evolutionary developments which led to the production of penitrem A by the closely related species *P. crustosum* and *P. glandicola*, on one hand, and *P. janczewskii* and *P. canescens*, on the other, and not by other species related to either pair. Further evolutionary complexity is indicated by the fact that some toxins, e.g. ochratoxin A and cyclopiazonic acid, are also made by a few *Aspergillus* species. One evolutionary point is clear: toxins made by species in subgenus *Biverticillium* are quite distinct from those in the other subgenera.

In this paper, the biology and ecology of toxigenic *Penicillium* species is addressed by looking in turn at each subgenus, and the teleomorph genera associated with *Penicillium*.

TOXIGENIC SPECIES IN *PENICILLIUM* SUBGENUS *PENICILLIUM*

As noted in Table 1, a high proportion of species in *Penicillium* subgenus *Penicillium* are capable of producing one or more mycotoxins. The species with positive links to compounds of definite toxicity are shown in Table 2. An indication of toxicity is also given: to provide some comparison, toxicities in standard animals and by standard routes of administration have been given where possible. The table is conservative, and other links may exist. Notes on these species and toxins follow.

Table 2. Toxigenic species in *Penicillium* subgenus *Penicillium* and associated toxins [1]

Species	Toxin	Toxicity, LD_{50}
P. aethiopicum	Griseofulvin Viridicatumtoxin	Rats, 500 mg kg^{-1} i.p. Mice, 70 mg kg^{-1} i.p.
P. aurantiogriseum	Penicillic acid Viomellein Xanthomegnin	Mice, 250 mg kg^{-1} i.v.; rats, 600 mg kg^{-1} oral Mice, 450 mg kg^{-1} diet, illness only Mice, 450 mg kg^{-1} diet
P. brevicompactum	Mycophenolic acid	Rats, 700 mg kg^{-1} oral
P. camemberti	Cyclopiazonic acid	Rats, 2.3 mg kg^{-1} i.p.; male rats, 36 mg kg^{-1} oral; female rats, 63 mg kg^{-1} oral
P. chrysogenum	Cyclopiazonic acid (occasional isolates)	
P. commune	Cyclopiazonic acid	
P. coprophilum	Griseofulvin	
P. crustosum	Penitrem A Cyclopiazonic acid	Mice, 1 mg kg^{-1} i.p.
P. fennelliae	Penicillic acid	Mice, 250 mg kg^{-1} i.v.; mice, 600 mg kg^{-1} oral
P. expansum	Patulin Citrinin	Mice, 5 mg kg^{-1} i.p.; mice, 35 mg kg^{-1} oral Mice, 35 mg kg^{-1} i.p.; mice,110 mg kg^{-1} oral
P. glandicola	Penitrem A	
P. griseofulvum	Cyclopiazonic acid, griseofulvin, patulin	
P. hirsutum	Cyclopiazonic acid	
P. roqueforti var. *roqueforti*	PR toxin	Mice, 6 mg kg^{-1} i.p., rats, 115 mg kg^{-1} oral
P. roqueforti var. *carneum*	Patulin, penicillic acid	
P. verrucosum	Ochratoxin A Citrinin	Dogs, 0.2 mg kg^{-1} oral; mice, 50 mg kg^{-1} oral
P. viridicatum	Cyclopiazonic acid, viomellein, xathomegnin	
P. vulpinum	Patulin	

[1] From Frisvad and Filtenborg (1989); Pitt and Leistner (1991), Pitt and Hocking (1997)

Penicillium verrucosum and Ochratoxin A

On the world scale, five mycotoxins or groups of mycotoxins are considered to be important in human health (GASGA,1993; Miller 1996). These are aflatoxins, ochratoxin A, fumonisins, trichothecenes and zearalenone. Ochratoxin A is the only toxin on this list produced by a *Penicillium* species, and by only a single species, *P. verrucosum*. This toxin is also produced by several *Aspergillus* species, notably *A. ochraceus* and *A. carbonarius* (Ciegler, 1972; Téren et al., 1996). These three major producing species have quite different ecologies, making it relatively easy to determine which species is responsible for ochratoxin A formation in a particular food or geographical location. In brief, *A. ochraceus* and closely related species grow at low water activities, and at moderate temperatures. They are mostly associated with dried and stored foods, especially cereals. Although many texts describe *A. ochraceus* as the main producer of ochratoxin A, production by this and related species has not often been reported, and their importance appears to have been overstated. The second producing *Aspergillus* species, *A. carbonarius* (and the closely related *A. niger* which produces toxin more rarely) grows well at high temperatures and produces pigmented hyphae and spores, making it resistant to UV light. Consequently this species is commonly found in grapes and similar fruit that mature in sunlight and at high temperatures.

The formation of ochratoxin A (and citrinin) in Danish cereals and pig meats was reported more than 25 years ago (Krogh et al., 1973), and the fungus responsible was believed to be *Penicillium viridicatum*, a species earlier reported to produce this toxin (van Walbeek et al., 1969). After some confusion, it became clear that isolates regarded as *P. viridicatum* but producing ochratoxin and citrinin were more correctly classified in a separate species, *P. verrucosum* (Pitt, 1987; Frisvad and Filtenborg, 1989).

Ochratoxin A is derived from isocoumarin linked to phenylalanine. The International Agency for Research on Cancer has classified this compound as a possible human carcinogen (Group 2B), based on sufficient evidence of carcinogenicity in experimental animal studies and inadequate evidence in humans (IARC, 1994). The target organ(s) of toxicity in all mammalian species tested are the kidneys, in which lesions can be produced by both acute and chronic exposure (Harwig et al., 1983).

Biology and Ecology. *Penicillium verrucosum* is a slowly growing species with an unusually low optimum temperature for growth, near 20°C. Growth occurs from 0-31°C, and down to 0.80 a_w (Pitt and Hocking, 1997). One consequence is that *P. verrucosum* appears to be confined to cool temperate zones: it is uncommon, indeed almost unknown, in the warm temperate or tropical zones.

Ochratoxin A is produced over the whole temperature range for growth of *P. verrucosum*, and down to 0.86 a_w (Northolt et al., 1979). The optimum conditions on a bread dough were reported to be 0.92 a_w at pH 5.6 (Patterson and Damoglou, 1986). Higher amounts of ochratoxin A were produced on wheat than on other substrates including maize, peanuts, rapeseed and soybeans (Madhyastha et al., 1990).

The main food habitat for *P. verrucosum* appears to be cereals grown in the cool temperate zones, ranging across Northern and Central Europe and Canada. In an extensive survey on 70 Danish barley samples from farms where pigs were suffering from nephritis, 67 contained high populations of *P. verrucosum* and 66 contained ochratoxin A (Frisvad, 1986; Frisvad and Vuif, 1986). It seems certain that growth of *P. verrucosum* in cereals is the major source of ochratoxin A in northern Europe and other cool temperate zone areas.

The occurrence of *P. verrucosum* in European cereals has two consequences: ochratoxin A is present in European bread and flour-based foods, and in the meat of animals that eat cereals as a major dietary component. In consequence, ochratoxin A has been found to be widespread in Europeans' blood (Breitholtz et al., 1991; Hald, 1991) and breast milk (Miraglia et al., 1996). The implications of this finding for human health in Europe should not be underestimated.

Ochratoxin A has been implicated in the etiology of Balkan Endemic Nephropathy, a kidney disease which occurs in parts of Bulgaria, Roumania and the former Yugoslavia. It has been sufficiently serious to result in the closing down of villages and moving of all surviving residents to other locations. The symptoms in people are sufficiently similar to those of ochratoxin A toxicity in animals that a role for ochratoxin A in this disease has been postulated on a number of occasions (Krogh, 1974; Pavlović et al., 1979; Austwick, 1981). However, no conclusive epidemiological evidence has emerged to support this hypothesis, and its relevance to studies on ochratoxin A toxicity remains doubtful (Mantle and McHugh, 1993). Other toxins or contributing factors are probably involved (Tatu et al., 1998).

Penicillium crustosum and Penitrem A

Penitrem A is a powerful neurotoxin, probably the most toxic compound produced by any species in *Penicillium* subgenus *Penicillium*. It is a complicated molecule, a polycyclic ether with 10 rings. It has a well-documented history of causing death and debility to farm and domestic animals (Wilson et al., 1968; Ciegler et al., 1976; Arp and Richard, 1979; Hocking et al., 1988), but the effect on humans is much less well understood (Cole et al., 1983). After much early confusion about species names, it is now clear that *Penicillium crustosum* is the major producer of penitrem A (Pitt, 1979; El-Banna et al., 1987; Frisvad and Filtenborg, 1989).

Penicillium crustosum is capable of growth from ca -2 - 30°C, with an optimum near 25°C. The pH limits for growth are from less than 2.2 to greater than 10.0, optimally 4.5 - 9.0 (Pitt and Hocking, 1997). *P. crustosum* has been responsible for spoilage of maize, processed meats, cheese, biscuits, cakes and fruit juices (Pitt and Hocking, 1997), and has been isolated as a weak pathogen from citrus fruits and melons (Snowdon, 1990) as well as apples and pears (Barkai-Golan, 1974; Sanderson and Spotts, 1995).

Nearly all isolates of *P. crustosum* produce penitrem A at high levels, so the presence of this species in foods (or feeds) is a warning signal (Pitt, 1979; El-Banna et al., 1987). Penitrem A is produced only at high moisture levels, above about 0.92 a_w, with an optimum around 0.995 a_w (ICMSF, 1996). This probably explains the relatively low number of reports of poisoning from a very toxic compound produced by a very common fungus.

Penicillium expansum and Patulin

Penicillium expansum is an important producer of patulin (and citrinin)(Harwig et al., 1973; Ciegler et al., 1977). Patulin is a lactone and was once considered to be a relatively important mycotoxin (Stott and Bullerman, 1975), but early reports of carcinogenicity (Dickens and Jones, 1961) have not been substantiated. Its significance is that *P. expansum* produces patulin as it rots apples and pears, and it therefore occurs in a product, apple juice, which is widely consumed, by children as well as adults. In consequence, a number of countries have set an upper limit of 50 μg kg^{-1} for patulin in apple juice and other apple products (van Egmond, 1989). Poor quality control, i.e. the use of rotting fruit in juice or cider manufacture, can result in unacceptable concentrations of patulin in juice, which may exceed 1000 μg kg^{-1} (Lindroth and Niskanen, 1978; Wheeler et al., 1987; Burda, 1992). Removal of rots by water flumes or, better, by high pressure water jets before apples are crushed, provides effective control.

Given that patulin appears to lack chronic toxic effects in humans, low levels in juices (less than 50 μg kg^{-1}) are of little concern. Patulin is certainly important, however, as an indicator of the use of poor quality raw materials in juice manufacture.

P. expansum has its major habitat as a destructive pathogen of pome fruits, but it is in reality a broad spectrum pathogen on fruits, including avocados, grapes, tomatoes and mangoes (Snowdon, 1990). Growth occurs from ca –3-35°C with an optimum near 25°C, and down to 0.82 a_w (Pitt and Hocking, 1997). Patulin can be produced over the range 0-25°C, but not at 31°C, the optimum being at 25°C. The minimum a_w for patulin production by *Penicillium expansum* is 0.95 at 25°C. Patulin is produced over the narrow pH range of 3.2-3.8 in apple juice. A modified atmosphere of 3% CO_2 and 2% O_2 completely inhibited patulin production at 25°C, but production occurred in atmospheres of 2% CO_2 and 10 or 20% O_2. Patulin is quite stable in apple juice during storage; pasteurisation at 90°C for 10 seconds caused less than 20% reduction (Pitt and Hocking, 1997).

Cyclopiazonic Acid Producers

As noted above, cyclopiazonic acid is produced by at least seven species of *Penicillium*, all classified in subgenus *Penicillium* (Table 2). The most interesting of these from the point of view of human health is *P. camemberti*, which is used in the production of Camembert and similar cheeses. Screening of *P. camemberti* isolates for cyclopiazonic acid production failed to identify toxin free isolates (Leistner and Eckardt, 1979). Although most observers agreed that the toxin was not produced in cheese (e.g. Orth, 1981; Schoch et al., 1984), not all were in agreement (e.g. Le Bars, 1979: Scott, 1981). Cyclopiazonic acid is a toxic indole-tetramic acid causing liver and kidney damage in animals (Dorner et al., 1985). A role in "Turkey X" disease along with aflatoxin appears likely (Cole, 1986). Although no direct evidence of toxicity to humans has been reported, absence of toxin in cheeses cannot be taken for granted. As searches for naturally nontoxigenic stains have been unsuccessful (Leistner, 1990), work on development of mutants unable to produce cyclopiazonic acid continues (Leistner et al., 1991).

P. camemberti grows rapidly at refrigeration temperatures, and up to about 35°C. Cyclopiazonic acid is produced by *Penicillium camemberti* in cheese after 5 days at 25°C, but not during normal storage at refrigeration temperatures (Still et al., 1978).

TOXIGENIC SPECIES IN *PENICILLIUM* SUBGENUS *FURCATUM*

A number of species in *Penicillium* subgenus *Furcatum* produce mycotoxins (Table 3). However, only *P. citrinum* and *P. oxalicum* are of common occurrence in foods. The other species listed in Table 3 are all found either entirely associated with soil or with soil and decaying vegetation. The presence of these species in foods is almost always adventitious, and due to contamination with spores from soil. The possibility exists of sufficient growth to cause mycotoxin production on a rare occasion.

Penicillium citrinum and Citrinin

Citrinin is produced by several *Penicillium* species, the best known and most studied of which is *P. citrinum.* Citrinin is moderately toxic, a significant renal toxin to monogastric domestic animals, including pigs (Friis et al., 1969) and dogs (Carlton et al., 1974). Domestic birds are also susceptible: citrinin causes watery diarrhoea, increased food consumption and reduced weight gain due to kidney degeneration in chickens, ducklings and turkeys (Mehdi et al., 1981, 1984). The effect of citrinin on humans remains undocumented. However, kidney damage appears to be a likely result of prolonged ingestion. Citrinin occurs along with patulin when *P. expansum* grows, and sometimes with ochratoxin A production by *P. verrucosum*, but nothing is known about potential synergies.

P. citrinum grows between ca 5°C and 37°C, optimally at 26-30°C. At 25°C, the minimum a_w for growth has been reported as 0.80-0.84. *P. citrinum* grows over the pH range 2-10 (Pitt and Hocking, 1997). This is one of the commonest fungi on earth, occurring in all kinds of foods and feeds, in almost all climates. Citrinin is produced over the range of 15-30°C at least, and optimally at 30°C (ICMSF, 1996) and is probably a low level contaminant in many foods.

Table 3. Toxigenic species in *Penicillium* subgenus *Furcatum* and associated toxins [1]

Species	Mycotoxin	Toxicity, LD_{50}
P. canescens	Penitrem A, griseofulvin	
P. citrinum	Citrinin	Mice, 35 mg kg^{-1} i.p.; mice, 110 mg kg^{-1} oral
P. janczewskii	Penitrem A, griseofulvin	
P. janthinellum	Janthitrems	Not available
P. miczynskii	Citreoviridin	Mice, 7.5 mg kg^{-1} i.p.; mice 20 mg kg^{-1} oral
P. oxalicum	Secalonic acid D	Mice, 42 mg kg^{-1} i.p.
P. paxilli	Paxilline Verruculogen	Mice, 35-225 mg kg^{-1}, tremors only Mice, 2.4 mg kg^{-1} i.p.; mice, 130 mg kg^{-1} oral
P. raistrickii	Griseofulvin	
P. simplicissimum	Fumitremorgen B Verruculogen, penicillic acid, viridicatumtoxin	Mice, 1-5 mg/mouse i.p.

[1] From Cole and Cox (1981); Frisvad and Filtenborg (1989); Pitt and Leistner (1991); Pitt and Hocking (1997). For other toxicity figures see Table 2.

P. oxalicum and Secalonic acid D

Secalonic acid D is produced as a major metabolite of *Penicillium oxalicum* and has significant animal toxicity (Ciegler et al., 1980). It has been found in nature, in grain dusts, at levels of up to 4.5 mg/kg (Ehrlich et al., 1982). The possibility that such levels can be toxic to grain handlers cannot be ignored. However, the role of secalonic acid D in human or animal disease remains a matter for speculation.

Growth of *Penicillium oxalicum* has been reported from 8-37°C, with an optimum near 30°C. Judged from rapid growth at 37°C, the maximum temperature for growth is in excess of 40°C. The minimum a_w for germination has been reported as 0.86, both in glucose media at 23 and 30°C and in NaCl media at 25°C (Pitt and Hocking, 1997).

As a result of its rapid growth at 37°C, *Penicillium oxalicum* is widespread in tropical commodities and foods, including maize, rice, peanuts, kemiri nuts, soybeans, mung beans, cowpeas and sorghum in Southeast Asia (Pitt et al., 1993, 1994, 1998). Nothing is known about the possible significance of secalonic acid D in any of these commodities.

TOXIGENIC SPECIES IN *PENICILLIUM* SUBGENUS *BIVERTICILLIUM*

A few species in *Penicillium* subgenus *Biverticillium* produce mycotoxins (Table 4). Although rugulosin has been reported as a quite toxic compound, disease due to its occurrence has not been reported and nothing is known of its occurrence in nature. The other toxins are discussed below.

Penicillium islandicum and Associated Toxins

Penicillium islandicum produces four mycotoxins, unique to the species. Cyclochlorotine and islanditoxin are very toxic cyclic peptides which have the same toxic moiety, a pyrrolidine ring with two attached chlorine atoms, and share a number of other physical and chemical properties (Scott, 1977). Luteoskyrin is a dimeric anthraquinone and erythroskyrin a heterocyclic red pigment. Both are liver and kidney toxins.

P. islandicum grows between about 10 and 40°C, down to 0.83 a_w at 31°C and 0.86 a_w at 25°C (Pitt and Hocking, 1997). Nothing is known about the factors influencing toxin production.

The toxic "yellow rice" syndrome, described in Japan last century, has resulted in Japanese scientists taking a particular interest in *Penicillium islandicum*, a species that can cause yellowing of rice. However, the practical importance of the toxins produced by this species remains unclear. *P. islandicum* was not commonly found in recent studies on Southeast Asian foods, including rice (Pitt et al., 1993, 1994, 1998).

Rubratoxins

Originally described as agents of animal disease (Burnside et al., 1957), rubratoxins were extensively studied, and are quite toxic. However, only a very few *Penicillium* isolates have been shown to be producers. Even the species producing these toxins remains in doubt. Rubratoxins appear to be a classic example of toxic compounds made by uncommon isolates which rarely cause disease in humans or animals.

Table 4. Toxigenic species in *Penicillium* subgenus *Biverticillium* and associated toxins [1]

Species	Mycotoxin	Toxicity, LD_{50}
P. islandicum	Cyclochlorotine	Mice, 6.5 mg kg^{-1} oral
	Erythroskyrin	Mice, 60 mg kg^{-1} i.p.
	Islanditoxin	Mice, 3 mg kg^{-1} i.p.
	Luteoskyrin	Mice, 40 mg kg^{-1} i.p.; mice, 220 mg kg^{-1} oral
P. purpurogenum	Rubratoxin A	Mice, 6.6 mg kg^{-1} i.p.
	Rubratoxin B	Chicks, 85 mg kg^{-1} oral
P. rugulosum	Rugulosin	Mice, 83 mg kg^{-1} i.p.
P. variabile	Rugulosin	

[1] From Pitt and Leistner (1991); Pitt and Hocking (1997).

SPECIES IN *PENICILLIUM* SUBGENUS *ASPERGILLOIDES* AND THEIR TOXINS

Only one mycotoxin of consequence is produced by species in *Penicillium* subgenus *Aspergilloides*: citreoviridin. This is produced by *P. citreonigrum*, formerly known as *P. citreoviride*. Curiously enough, it is also produced by *P. miczynskii* (in subgenus *Furcatum*) and *Eupenicillium ochrosalmoneum*. It is not clear whether the path to this toxin arose more than once, or whether these species are more closely related than current taxonomies suggest.

Penicillium citreonigrum and Citreoviridin

Penicillium citreonigrum is the major source of citreoviridin (El-Banna et al., 1987), the cause of the Oriental disease known in Japan as acute cardiac beriberi. Recognised for the past three centuries (Ueno and Ueno, 1972), this disease frequently affected young healthy adults, and death could occur within a few days. The work of Sakaki in the 1890s (Ueno and Ueno, 1972) implicated mouldy "yellow rice" as a probable cause, and led to a ban on the sale of yellow rice in Japan in 1910. The disease in Japan is now only of historical interest. Uraguchi

(1969) and Ueno and Ueno (1972) showed that acute cardiac beriberi was a mycotoxicosis due to the growth of *P. citreonigrum* in rice.

P. citreonigrum grows from less than 5°C to ca 37°C, and is probably a xerophile (Pitt and Hocking, 1997). Citreoviridin is a neurotoxin, an unusual molecule consisting of a lactone ring conjugated to a furan ring (Cole and Cox, 1981). It is produced by *P. citreonigrum* from 10-37°C, with a maximum near 20°C (ICMSF, 1996). Although not a commonly isolated species, *Penicillium citreonigrum* is widely distributed. It has been reported from rice and other cereals from time to time, but infrequently from other foods (Pitt and Hocking, 1997).

SPECIES IN *EUPENICILLIUM* AND *TALAROMYCES* AND THEIR TOXINS

Only a single mycotoxin of importance is produced by species classified in *Eupenicillium*. Citreoviridin is produced by *E. ochrosalmoneum*, and this toxin has been found naturally occurring in US maize, at levels up to 2.8 mg kg^{-1} (Wicklow and Cole, 1984; Wicklow et al., 1988). The importance of these reports is difficult to assess in the absence of further data.

Significant mycotoxin production has rarely been reported from *Talaromyces* species. Rugulosin is produced by *T. wortmannii* and wortmannin by an atypical isolate of *T. flavus* (Frisvad et al., 1990).

REFERENCES

Alsberg, C.L., and Black, O.F., 1913, Contributions to the study of maize deterioration: biochemical and toxicological investigations of *Penicillium puberulum* and *Penicillium stoloniferum*, *Bull. Bur. Anim. Ind. U.S. Dep. Agric.* 270: 1-47.

Arp, L.H., and Richard, J.L., 1979, Intoxication of dogs with the mycotoxin penitrem A, *J. Am. Vet. Med. Assoc.* 175: 565-566.

Austwick, P.K.C., 1981, Balkan nephropathy, *Practitioner* 225: 1031-1038.

Barkai-Golan, R., 1974, Species of *Penicillium* causing decay of stored fruits and vegetables in Israel, *Mycopathol. Mycol. Appl.* 54: 141-145.

Breitholtz, A., Olsen, M., Dahlbäck, A., and Hult, K., 1991, Plasma ochratoxin A levels in three Swedish populations surveyed using an ion-pair HPLC technique, *Food Addit. Contam.* 8: 183-92.

Burda, K., 1992, Incidence of patulin in apple, pear, and mixed fruit products marketed in New South Wales, *J. Food Prot.* 55: 796-798.

Burnside, J.E., Sippel, W.L., Forgacs, J., Carll, W.L., Atwood, M.B., and Doel, E.R., 1957, A disease in swine and cattle caused by eating moldy corn. II. Experimental production with pure cultures of moulds, *Am. J. Vet. Res.* 18: 817-824.

Carlton, W.W., Sansing, G., Szczech, G.M., and Tuite, J., 1974, Citrinin mycotoxicosis in beagle dogs, *Food Cosmet. Toxicol.* 12: 479-490.

Ciegler, A., 1972, Bioproduction of ochratoxin A and penicillic acid by members of the *Aspergillus ochraceus* group, *Can. J. Microbiol.* 18: 631-636.

Ciegler, A., Vesonder, R.F., and Cole, R.J., 1976, Tremorgenic mycotoxins, *Adv. Chem.* 149: 163-177.

Ciegler, A., Vesonder, R.F., and Jackson, L.K., 1977, Production and biological activity of patulin and citrinin from *Penicillium expansum*, *Appl. Environ. Microbiol.* 33: 1004-1006.

Ciegler, A., Hayes, A.W., and Vesonder, R.F., 1980, Production and biological activity of secalonic acid D, *Appl. Environ. Microbiol.* 39: 285-287.

Cole, R.J., 1986, Etiology of Turkey "X" disease in retrospect: a case for the involvement of cyclopiazonic acid, *Mycotoxin Res.* 2: 3-7.

Cole, R.J., and Cox, R.H., 1981, *Handbook of Toxic Fungal Metabolites,* Academic Press, New York.

Cole, R.J., Dorner, J.W., Cox, R.H., and Raymond, L.W., 1983, Two classes of alkaloid mycotoxins produced by *Penicillium crustosum* Thom isolated from contaminated beer, *J. Agric. Food Chem.* 31: 655-657.

Dickens, F., and Jones, H.E.H., 1961, Carcinogenic activity of a series of reactive lactones and related substances, *Br. J. Cancer* 15: 85-100.

Dorner, J., Cole, R.H., and Lomax, L.G., 1985, The toxicity of cyclopiazonic acid, in: *Trichothecenes and Other Mycotoxins*, J. Lacey, ed., John Wiley, Chichester, UK. pp. 529-535.

Ehrlich, K.C., Lee, L.S., Ciegler, A., and Palmgren, M.S., 1982, Secalonic acid D: natural contaminant of corn dust, *Appl. Environ. Microbiol.* 44: 1007-1008.

El-Banna, A.A., Pitt, J.I., and Leistner, L., 1987, Production of mycotoxins by *Penicillium* species, *System. Appl. Microbiol.* 10: 42-46.
Friis, P., Hasselager, E., and Krogh, P., 1969, Isolation of citrinin and oxalic acid from *Penicillium viridicatum* Westling and their nephrotoxicity in rats and pigs, *Acta Pathol. Microbiol. Scand.* 77: 559-560.
Frisvad, J.C., 1986, Taxonomic approaches to mycotoxin identification, in: *Modern Methods in the Analysis and Structural Elucidation of Mycotoxins,* R.J. Cole, ed., Academic Press, Orlando, Florida. pp. 415-457.
Frisvad, J.C., 1987, High performance liquid chromatographic determination of profiles of mycotoxins and other secondary metabolites, *J. Chromatogr.* 392: 333-347.
Frisvad, J.C., and Filtenborg, O., 1989, Terverticillate Penicillia: chemotaxonomy and mycotoxin production, *Mycologia* 81: 837-861.
Frisvad, J.C., and Viuf, B.T., 1986, Comparison of direct and dilution plating for detection of *Penicillium viridicatum* in barley containing ochratoxin, in: *Methods for the Mycological Examination of Food*, A.D. King, J.I. Pitt, L.R. Beuchat and J.E.L. Corry, eds, Plenum Press, New York. pp. 45-47.
Frisvad, J.C., Filtenborg, O., Samson, R.A., and Stolk, A.C., 1990, Chemotaxonomy of the genus *Talaromyces*, *Antonie van Leeuwenhoek* 57: 179-189.
GASGA (Group for Assistance on Systems relating to Grain After-harvest), 1993, *Mycotoxins in Food and Feedstuffs. Report of GASGA Working Party on Fungi and Mycotoxins in Asian Food and Feedstuffs,* Australian Centre for International Agricultural Research, Canberra, ACT.
Hald, B., 1991, Ochratoxin A in human blood in European countries, in: *Mycotoxins, Endemic Nephropathy and Urinary Tract Tumours*, F. Castegnaro, R. Pleština, G. Dirheimer, I.N. Chernozemsky, and H. Bartsch, eds, International Agency for Research on Cancer, Lyon, France. pp. 159-164.
Harwig, J., Chen, Y.-K., Kennedy, B.P.C., and Scott, P.M., 1973, Occurrence of patulin and patulin-producing strains of *Penicillium expansum* in natural rots of apple in Canada, *Can. Inst. Food Sci. Technol. J.* 6: 22-25.
Harwig, J., Kuiper-Goodman, T., and Scott, P.M., 1983, Microbial food toxicants: ochratoxins, in: *Handbook of Foodborne Diseases of Biological Origin, M.* Reichcigl, ed., CRC Press, Boca Raton, Florida. pp. 193-238.
Hocking, A.D., Holds, K., and Tobin, N.F., 1988, Intoxication by tremorgenic mycotoxin (penitrem A) in a dog, *Aust. Vet. J.* 65: 82-85.
Krogh, P. ,1974, Mycotoxic porcine nephropathy: a possible model for Balkan (endemic) nephropathy, in: *Endemic nephropathy. Proceedings of the 2nd International Symposium on Endemic Nephropathy*, A. Puchlev, ed., Bulgarian Academy of Sciences, Sofia, Bulgaria. pp. 266-270.
Krogh, P., Hald, B., and Pedersen, E.J., 1973, Occurrence of ochratoxin and citrinin in cereals associated with mycotoxic porcine nephropathy, *Acta Pathol. Microbiol. Scand., Sect. B*, 81: 689-695.
IARC (International Agency for Research on Cancer), 1994, Ochratoxin A, in *IARC Monographs on the Evaluation of Carcinogenic Risks to Humans, Vol. 56. Some Naturally Occurring Substances: Food Items and Constitutents, Heterocyclic Aromatic Amines and Mycotoxins*. International Agency for Research on Cancer, Lyon, France. pp. 489-521.
ICMSF (International Commission on Microbiological Specifications for Foods), 1996, Toxigenic fungi: *Penicillium*, in: *Microorganisms in Foods. 5. Characteristics of Food Pathogens*, Blackie Academic and Professional, London. pp. 397-413.
Le Bars, J., 1979, Cyclopiazonic acid production by *Penicillium camemberti* Thom and natural occurrence of this mycotoxin in cheese, *Appl. Environ. Microbiol.* 38: 1052-1055.
Leistner, L., 1990, Mould-fermented foods: recent developments, *Food Biotechnol.* 4: 433-441.
Leistner, L., and Eckardt, C., 1979, Vorkommen toxinogener Penicillien bei Fleischerzeugnissen, *Fleischwirtschaft* 59: 1892-1896.
Leistner, L., and Pitt, J.I., 1977, Miscellaneous *Penicillium* toxins, in: *Mycotoxins in Human and Animal Health*, J.V. Rodricks, C.W. Hesseltine and M.A. Mehlman, eds, Pathotox Publishers, Park Forest South, Illinois. pp. 639-653.
Leistner, L., Geisen, R., and Böckle, B., 1991, Possibilities of and limits to genetic change in starter cultures and protective cultures, *Fleischwirtschaft* 71: 682-683.
Lindroth, S., and Niskanen, A., 1978, Comparison of potential patulin hazard in home-made and commercial apple products, *J. Food Sci.* 43: 446-448.
Madhyastha, S.M., Marquardt, R.R., Frohlich, A.A., Platford, G., and Abramson, D., 1990, Effects of different cereal and oilseed substrates on the growth and production of toxins by *Aspergillus alutaceus* and *Penicillium verrucosum*, *J. Agric. Food Chem.* 38: 1506-1510.
Mantle, P.G., and McHugh, K.M., 1993, Nephrotoxic fungi in foods from nephropathy households in Bulgaria, *Mycol. Res.* 97: 205-212.
Mehdi, N.A.Q., Carlton, W.W., and Tuite, J., 1981, Citrinin mycotoxicosis in broiler chickens, *Food Cosmet. Toxicol.* 19: 723-733.
Mehdi, N.A.Q., Carlton, W.W., and Tuite, J., 1984, Mycotoxicoses produced in ducklings and turkeys by dietary and multiple doses of citrinin, *Avian Pathol.* 13: 37-50.

Miller, J.D., 1996, Food-borne natural carcinogens: issues and priorities, *Afr. Newslett. Occup. Health Safety* 6(Suppl. 1): S22-8.

Miraglia, M., Brera, C., and Colatosti, M., 1996, Application of biomarkers to assessment of risk to human health from exposure to mycotoxins, *Microchem. J.* 54: 472-7.

Miyake, I., Naito, H., and Sumeda, H., 1940, (Japanese title), *Rept Res. Inst. Rice Improvement* 1: 1.

Northolt, M.D., van Egmond, H.P., and Paulsch, W.E., 1979, Ochratoxin A production by some fungal species in relation to water activity and temperature, *J. Food Prot.* 42: 485-490.

Orth, R., 1981, Mykotoxine von Pilzen der Käseherstellung, in: *Mykotoxine in Lebensmitteln*, J. Reiss, ed., Gustaf Fisher Verla, Stuttgart, Germany. pp. 273-296.

Patterson, M., and Damoglou, A.P., 1986, The effect of water activity and pH on the production of mycotoxins by fungi growing on a bread analogue, *Lett. Appl. Microbiol.* 3: 123-125.

Pavlovi*f*, M., Plestina, R., and Krogh, P., 1979, Ochratoxin A contamination of foodstuffs in an area with Balkan (endemic) nephropathy, *Acta Pathol. Microbiol. Scand., Sect. B*, 87: 243-6.

Pitt, J.I., 1979, *Penicillium crustosum* and *P. simplicissimum*, the correct names for two common species producing tremorgenic mycotoxins, *Mycologia* 71: 1166-1177.

Pitt, J.I., 1987, *Penicillium viridicatum*, *Penicillium verrucosum* and production of ochratoxin A, *Appl. Environ. Microbiol.* 53: 266-269.

Pitt, J.I., and Hocking, A.D., 1997, *Fungi and Food Spoilage*. 2nd ed. London: Blackie Academic and Professional. 592 pp.

Pitt, J.I., and Leistner, L., 1991, Toxigenic *Penicillium* species, in: *Mycotoxins and Animal Foods*, J.E. Smith and R.S. Henderson, eds, CRC Press, Boca Raton, Florida. pp. 91-99.

Pitt, J.I., and Samson, R.A., 1990, Approaches to *Penicillium* and *Aspergillus* systematics, *Stud. Mycol., Baarn* 32: 77-90.

Pitt, J.I., and Samson, R.A., 1993, Species names in current use in the *Trichocomaceae* (Fungi, Eurotiales), in: *Names in Current Use in the Families* Trichocomaceae, Cladoniaceae, Pinaceae, *and* Lemnaceae, W. Greuter, ed., Koeltz Scientific Books, Königstein, Germany. *Regnum Vegetabile* 128: 13-57.

Pitt, J.I., Hocking, A.D., Bhudhasamai, K., Miscamble, B.F., Wheeler, K.A., and Tanboon-Ek, P., 1993, The normal mycoflora of commodities from Thailand. 1. Nuts and oilseeds, *Int. J. Food Microbiol.* 20: 211-226.

Pitt, J.I., Hocking, A.D., Bhudhasamai, K., Miscamble, B.F., Wheeler, K.A., and Tanboon-Ek, P., 1994, The normal mycoflora of commodities from Thailand. 2. Beans, rice, small grains and other commodities, *Int. J. Food Microbiol.* 23: 35-53.

Pitt, J.I., Hocking, A.D., Miscamble, B.F., Dharmaputra, O.S., Kuswanto, K.R., Rahayu, E.S., and Sardjono, 1998, The mycoflora of food commodities from Indonesia, *J. Food Mycol.* 1: 41-60.

Sanderson, P.G., and Spotts, R.A., 1995, Postharvest decay of winter pear and apple fruit caused by species of *Penicillium*, *Phytopathology* 85: 103-110.

Schoch, U., Lüthy, J., and Schlatter, C., 1984, Mutagenitätsprüfung industriell verwendeter *Penicillium camemberti*- und *P. roqueforti* -Stämme, *Z. Lebensm.-Unters. Forsch.* 178: 351-355.

Scott, P.M., 1977, *Penicillium* mycotoxins, in: *Mycotoxic Fungi, Mycotoxins, Mycotoxicoses, an Encyclopedic Handbook. Vol. 1. Mycotoxigenic Fungi*, T.D. Wyllie and L.G. Morehouse, eds, Marcel Dekker, New York. pp. 283-356.

Scott, P.M., 1981, Toxins of *Penicillium* species used in cheese manufacture, *J. Food Prot.* 44: 702-710.

Snowdon, A.L., 1990, *A Colour Atlas of Post-harvest Diseases and Disorders of Fruits and Vegetables. 1. General Introduction and Fruits*, Wolfe Scientific, London.

Still, P., Eckardt, C., and Leistner, L., 1978, Bildung von Cyclopazonsaure durch *Penicillium camembertii*-isolate von Kase, *Fleischwirtschaft* 58: 876-878.

Stott, W. T., and Bullerman, L.B., 1975, Patulin: a mycotoxin of potential concern in foods, *J. Milk Food Technol.* 38: 695-705.

Svendsen, A., and Frisvad, J.C., 1994, A chemotaxonomic study of the terverticillate Penicillia based on high performance liquid chromatography of secondary metabolites, *Mycol. Res.* 98: 1317-1328.

Tatu, C.A., Orem, W.H., Finkelman, R.B., and Feder, G.L., 1998, The etiology of Baltic Endemic Nephropathy: still more questions than answers, *Environ. Health Perspect.* 106: 689-700.

Téren, J., Varga, J., Hamari, Z., Rinyu, E., and Kevei, É., 1996, Immunochemical detection of ochratoxin A in black *Aspergillus* strains, *Mycopathologia* 134: 171-176.

Ueno, Y., and Ueno, I., 1972, Isolation and acute toxicity of citreoviridin, a neurotoxic mycotoxin of *Penicillium citreo-viride* Biourge, *Jpn. J. Exp. Med.* 42: 91-105.

Uraguchi, K., 1969, Mycotoxic origin of cardiac beriberi, *J. Stored Prod. Res.* 5: 227-236.

Van Egmond, H.P., 1989, Current situation on regulations for mycotoxins. Overview of tolerances and status of standard methods of sampling and analysis, *Food Addit. Contam.* 6: 139-188.

Van Walbeek, W., Scott, P.M., Harwig, J., and Lawrence, J.W., 1969, *Penicillium viridicatum* Westling: a new source of ochratoxin A, *Can. J. Microbiol.* 15: 1281-1285.

Wheeler, J.L., Harrison, M.A., and Koehler, P.E., 1987, Presence and stability of patulin in pasteurized apple cider, *J. Food Sci.* 52: 479-480.

Wicklow, D.T., and Cole, R.J., 1984, Citreoviridin in standing corn infested by *Eupenicillium ochrosalmoneum*, *Mycologia* 76: 959-961.
Wicklow, D.T., Stubblefield, R.D., Horn, B.W.. and Shotwell, O.L,, 1988, Citreoviridin levels in *Eupenicillium ochrosalmoneum*-infested maize kernels at harvest, *Appl. Environ. Microbiol.* 54: 1098-1098.
Wilson, B.J., Wilson, C.H., and Hayes, A.W., 1968, Tremorgenic toxin from *Penicillium cyclopium* grown on food materials, *Nature (London)* 220: 77-78.

CHEMISTRY AND TOXICOLOGY OF MOLDS ISOLATED FROM WATER-DAMAGED BUILDINGS

Bruce B. Jarvis

Department of Chemistry and Biochemistry
University of Maryland
College Park, MD 20742

ABSTRACT

There is increasing evidence of health risks associated with damp buildings and homes in which high levels of microbes are found. Although concerns have traditionally centered on microbial pathogens and allergenic effects, recent work has suggested that fungi pose the more serious risk. Evidence is accumulating that certain toxigenic molds are particularly a risk for human health through exposure, via inhalation, of fungal spores. Many of these fungi produce toxins (mycotoxins) some of which have been shown to cause animal and human intoxications, usually in an agricultural setting. The fungus, *Stachybotrys chartarum* (*S. atra*) is considered to be one of the more serious threats to people living and working in water-damaged buildings. This mold has a long history of being responsible for animal toxicoses, and in recent years, being associated with infant pulmonary hemosiderosis (bleeding in the lungs) of infants exposed to spores of this fungus in their homes. *S. atra* produces a variety of potent toxins and immunosuppressant agents, including a novel class of diterpenes (atranones) of unusual structure. More research is needed to determine the impact to health resulting from inhalation of toxigenic mold spores.

INTRODUCTION

In the past twenty years, there is increasing recognition that an important factor in the health of people in indoor environments is the dampness of the buildings in which they live and work (Miller, 1995). Furthermore, it is now appreciated that the principal non-pathogenic biologics responsible for the health problems in such buildings are usually fungi rather than bacteria or viruses (Miller, 1992; 1993). Although fungi in this context have traditionally been viewed as allergens (and in unusual circumstances, pathogens), data have accumulated to show that the adverse health effects resulting from inhalation of fungal spores are due to multiple factors. One factor associated with certain fungi is the production of small molecular weight toxins (mycotoxins) by these fungi (Miller, 1995). Traditionally, mycotoxins are held to be

Mycotoxins and Food Safety, edited by Trucksess *et al.*
Kluwer Academic/Plenum Publishers, 2002

important in human and animal health because of their production by toxigenic fungi associated with food and feed (Betina, 1989). However, mycotoxins tend to concentrate in fungal spores (Sorenson et al., 1987: Larsen and Frisvad, 1994), and thus present a potential hazard to those inhaling airborne spores. There are well documented cases of mycotoxicosis resulting from inhalation of toxigenic spores by agriculture workers handling moldy farm material (Autrup et al., 1991; Autrup et al., 1993), but until recently there have been few reports of such toxicoses in an urban setting (Jarvis, 1990).

From an agricultural view, the three most important genera of toxigenic fungi are *Fusarium*, *Aspergillus*, and *Penicillium*. The principal classes of mycotoxins of concern from ingestion are (fungal genera): the trichothecenes and fumonisins (*Fusarium*), aflatoxins (*Aspergillus*), and the ochratoxins (*Aspergillus* and *Penicillium*). There also are numerous other mycotoxins that intermittently cause problems. Some trichothecenes are acutely toxic while others are immunosuppressant and cause feed refusal in livestock. Fumonisins and especially some of the aflatoxins (e.g. B_1) are carcinogenic (Peraica et al., 1999). There is now overwhelming epidemiological evidence that aflatoxin B_1 in the diet contributes significantly to the high incidence of liver cancer in many third world countries. Ochratoxin A is nephrotoxic and a possible cause of urinary tract tumors and Balkan-endemic nephropathy (Peraica et al., 1999).

In an indoor environment, the toxigenic molds of most concern are *Penicillium, Aspergillus*, *Chaetomium*, and *Stachybotrys*. However, the specific species of *Penicillium* and *Aspergillus* in the indoor environments differ from those of an agricultural concern. For agricultural-based mycotoxicoses, *P. verrucosum*, *A. flavus*, and *A. parasiticus* (the latter two are aflatoxin producers) are of most concern; whereas, indoors, other species such as *P. aurantiogriseum, P. brevicompactum,* and *A. versicolor* are the most important toxigenic species (Samson, 1999). Often, the picture is confused by the difficulty of species identification, a situation common in the identification of *Penicillium* and to a lesser extent of *Aspergillus*. Another confounding factor in the evaluation of mold-contaminated buildings is the fact that levels of airborne spores can not be easily assessed by measuring the colony-forming units (CFU's)/m^3 since such measurements will detect only viable spores (and both viable and non-viable spores contain mycotoxins), and even there, the number and type of fungal colonies observed will depend very much upon the media employed among other factors (Miller, 1993; Tsai et al., 1999).

Stachybotrys chartarum (Ehrenb. ex Link) Hughes (*S. atra* Corda) has been a source of focus in recent years due to its association with idiopathic pulmonary hemosiderosis (IPH) in children, several of whom have died (Etzel et al., 1998; Dearborn et al., 1999). However, the *Penicillium* spp. and *A. versicolor* are far more commonly encountered in water-damaged buildings than is *S. chartarum*. The basic reason for this is that the former species grow well on building material with water activities (a_W) in the range of 0.8; whereas, *S. chartarum* requires a_W of 0.9 and higher and >0.96 on building material to grow well (Nielsen et al., 1999). Generally, moisture levels this high are the result of local leaking (e. g. from a broken pipe) and are confined to a small area, but with leaking roofs, water intrusion can be quite extensive. There is nothing special about toxigenic fungi in that like all fungi, their growth properties depend upon substrate, temperature and water activity (Samson, 1999).

SECONDARY METABOLITES OF THE INDOOR MOLDS

Table 1 lists the more common fungi (and their associated mycotoxins) encountered in damp indoor environments. For the most part, the mycotoxins in Table 1 have been found to be produced by these fungi in laboratory cultures, and in only a few cases have these mycotoxins actually been shown to be produced when cultured on building materials (Nielsen et al., 1998a; 1999) or been isolated from mold-contaminated buildings (Andersson et al., 1997; Nielsen et al., 1999). Figure 1 illustrates a number of mycotoxins produced by fungi isolated from indoor environments, although several (e. g. ochratoxin A, deoxynivalenol, and roquefortine C) are far more likely to be associated with mold-contaminated food and feed.

Table 1. Common toxigenic fungi isolated from water-damaged buildings.[1]

Fungus	Mycotoxins
Alternaria alternata	tenuazonic acid, alternatiol, alternatiol monomethyl ether, altertoxins
Aspergillus fumigatus	**gliotoxin**, verrucologen, fumitremorgceusins, fumitoxins, tryptoquivalins
A. niger	naphthopyrones, malformins, nigragillin, orlandin
A. nidulans	**sterigmatocystin**, nidulotoxin
A. ustus	austamide, austdiol, austins, austocystins, kotanin
A. versicolor	**sterigmatocystin**, **5-methoxystermatocystin**, versicolorins
Chaetomium globosum	**chaetoglobosins, chetomin**
Memnoniella echinata	trichodermol, trichodermin, dechlorogriseofulvins, memnobotrins A and B, memnoconol, memnoconone
Penicillium aurantiogriseum	auranthine, penicillic acid, verrucosidin, **nephrotoxic glycopeptides**
P. brevicompactum	mycophenolic acid
P. chrysogenum	roquefortine C, meleagrin, chrysogin
P. polonicum	3-methoxyviridicatin, verrucosidin, verrucofortine
Stachybotrys chartarum	**satratoxins**, **verrucarins**, **roridins**, atranones, dolabellanes stachybotrylactones and lactams, stachybotrydials
Trichoderma harzianum	**alamethicins**, emodin, suzukacillin, trichodermin
Wallemia sebi	walleminols A and B

[1]Taken in part from Samson, 1999. Toxins in boldface are of high potency

The mycotoxins found in indoor air are usually concentrated in the aerosolized spores of the toxigenic fungi (Sorenson, 1999); high levels can also be found in mycelia as well (Nielsen et al., 1998a; Gravesen et al., 1999). In any event, levels of exposure to mycotoxins in airborne particulate matter are unlikely to approach those encountered in animal feed intoxications. However, Creasia et al. (1987 and 1990) reported that T-2 toxin is considerably more toxic when inhaled than when ingested, although others have reported lower toxicities for T-2 inhaled compared with toxicities resulting from subcutaneous exposure (Marrs et al., 1986) or intravenous exposure (Pang et al., 1988). Comparisons of these studies is made difficult by the workers using not only different animals but by employing different toxin administration regimes.

The effects of mycotoxins are multiple, and in particular, the immunosuppressant effects of these toxins on alveolar macrophage function are very pronounced (Sorenson, 1990; Flannigan and Miller, 1994). Studies of the effects of fungal spores on rat pulmonary alveolar macrophage cells have shown variability probably mediated by spore-borne toxins (Sorenson, 1999). Spores of five *Aspergillus, Penicillium spinulosum*, and *Cladosporium cladosporioides* caused the release of leukotriene B4 and increased superoxide anion production and activation of complement. Several species of *Aspergillus* inhibited LPS-stimulated IL-1; whereas, the other species had no effect (Sorenson et al., 1994). Jakab et al. (1994) demonstrated that inhalation exposure of aflatoxin to rats led to persistent reductions in phagocytosis, among other effects. In rats, inhalation exposure to aflatoxin-containing spores of *Aspergillus* has been demonstrated to be an effective route of exposure (Zarba et al., 1992). Damage to clearance mechanisms would affect processing of antigen and lead to accumulation of material in granulomatous matter leading to an immunotoxic effect (Donaldson and Brown, 1990; Driscoll et al., 1990; Holt, 1990). Low exposure to spores of toxigenic *S. chartarum* affects lung surfactant production in lab animals (Mason et al., 1998).

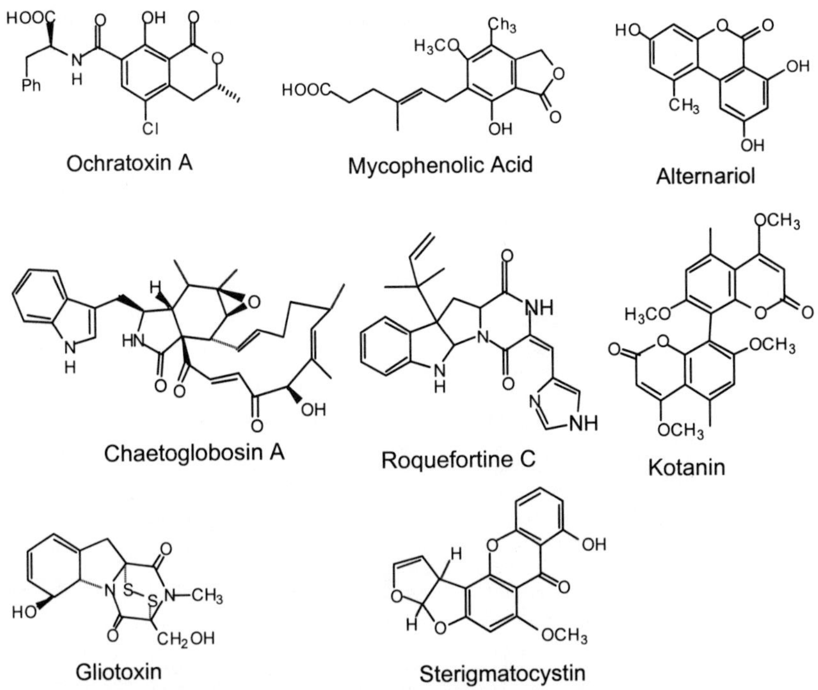

Figure 1. Selected mycotoxins reported produced by fungi isolated from water-damaged buildings.

Toxigenic spores that strongly affect alveolar macrophage function pose a threat to workers handling mycotoxin-contaminated material. Occupational exposures to spores of *Aspergillus flavus*, a producer of the potent carcinogenic aflatoxins have been shown to correlate to increased risk of liver cancer (Autrup et al., 1991; Autrup et al., 1993). A female agricultural worker suffered acute renal failure following exposure to grain dust in an enclosed granary, whose wheat was shown to be contaminated by *Aspergillus ochraceous* and its mycotoxin metabolite, ochratoxin A (see Figure 1). Although ochratoxin A was not demonstrated to be in air samples from the granary, caged experimental animals (guinea pigs and rabbits) experienced acute renal failure when exposed for 8 hours to aerosols generated by their movement on moldy wheat (Di Paolo et al., 1993). In this regard, a recent report of high concentrations (1500 ppb) of ochratoxin A in dust from a home air handling system in which occupants had complained of recurring health problems, is note worthy (Richard et al., 1999). Other reports have indicated that *Stachybotrys chartarum* (Croft et al., 19986; Johanning et al., 1996; Johanning et al., 1998), *A. versicolor*, and several toxigenic species of *Penicillium* are potentially hazardous, especially when the air-handling systems have become heavily contaminated (Flannigan and Miller, 1994).

Epidemiological studies have connected the increased incidence of early births in Norwegian farmers to their exposure to grain-borne toxigenic fungi (Kristensen, 1997). Researchers have shown that Finnish farmers are commonly exposed to relatively high levels of airborne spores (10^3-10^6 CFU/m^3) and that low to moderate (0.00-11 ppm) levels of *Fusarium* toxins (e.g. deoxynivalenol, see Figure 1) are commonly found in the grain, although less commonly in air samples (Lappalainen et al., 1996).

The literature on the growth characteristics of fungi in damp buildings is extensive (Miller, 1992; Miller, 1993; Miller, 1995; Flannigan and Miller, 1994; Samson, 1999).

However, with respect to the production of organic compounds by these fungi, most of the literature has focused on the microbial-generated microbial volatile organic compounds (MVOCs) (Godish, 1996). Although some of the MVOCs might be of health concern from a chronic perspective (e.g. styrene), MVOCs are not considered an acute toxic hazard (although the associated odors can present problems) (Godish, 1996). However, confounding factors such as the co-occurrence of ozone with MVOCs may result in a significant increase in volatile irritants (Wolkoff et al., 1999).

Although a great deal is known about the spectrum of mycotoxins produced by the relevant toxigenic fungi found indoors (Etzel et al., 1998; Dearborn et al., 1999; Neilsen et al., 1999), there are few data available on the presence of the mycotoxins in mold-contaminated buildings. In those few cases where mycotoxins have been found in mold-contaminated buildings (Croft et al., 1986; Johanning et al., 1996; Nielsen et al., 1998a; Anderson et al., 1997; Johanning et al., 1998; Hodson et al., 1998; Flappen et al., 1999), they have been found in bulk samples rather than in air samples. Following from this, there remains the critical question of the exposure of people in these buildings to the airborne toxigenic fungal spores, a problem that merits further discussion (*infra vide*).

As noted above, many studies report the presence of toxigenic fungi in mold-contaminated buildings, but few actually determine the presence of the mycotoxins in the buildings. Nielsen et al. (1999) showed that bulk samples from naturally mold-contaminated building materials from water-damaged domestic residences contained both toxigenic fungi as well as the specific mycotoxins produced by these fungi. Although their data are complicated by these natural cultures being mixtures of two or more fungal species growing together (mixed cultures are the rule rather than the exception under natural conditions), the mycotoxins found are those expected from the fungi observed. Thus, the sterigmatocystin (an IARC 2A carcinogen) and 5-methoxysterigmatocystin found in the bulk samples most certainly came from *A. versicolor* (which in fact produces these mycotoxins when grown on building materials) (Nielsen, 1998b) and the chaetoglobosins from the *Chaetomium* sp. *P. chrysogenum* was commonly observed in these bulk samples, but no mycotoxins were found that could be ascribed to this fungus.

Our recent work has been centered on the chemistry of *Stachybotrys chartarum* and the closely related fungus (Jong and Davis, 1976), *Memnoniella echinata*. The similarities of the metabolites produced by these fungi (Figure 2) supports their being classified as closely related species. Both produce trichothecenes and the phenolic drimanes (e.g. stachybotrylactones/lactams and memnobotrins) (Jarvis et al., 1995; Hinkley et al., 1999a), but to date, we have found griseofulvins to be produced only by *M. echinata* (Jarvis et al., 1996) and the atranones (Hinkley et al., 1999b; Hinkley et al., 2000) to be produced only by *S. chartarum*. Furthermore, both *S. chartarum* and *M. echinata* produce the simple trichothecenes, trichodermol and trichodermin, but *S. chartarum* appears to produce the macrocyclic trichothecenes, e. g. satratoxins and roridins (Figure 2). Initially, we reported that *S. chartarum* occurs as two chemotypes, one that produces trichothecenes and one that produces atranones (Hinkley et al., 1999b). However, we have now examined many *S. chartarum* isolates from the U.S. and Europe and find that about one-third of the isolates produce the macrocyclic trichothecenes, and the remaining isolates produce both the atranones and the simple trichothecenes, trichodermol and trichodermin. Culture extracts from the former chemotypes are considerably more cytotoxic than are culture extracts from the latter chemotypes.

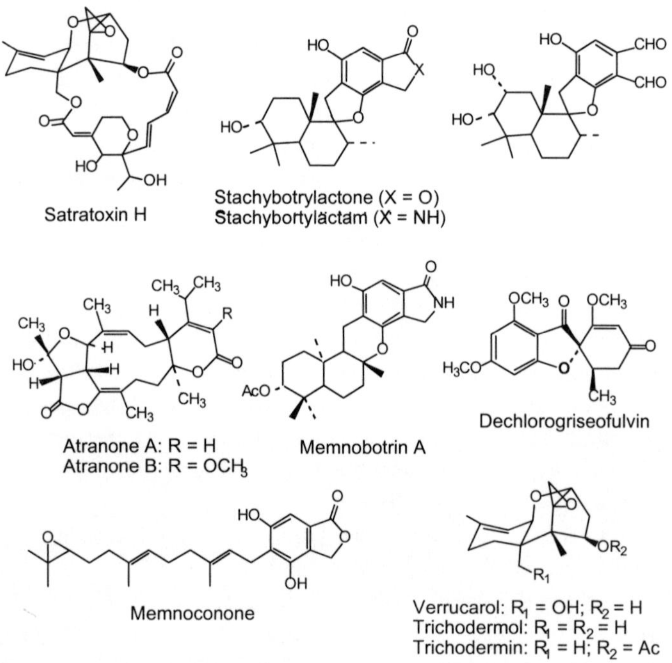

Figure 2. Mycotoxins isolated from *S. chartarum* and *M. echinata*

Many isolates of *S. chartarum* were obtained from both case and control homes in the Cleveland idiopathic pulmonary hemosiderosis (IPH) investigation. Of the original cultures of 16 isolates of *S. chartarum* from the Cleveland homes, six proved to be highly cytotoxic and to produce appreciable levels of macrocyclic trichothecenes (Jarvis et al., 1998). Four other isolates were reported to produce low levels of macrocyclic trichothecenes, but upon reanalysis, employing a more sensitive and reliable method (Jarvis and Hinkley, 1999), only one of these four isolates was confirmed to produce macrocyclic trichothecenes. Thus of these 16 isolates, seven produce macrocyclic trichothecenes and the remaining nine produce atranones (unpublished work); none produces both macrocyclic trichothecenes and atranones. We have examined an additional 20 isolates from the Cleveland study and found similar numbers. Rice cultures of six of these latter isolates produce high levels (70-145 ppm) of macrocyclic trichothecenes and correspondingly, the culture extracts are highly cytotoxic. Extracts of the 14 remaining isolates are $\sim 10^5$ less cytotoxic. Of these 14 isolates, 12 produce the atranones. Most of these latter 14 isolates also produce trichodermin and trichodermol. All *S. chartarum* isolates examined in our laboratory produce the stachybotrylactones/lactams at levels considerably higher than those observed for either the macrocyclic trichothecenes or the atranones.

These data suggest that *S. chartarum* comes in two chemotypes: one that produces the macrocyclic trichothecenes (and is highly cytotoxic) and one that produces the atranones (and is considerably less cytotoxic). It is important to note however, the remarkable fact that both chemotypes can occur together and hence those residents in damp *Stachybotrys*-contaminated buildings can be exposed to both fungal chemotypes. Attempts to correlate these chemotypes with taxonomic characteristics, either by growth patterns or through molecular biological techniques (Vesper et al., 1999) to date have not been successful. Inhalation studies in mice have shown that the spores of toxigenic *S. chartarum* isolates are considerably more destructive to lung tissue than are the spores of the low toxic isolates (Nikulin et al., 1996; Nikulin et al., 1997). Furthermore, extraction of highly toxic spores of *S. chartarum* with methanol prior to instillation in rats renders the spores considerably less damaging to lung tissue of the animals (Rao et al., 2000).

CONCLUSIONS

Human and animal exposures to toxigenic molds occur most often through ingestion of contaminated foods and feed. There is no question that such exposures are of concern from both health and economic perspectives. With respect to human exposures to molds as a result of breathing the air in moldy buildings, epidemiological data indicate that such exposures do adversely affect the people in living and working in such buildings. Although these effects are more than allergenic in nature, the role played by mycotoxins is by no means clear. There are few cases of mycotoxins being detected in these buildings and fewer still, if any, of actual measurements of mycotoxins in air samples. The reports of the association of IPH in the Cleveland infants (Dearborn et al., 1999) is controversial (Fung, 1998; Burge, 1999; Sudakin, 2000; Anon, 2000), but all agree that more data are needed before any definite conclusions can be made with respect to the health effects resulting from inhalation of toxigenic mold spores.

At this time, we have no reliable way to determine human exposure to these indoor molds (Dillon et al., 1999). The problem is extremely complex, and what are needed are biological markers that can be employed to measure human exposure to these molds. The difficulty of this problem can be appreciated by noting the enormous amount of time, energy, and expense that has gone into establishing the carcinogenic properties of alfatoxin B_1, and assessing the role that this mycotoxin has played in human cancers as a result of environmental exposures. A similar thorough evaluation of the effects of the inhalation of toxigenic spores (e.g. those of *Stachybotrys chartarum*) will be considerably more difficult for a variety of reasons, *inter alia* real life exposure is always to more than just one fungal species, co-occurring isolates of the same species may have considerably different toxigenic potential; we know virtually nothing about the metabolic fate of the vast majority of mycotoxins produced by indoor molds and thus have little basis at the moment to suggest metabolic markers for exposure.

REFERENCES

Andersson, M.A., Nikulin, M., Köljalg, U., Andersson, M.C., Rainey, F., Reijula, K., Hintikka, E. L., and Salkinoja-Salonen, M., 1997, Bacteria, molds, and toxins in water-damaged building materials, *Appl. Environ. Microbiol.*, 63:387-393.

Anon., 2000, *MMWR Weekly*, 49:180-184.

Autrup, J.L., Schmidt, J., Seremet, T., and Autrup, H., 1991, Determination of exposure to aflatoxins among Danish workers in animal feed production through the analysis of aflatoxin B_1 adducts to serum albumin. *Scand. J. Work Environ. Health*, 17:436-440.

Autrup, J.L., Schmidt, J., Seremet, T., and Autrup, H., 1993, Exposure to aflatoxin B in animal feed production plant workers, *Scand. Environmental Health Perspectives*, 99:195-197.

Betina, V., 1989, *Mycotoxins: Chemical, Biological and Environmental Aspects*. Elsevier, New York, NY.

Burge, H.A., 1999, Fungal growth in buildings: the aerobiological perspective, in: *Bioaerosols, Fungi, and Mycotoxins: Health Effects, Assessment, Prevention, and Control*, E. Johanning and C.S. Yang, eds., Boyd Printing, Albany, NY, pp. 306-312.

Creasia, D.A., Thurman, J.D., Jones, L.J., III, Nealley, M.L., York, C.G., Wannemacher, R.W., and Bunner, D.L., 1987, Acute inhalation toxicity of T-2 mycotoxin in mice, *Fundam. Appl. Toxicol.*, 8:230-235.

Creasia, D.A., Thurman, J.D., Wannemacher, R.W., and Bunner, D.L., 1990, Acute inhalation toxicity of T-2 mycotoxin in the rat and guinea pig, *Fundam. Appl. Toxicol.*, 14:54-59.

Croft, W.A., Jarvis, B.B., and Yatawara, C.S., 1986, Airborne outbreak of trichothecene toxicosis. *Atmospheric Environment*, 20:549-552.

Dearborn, D.G., Yike, I., Sorenson, W.G., Miller, M. J., and Etzel, R.A., 1999. Overview of investigation into pulmonary hemorrhage among infants in Cleveland, Ohio, *Environ. Health Persp.* Suppl. 3, 107:495-499.

Dillon, H.K., Miller, J.D., Sorenson, W.G., Douwes, J., and Jacobs, R.R., 1999, Review of methods applicable to the assessment of mold exposure to children, *Environ. Health Persp.*, Suppl. 3, 107:473-480.

Di Paolo, N., Guarnieri, A., Loi, F., Sacchi, G., Mangiarotti, A.M., and Di Paolo, M., 1993, Acute renal failure frominhalation of mycotoxins, *Nephron*, 64:621-625.

Donaldson, K. and Brown, D.M., 1990, Tumor necrosis factor production by rat alveolar macrophages exposed to organic and inorganic dust, *Respiratory Med.*, 84:432-434.

Driscoll, K.E., Maurer, J.K., and Crosby, L.L., 1990, Overload of lung clearance is associated with activation of alveolar macrophage tumor-necrosis-factor and fibronectin release, *J. Aerosol Med.*, 3:83-91.

Etzel, R.A., Montana, E., Sorenson, W.G., Kullman, G.J., Allen, T.M., Olson, D.R., Jarvis, B.B., Miller, J.D., and Dearborn, D.G., 1998, Acute pulmonary hemorrhage in infants associated with exposure to *Stachybotrys atra* and other fungi. *Arch. Pediatr. Adolesc. Med.*, 152:757-762.

Flannigan, B. and Miller, J.D., 1994, Health implications of fungi in indoor environments: an overview, in: *Health Implication of Fungi in Indoor Environments*, R. Samson, B. Flannigan, B.M. Flannigan, and S. Graveson, eds., Elsevier, Amsterdam, pp. 3-28.

Flappan, S.M., Portnoy, J., Jones, P., and Barnes, C., 1999, Infant pulmonary hemorrhage in a suburban home with water damage and mold (*Stachybotrys atra*). *Environmental Health Perspectives*, 107:927-930.

Fung, F., Clark, R., and Williams, S., 1998, *Stachybotrys*, a mycotoxin-producing fungus of increasing toxicologic importance. *J. Toxicol. Clin. Toxicol.*, 36:79-86.

Godish, T., 1996, *Sick Buildings: Definition, Diagnosis and Mitigation*, Lewis Publishers, Boca Raton, FL

Gravesen, S., Nielsen, P.A., Iversen, R., and Nielsen, K. F., 1999, Micofungal contamination of damp buildings-examples of risk constructions and risk materials, *Environ. Health Persp.* Suppl. 3, 107:505-508.

Hinkley, S.F., Fettinger, J.C., Dudley, K., and Jarvis, B.B., 1999, Memnobotrins and memnoconols: novel metabolites from *Memnoniella echinata*, *J. Antibiotics*, 52:988-997.

Hinkley, S.F., Jaing, J., Mazzola, E.P., and Jarvis, B.B., 1999, Atranones: novel diterpenoids from the toxigenic mold *Stachybotrys atra*, *Tetrahedron Lett.*, 40:2726-2728.

Hinkley, F.F., Mazzola, E. P., Fettinger, J. C., Lam, Y.-F., and Jarvis, B.B., 2000, Atranones A-G: a unique series of metabolites from the toxigenic mold *Stachybotrys chartarum, Phytochemistry*, 55:663-673.

Hodgson, M.J., Morey, P., Leung, W.-Y., Morrow, L., Miller, J.D., Jarvis, B.B., Robbins, H., Halsey, J.F. and Storey, E., 1998, Building-associated pulmonary disease from exposure to *Stachybotrys chartarum* and *Aspergillus versicolor*. *J. Occup. Environ. Med.*, 40:241-249.

Holt, P.G., 1990, Inflammation in organic dust-induced lung-disease - new approaches for research into underlying mechanisms, *Am. J. Indust. Med.*, 47:47-54.

Jakab, G.J., Hmieleski, R., Zarba, A., Hemenway, D.R., and Groopman, J.D., 1994, Respiratory aflatoxicosis: supression of pulmonary and systemic host defenses in rats and mice, *Toxicol. Appl. Pharmacol.*, 125:198-205.

Jarvis, B.B., 1990, Mycotoxins and indoor air quality, in: *Biological Contaminants in Indoor Environments*, AST STP 1071, P.R. Morey, J.C. Feely, Sr., and J.A. Otten, eds., American Society for Testing and Materials, Philadelphia, pp. 201-214.

Jarvis, B.B. and Hinkley, S. 1999, Analysis for *Stachybotrys* toxins, in: *Bioaerosols, Fungi, and Mycotoxins*, E. Johanning and C.S. Yang (eds.), Boyd Printing, Albany, NY, pp. 232-239.

Jarvis, B.B., Salemme, J., and Morais, A., 1995, *Stachybotrys* toxins. 1. *Nat. Toxins* 3:10-16.

Jarvis, B.B., Zhou, Y., Jiang, J.,Wang, S., Sorenson, W.G., Hintikka, E.-L., Nikulin, M., Parikka, P., Etzel, R.A., and Dearborn, D.G., 1996, Toxigenic molds in water-damaged buildings: dechlorogriseofulvins from *Memnoniella echinata, J. Nat. Prod.*, 59:553-554.

Jarvis, B.B., Sorensen, W.G., Hintikka, E.-L., Nikulin, M., Zhou, Y., Jiang, J., Wang, S., Hinkley, S., Etzel, R.A., and Dearborn, D.G., 1998, Study of toxin production by isolates of *Stachybotrys chartarum* and *Memnionella echinata* isolated during a study of pulmonary hemosiderosis in infants, *Appl. Environ. Microbiol.*, 64:3620-3625.

Johanning, E., Biagini, R., Hull, D.L., Morey, P., Jarvis, B., Landsbegis, P., 1996, Health and immunology study following exposure to toxigenic fungi (*Stachybotrys chartarum*) in a water-damaged office environment. *Int. Arch. Occup. Environ. Health*, 68:207-218.

Johanning, E., Gareis, M., Hintikka, E.-L., Yang, C.S., Nikulin, M., Jarvis, B.B., and Dietrich, R., 1998, Toxicity screening of fungal samples: results of sentinel health investigations related to indoor *Stachybotrys atra* (*chartarum*) exposure. *Mycotoxin Research,* 14:60-73.

Jong, S.C. and Davis E.E., 1976, Contribution to the knowledge of *Stachybotrys* and *Memnoniella* in culture *Mycotaxon* 3:409-485.

Kristensen, P., Irgens, L.M., Andersen, A., Bye, A.S., and Sundheim, L., 1997, Gestational age, birth weight, and perinatal death among births to Norwegian farmers, 1967-1991, *Am. J. Epidemiol.*, 146:329-338.

Lappalainen, S., Nikulin, M., Berg, S., Parikka, P., Hintikka, E.-L., and Pasanen, A.L., 1996, Fusarium toxins and fungi associated with handling of grain on eight Finnish farms *Atmospheric Environment*, 30:3059-3065.

Larsen, T.O. and Frisvad, J.C., 1994, Production of volatiles and mycotoxins in conidia of common indoor *Penicillium* and *Aspergillus* species, in: *Health Implication of Fungi in Indoor Environments*, R. Samson, B. Flannigan, M. Flannigan and S. Graveson, eds., Elsevier, Amsterdam, pp. 251-279.

Marrs, T.C., Edginton, J.A.G., Price, P.N., and Upshall, D.G., 1986, Acute toxicity of T-2 mycotoxin to the guinea pig by inhalation and subcutaneous routes, *Br. J. Exp. Path.*, 67:259-268.

Mason, C.D., Rand, T.G., Oulton, M., MacDonald, J.M., and Scott, J.E., 1998, Effects of *Stachybotrys chartarum* conidia and isolated toxin on lung surfactant production and homeostasis, *Natural Toxins*, 6:27-33.

Miller, J.D., 1992, Fungi as contaminants in indoor air, *Atmospheric Environment* 26:2163-2172.

Miller, J.D., 1993, Fungi and the building engineer, in: *Approaches to Assessment of the Microbial Flora of Buildings*. ASHRAE IAO 92, Environments for Healthy People. ASHRAE, Atlanta, Georgia, pp. 147-159.

Miller, J.D., 1995, Quantification of health effects of combined exposures: a new beginning. in: *Indoor Air Quality-an Integrated Approach*, L. Morawska, ed., lsevier, Amsterdam, pp. 159-168.

Nielsen, K.F., Hansen, M.O., Larsen, T.O., and Thrane, U., 1998a, Production of trichothecene mycotoxins on water damaged gypsum boards in Danish buildings, *Internat. Bioterior. Biodegrad.*, 42:1-7.

Nielsen, K.F., Thrane, U., Larsen, T.O., Nielsen, P.A., and Gravesen, S., 1998b, Production of mycotoxins on artificially inoculated building materials, *Internat. Bioterior. Biodegrad.*, 42:9-16.

Nielsen, K.F., Gravesen, S., Nielsen, P.A., Andcrsen, B., Thrane, U., and Frisvad, J.C., 1999, Production of mycotoxins on artificially and naturally infested building materials, *Mycopathogia*, 145:43-56.

Nikulin M., Reijula K., Jarvis B.B., Hintikka E.-L., 1996, Experimental lung mycotoxicosis in mice induced by *Stachybotrys atra*, *Int. J. Exp. Pathol.*, 77:213-218.

Nikulin M., Reijula K., Jarvis B.B., Veijalainen P., Hintikka E.-L., 1997, Effects of intranasal exposure to spores of *Stachybotrys atra* in mice. *Fundam. Appl. Toxicol.*, 35:182-188.

Pang, V.F., Lambert, R.J., Felsburg, P.J., Beasley, V.R., Buck, W.B., and Halchek, W.M., 1988, Experimental T-2 toxicosis in swine following inhalation exposure: clinical signs and effects on hematology, serum biochemistry, and immune response, *Fund. Appl. Toxicol.*, 11:100-109.

Peraica, M., Radic, B., Lucic, A. and Pavlovic, M., 1999, Toxic effects of mycotoxins in human health, *Bull. WHO*, 77:754-766.

Rao, C.Y., Brain, J.D., and Burge, H.A., 2000, Reduction of pulmonary toxicity of *Stachybotrys chartarum* spores by methanol extraction of mycotoxins, *Appl. Environ. Microbiol.*, 66:2817-2821.

Richard, J.L., Plattner, R. D., May, J. and Liska, S. L., 1999, The occurrence of ochratoxin A in dust collected from a problem home, *Mycopathologia*, 146:99-103.

Samson, R.A., 1999, Ecology, detection and identification problems of moulds in indoor environments, in: *Bioaerosols, Fungi, and Mycotoxins: Health Effects, Assessment, Prevention, and Control*, E. Johanning and C.S. Yang, eds., Boyd Printing, Albany, NY, pp. 33-37.

Sorenson, W.G., 1990, Mycotoxins as potential occupational hazards, *Dev. Industrial Microbiology*, 31:205-211.

Sorenson, W.G., Shahan, T.A., and Lewis, D.M. 1994, Activation of alveolar macrophages by conidia of common ODTS fungi, in: *Health Implication of Fungi in Indoor Environments*, R. Samson, B. Flannigan, M. Flannigan and S. Graveson, eds., Elsevier, Amsterdam, pp. 325-343.

Sorenson, W.G., 1999, Fungal spores: hazardous to health? *Environ. Health Persp.* Suppl. 3, 107:469-472.

Sorenson, W.G., Frazer, D.G., and Jarvis, B.B., 1987, Trichothecene mycotoxins in airborne conidia of *Stachybotrys atra.*, *Appl. Environ. Microbiol.*, 53:1370-1375.

Sudakin, D.L., 2000, *Stachybotrys chartarum*: current knowledgeof its role in disease, *Medscape Gen. Med.*, February 29.

Tsai, S.M., Yang, C.S., and Heinsohn, P., 1999, Comparative studies of fungal media for recovery of *Stachybotrys chartarum* from environmental samples, in: *Bioaerosols, Fungi, and Mycotoxins: Health Effects, Assessment, Prevention, and Control*, E. Johanning and C.S. Yang, eds., Boyd Printing, Albany, NY, pp. 330-334.

Vesper, S.J., Dearborn, D.G., Yike, I., Sorenson, W.G., and Haugland, R.A., 1999, Hemolysis, toxicity, and randomly amplified polymorphic DNA analysis of *Stachybotrys chartarum* strains *Appl. Environ. Microbiol.*, 65:3175-3181.

Wolkoff, P., Clausen, P.A., Wilkins, C.K., Hougaard, K.S., and Nielsen, G. D., 1999, formation of strong airway irritants in a model mixture of (+)-α-pinen/ozone, *Atmospheric Environment*, 33:693-698.

Zarba, A., Hmieleski, R., Hemenway, D.R., Jakab, G.J., and Groopman, J.D., 1992, Aflatoxin B_1-DNA adduct formation in rat liver following exposure by aerosol inhalation. *Carcinogenesis*, 13:1031-1033.

BIOLOGICAL CONTROL OF FUSARIUM HEAD BLIGHT OF WHEAT AND DEOXYNIVALENOL LEVELS IN GRAIN VIA USE OF MICROBIAL ANTAGONISTS

David A. Schisler,[1] Naseem I. Khan,[2] and Michael J. Boehm[2]

[1]USDA-ARS, National Center for Agricultural Utilization Research, Peoria, IL 61604
[2]Department of Plant Pathology, Ohio State University, Columbus, OH 43210

ABSTRACT

Efforts to reduce mycotoxin contamination in food logically start with minimizing plant infection by mycotoxin producing pathogens. *Fusarium graminearum* (perfect state, *Gibberella zeae*) infects wheat heads at flowering, causing the disease Fusarium head blight (FHB) and losses of over 2.6 billion dollars in the U.S. during the last 10 years. The pathogen often produces deoxynivalenol (DON) resulting in grain size and quality reduction. Highly resistant wheat cultivars currently are not available for reducing FHB, and labeled fungicides are not consistently effective. The feasibility of biologically controlling FHB is currently being evaluated. Microbial isolates obtained from wheat anthers were screened for their ability to utilize tartaric acid, a compound that is poorly utilized by *F. graminearum* and could be utilized in formulations of biological control agents. Four strains that utilized tartaric acid and three that did not were effective in reducing FHB disease severity by up to 95% in greenhouse and 56% in field trials. Additional research programs around the globe have identified other antagonist strains with potential for biologically controlling FHB. Though a considerable body of research remains to be completed, strategies and microorganisms for biologically controlling FHB have reached an advanced stage of development and offer the promise of being an effective tool that could soon contribute to the reduction of FHB severity and DON contamination of grain in commercial agriculture.

INTRODUCTION

In natural ecosystems where plant pathogens have co-evolved with the plant community, plant disease rarely occurs on an epidemic scale. In part, this is because such ecosystems are comprised of a wide diversity of plant species that are colonized by an even greater diversity of nonpathogenic microorganisms, some of which are superior competitors for the nutrients, minerals and physical space that the pathogen needs to gain entry to a plant. Other microorganisms may directly parasitize the plant pathogen, produce compounds that inhibit the growth of the pathogen or trigger the activation of plant defense systems (Hammerschmidt, 1999) prior to pathogen inoculum arriving on the infection court. Harvesting, understanding and applying this naturally occurring, microbially led process that results in reduced plant pathogen activity has been the focus of biological control research in the field of plant pathology. Definitions of "biological control" as it pertains to plant pathogens are numerous (Baker and Cook, 1974; Campbell, 1989; Wilson, 1997). For purposes of this paper, the definition of Cook and Baker (1983) describes the topic well: "Biological control is the reduction of the amount of inoculum or disease-producing activity of a pathogen accomplished by or through one or more organisms other than man." Interest in using biological control as a tool against plant disease has grown exponentially since the early 1960's (Baker, 1987). A portion of this increased interest has followed the public's increasing preference for agricultural products that are produced using fewer pesticides. Also contributing to research interest in biological control has been the loss of fungicide registrations for some agricultural uses and the loss of pesticide efficacy in cases where pathogen populations have acquired resistance (Desjardins et al., 1993; Hanson and Loria, 1994).

Many plant pathogenic fungi have been the target of research designed to discover microorganisms and develop methods for biologically reducing disease. Of these pathogenic fungi, several selected for investigation are producers of potent mycotoxins. These include, but are not limited to studies on biologically controlling *Aspergillus flavus* and *A. parasiticus* (aflatoxin producers), *Fusarium verticillioides (*synonym *F. moniliforme*) (fumonisin producer), *F. sambucinum* (diacetoxyscirpenol producer) and *F. graminearum* (deoxynivalenol and zearalenone producer). For instance, Cotty (1994) demonstrated in field studies on cotton that the application of an atoxigenic isolate of *A. flavus* to soil could exclude aflatoxin-producing isolates during cotton boll infection and thereby reduce aflatoxin contamination of the cottonseed. Other investigations have concentrated on controlling the pathogen in other crops. *A. flavus* is capable of invading maturing ears of maize in the field and producing aflatoxin during the process. Wicklow et al. (1988) documented that the normal progression of the infection process could be altered by the presence of competing fungi in preharvest maize. *Fusarium verticillioides*, and to a lessor extent *Acremonium strictum*, and *Aspergillus niger* were shown to inhibit *A. flavus* and aflatoxin contamination from spreading beyond wounded and immediately adjacent healthy kernels in ears of maize. A reduction in germination of sclerotia of *A. parasiticus*, also capable of producing aflatoxin, was described for studies in soil artificially infested with the fungus *Paecilomyces lilacinus* (Will et al., 1994). In peanuts, Dorner et al. (1998) discovered that nontoxigenic color mutants of *A. flavus* and *A. parasiticus* applied as colonized rice at increasing concentrations to soil lowered preharvest aflatoxin contamination of peanuts. Applied at 2, 10 and 50 g of colonized rice per meter of soil, these antagonists reduced aflatoxin by 74.3, 95, and 99.9% compared to the untreated control, respectively. Wilson (1988) reviewed additional work on biologically controlling *A. flavus*.

The most consistent producer of fumonisins in corn is *F. verticillioides* (perfect state, *Gibberella fujikuroi* mating population A) (Munkvold and Desjardins, 1997). There has been relatively little work on developing biological control microorganisms and methodologies for reducing *F. verticillioides* infection and fumonisin contamination of corn. Although competitive exclusion of mycotoxin producing isolates of *F. verticillioides* by atoxigenic isolates hypothetically could be a successful approach in reducing fumonisin contamination in corn, results to date indicate that the pathogenicity of *F. verticillioides* is not tied to its ability to produce fumonisins (Desjardins and Plattner, 2000). Therefore, isolates of *F. verticillioides* that are both nonpathogenic and atoxigenic would have to be selected or engineered to be considered as candidate biocontrol agents. The colonization of maize roots by microorganisms that exhibit *in vitro* antagonism towards *F. verticillioides* has been documented (Hebbar et al., 1992). While of interest academically, it seems unlikely that laboratory demonstrations of reducing *F. verticillioides* inoculum in soil via the introduction of the mycophagous nematode *Aphelenchus avenae* (Gupta, 1986) will result in an applied method for reducing *F. verticillioides* in the field.

Trichothecenes, such as diacetoxyscirpenol, can be produced by *Fusarium sambucinum* (perfect state, *Gibberella pulicaris*) the primary causal agent of Fusarium dry rot on stored potato tubers in North America. Research on biologically controlling this disease has been extensive with successful laboratory studies using bacterial antagonists (Schisler and Slininger, 1994; Slininger et al., 1994; Schisler et al., 1997; Schisler et al., 2000a), yeasts (Schisler et al., 1995) and arbuscular mycorrhizae (Niemira et al., 1996). In the studies of Schisler and Slininger (1994), 18 bacterial antagonists were selected that substantially reduced disease in laboratory studies, sometimes to undetectable levels compared to controls. Studies that ranked the commercial development potential of these strains based on their efficacy and amenability to production in liquid media (Slininger et al., 1994; Schisler and Slininger, 1997) have been completed. Additional investigations have determined antagonist dose-response curves, demonstrated biocontrol efficacy enhancement when certain pairs of antagonists were applied to potato tubers prior to storage (Schisler et al., 1997), and documented the biological inhibition of tuber sprouting when tubers were treated with several of the dry rot antagonists (Slininger et al., 2000). Field trials in commercial storage settings have demonstrated significant disease reduction using bacterial antagonists against *F. sambucinum* (Schisler et al., 2000b).

The biological control of the mycotoxin producing fungus, *F. graminearum*, (perfect state, *Gibberella zeae*) has also received considerable research attention. The remainder of this chapter will concentrate on work that has been conducted on discovering and developing biological control microorganisms and processes for reducing the severity of FHB, primarily incited by *F. graminearum*. While emphasizing work conducted in the authors' laboratories in a cooperative research effort between the USDA-ARS, National Center for Agricultural Utilization Research in Peoria, Illinois, and The Ohio State University in Columbus, Ohio, additional work on biologically controlling this disease that has been conducted at other locations in North America and around the world will be mentioned.

FUSARIUM HEAD BLIGHT OF WHEAT

Cause and Importance

The primary causal agent of FHB of wheat and barley is *F. graminearum* Group 2 (Aoki and O'Donnell, 1999), although *F. culmorum*, *F. avenaceum*, *F. verticillioides*, *F. oxysporum* and *F. poae*, among others, have been identified as generally less important components that can contribute to a FHB disease complex (Bai and Shaner, 1994; Parry et al., 1995). Also known as scab of wheat, pink mold, whiteheads and tombstone scab, FHB is an important disease in humid and semi-humid regions of the world (McMullen et al., 1997) where it can cause extensive damage to grains, especially wheat and barley. *F. graminearum* has the potential to completely destroy an otherwise healthy crop only weeks before harvest by infecting developing heads at the time of flowering. Yield losses of 20-40% and higher are common when warm moist conditions exist during the time of wheat flowering and early grain fill (Martin and Johnston, 1982; Bottalico et al., 1989; Cook and Veseth, 1991; Luo, 1992; Sokolov et al., 1998). Epidemics of the disease are known to have occurred in the United States since the early 1900's (Atanasoff, 1920). Fusarium head blight in the U.S. has been particularly devastating in the 1990's (McMullen et al., 1997) with damage due to the disease estimated to exceed 3.0 billion dollars at the farm-gate alone (Anderson and Ward, 1999). Reports from around the globe have been made on the serious to severe effects of scab in recent times. These include reports from Canada (Sutton, 1982), Germany (Müller et al., 1997), Norway (Langseth et al., 1999), United Kingdom (Polley et al., 1991), Russia (Pavlova and

Izmalkova, 1995), China (Li et al., 1999), Argentina (Moschini and Fortugno, 1996; Rizzo et al., 1997), and Korea (Vestal, 1964).

Fusarium graminearum can produce potent toxins such as the estrogenic toxin zearalenone (F-2) (Hesseltine et al., 1978) and the trichothecene deoxynivalenol (DON, vomitoxin) (Snijders, 1990) during colonization of grain, and in some cases, during improper grain storage (Homdork et al., 2000). Grain that is heavily contaminated by DON is frequently unsuitable for human consumption and may be refused as feed (Forsyth et al, 1977; Vesonder and Hesseltine, 1980). Exposure to high levels of DON in the diet has been associated with severe gastrointestinal distress in humans and prolonged exposure has been associated with impaired immune systems in laboratory animals (Marasas, 1991; Beardall and Miller, 1994; Pestka and Bondy, 1994). Deleterious effects have also been noted when exposing various mammalian cell lines to DON *in vitro* (Knasmuller et al., 1997; Kasuga et al., 1998). Deoxynivalenol can inhibit amino acid incorporation and protein production in plant tissues (Casale and Hart, 1988; Miller and Ewen, 1997). In the case of *F. culmorum*, and likely *F. graminearum* as well, the toxin is associated not just with hyphae of the pathogen within host tissues but can be translocated from infected spikelets in advance of hyphae (Kang and Buchenauer, 1999). In both greenhouse and field tests, the virulence of *F. graminearum* mutants with disrupted ability to produce DON was less than the DON producing progenitor isolate or a mutant isolate restored to DON producing capability (Proctor et al., 1995; Desjardins et al., 1996). Methods for decontaminating DON infested cereals are expensive and labor intensive (Accerbi et al, 1999; Karlovsky, 1999). Therefore, cereals contaminated with high levels of these toxins are generally no longer suitable for human or animal consumption or for further processing. Infection of wheat kernels by *F. graminearum* reduces grain yield and affects grain quality (Clear and Patrick, 1990). The infection of seed reduces seed germination, seedling vigor and plant emergence (Bechtel et al., 1985).

Options for Reducing Fusarium Head Blight of Wheat

Several methods have been and are being pursued in an attempt to reduce the severity of FHB and the concomitant contamination of grain by DON and other mycotoxins. One approach is to treat wheat heads at flowering with fungicides active against *F. graminearum*. Though this approach can have some positive effect, (Wilcoxson, 1996; Suty and Mauler-Machnik, 1997; Jones 1999) concerns regarding residues when applying fungicides only a few weeks before harvest render this alternative less attractive. The development of resistant cultivars of wheat holds promise in reducing the severity of FHB. Considerable progress is being made in this arena but highly resistant cultivars that retain ideal agronomic traits have not yet been developed via either conventional breeding or genetic engineering approaches (Johnston, 1994; Bushnell et al., 1998; Bai et al., 2000). Although partially effective in reducing FHB, conventional tillage of fields after harvesting in order to bury plant residues and reduce available pathogen inoculum (Miller et al., 1998; Dill-Macky and Jones, 2000) is not compatible with the preferred agricultural practice of minimum tillage.

Biological control, though not a control option that is currently available commercially, holds considerable promise for reducing FHB (McMullen and Bergstrom, 1999). Wheat heads are most susceptible to infection by *F. graminearum* from the onset of flowering until the early stages of kernel filling (Fernando et al., 1997; Paulitz, 1999). Microorganisms could be used to reduce the amount of pathogen inoculum prior to its arrival on wheat heads, used to reduce the number of successful infections by the pathogen or used to activate plant host defenses prior to the arrival of pathogen inoculum. Many studies support the possibility of microbial inoculants achieving successful colonization of plant tissues minimally colonized by resident microorganisms and providing biological control of plant disease (Schisler et al., 1998; Benbow and Sugar, 1999; Droby et al., 1999; Stockwell et al., 1999b). In work in our laboratories, we chose to concentrate our research on developing strategies and microorganisms for protecting wheat heads during their most susceptible stage of development, flowering.

A Case Study: Discovery and Development of Microbial Antagonists Active Against *F. graminearum* on Wheat

Anthesis is a crucial time for the onset of FHB with anthers promoting infection of wheat heads by *F. graminearum* (Strange and Smith, 1971). Prior to initiating our investigation, we speculated that microorganisms isolated from wheat anthers would be a fruitful source of putative biocontrol agents. Logically, microorganisms isolated from wheat anthers would have an increased likelihood of being adept colonists of anthers compared to isolates from other sources. Choline and betaine are found in wheat anthers and stimulate the growth of conidial germ tubes of *F. graminearum* (Strange and Smith, 1978). We surmised that information on carbon utilization by anther colonists could prove useful for selecting and formulating biocontrol strains. During our study, HPLC analysis of culture broths containing individual anther colonists and choline in the form of choline bitartrate identified colonists capable of utilizing tartaric acid but was ineffective in identifying those capable of metabolizing choline. Fortuitously, a higher percentage of tartaric acid-utilizing microbes exhibited superior FHB biocontrol capabilities than did non-utilizing microbes. Tartaric acid is a compound that is poorly utilized by *F. graminearum* as a carbon source (D. Schisler, unpublished) and is a readily available byproduct in the processing of grapes and other fruits for juice and wine (Andrés et al., 1997). Populations of tartaric acid-utilizing antagonists may be selectively enriched on anthers if the compound was included in the formulations of such antagonists.

Isolation of Anther-Colonizing Microorganisms. Anthers were collected from flowering wheat plants across Illinois and Ohio, two states that had experienced FHB epidemics in the 1990's. Anthers were frozen in 10% glycerin at - 80°C in cryovials. Over 400 anther collections were obtained. Microorganisms were isolated from anthers by gradually thawing a vial, vigorously mixing the vial contents to liberate microorganisms from anther surfaces, and serially diluting vial contents onto a variety of solidified media (Khan et al., 2001). Single colonies of putative antagonists showing distinct growth morphology were streaked for purity with a total of 738 isolates obtained (Figure 1).

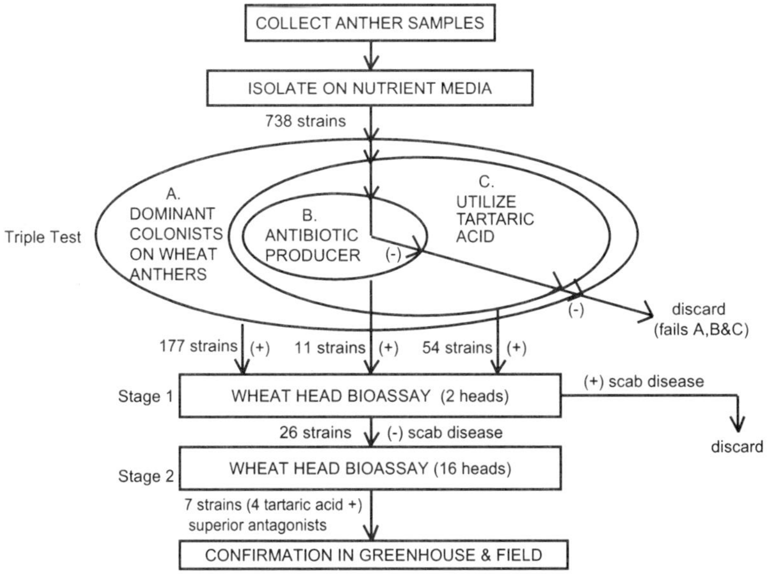

Figure 1. Selection strategy utilized to obtain microbial antagonists effective in reducing FHB on wheat.

Stage 1 Antagonist Selection: "Triple Test" Selection Procedure. The 738 microbial isolates were then subjected to a "triple test" selection procedure (Figure 1). Isolates had to pass at least one of the tests or else were eliminated from further consideration. Isolates had to be either randomly chosen from a pool of dominant colonists from wheat anthers (a "dominant colonist" being a microbial isolate that maintained a sufficient population on the anther to be recovered using traditional serial dilution techniques), produce antibiosis zones against *F. graminearum* in *in vitro* antibiosis tests on 1/5 strength tryptic soy broth agar (TSBA/5) or utilize tartaric acid when the compound was supplied as a carbon source in a liquid culture broth. Due to the ease of conducting utilization tests in liquid culture, all isolates were tested for tartaric acid utilization as determined by HPLC analysis of culture broths of each isolate. Isolates obtained from anthers were grown on TSBA/5 and used to inoculate 10 ml of a minimal salt medium that contained tartaric acid (as 1 g/L choline bitartrate) and urea (1.26 g/L) as a nitrogen source. After 48 h incubation at 25°C in a shaker incubator, spent culture broth was analyzed for the presence of tartaric acid using HPLC (Khan et al., 2001). Fifty-four of the original 738 isolates of microorganisms assayed used tartaric acid as a carbon source. These strains (54 strains), antibiotic producers (11) and randomly-selected, dominant anther colonists (177) were chosen for use in a two-head plant bioassay of biocontrol efficacy against FHB.

Stage 2 Antagonist Selection:Two-Head Plant Bioassay. Two seedlings of hard red spring wheat (cultivar "Norm") were grown in pasteurized potting mix in a growth chamber for approximately 8 weeks prior to use in bioassays. Conidial inoculum of *F. graminearum* isolate Z-3639 was produced in petri dishes of clarified V-8 juice agar. Two or three days before use, microbial strains that passed the triple test were recovered from storage in 10% glycerol at -80°C by streaking partially thawed glycerol onto TSBA/5 and restreaked for purity after 48 h. Inoculum for wheat heads was prepared that contained antagonist cells [final absorbance ~.2-.3 at 620 nm wavelength light (A620)], 5 x 10^5 conidia/ml and 0.04% Tween 80 (Sigma, St. Louis, MO). To initiate the two-head plant bioassay for each of the antagonists of *F. graminearum*, the middle florets of two wheat heads were co-inoculated with 10 Fl of a microbial suspension. Heads were inoculated within 2-4 days of flowering. After inoculation, wheat plants were misted with water, incubated in a plastic humidity chamber for 72 h at approximately 22°C, and then transferred to greenhouse benches. Fusarium head blight severity was visually estimated using a 0-100% scale (Stack and McMullen, 1995) 16 days after inoculation. Twenty-six strains, including nine that utilized tartaric acid, prevented any visible FHB disease development and were selected for stage 3 testing of efficacy against *F. graminearum*.

Stage 3 Antagonist Selection:Multiple-Head Plant Bioassay. Stage 3 plant bioassays were conducted as for stage 2 except that biomass of microbial strains was produced in a semi-defined complete liquid medium (SDCL) (Slininger et al., 1994) in 50-ml Erlenmeyer flasks (Khan et al., 2001). At anthesis, a mixture of cells of a putative antagonist (25% of a fully colonized broth; approximately 5 x 10^8 CFU/ml and 2.5 x 10^7 CFU/ml for bacteria and yeast strains, respectively) and *F. graminearum* conidia (5 x 10^5 conidia/ml) were used to point inoculate 16 wheat heads per antagonist strain. Seven of the 26 strains (Figure 1) assayed reduced FHB severity ($P \leq 0.05$, data not shown) and were selected for further studies in the greenhouse and field. Three of the strains were *Bacillus* spp. while four of the strains were yeasts (Table 1). Four of the seven selected strains were able to utilize tartaric acid in liquid culture.

Greenhouse Testing of Superior Antagonists Against *F. graminearum*. In repeated, multiple-head plant bioassays conducted against *F. graminearum* isolate Z-3639 (originally isolated in Kansas) in the greenhouse, all seven antagonists reduced FHB as indicated by increased 100-kernel weight of microbially treated wheat heads (Figure 2B; $P \leq 0.05$, Fisher Protected LSD; strain OH 72.4 data not shown). Compared to the pathogen inoculated control, 100-kernel weight was increased by over 100% in plants treated with *Bacillus* strains AS 43.3, AS 43.4 and *Cryptococcus* OH 182.9. These strains also decreased disease severity (Figure 2A). Interestingly, the content of deoxynivalenol in wheat inoculated with the pathogen was reduced by more than 10-fold for AS 43.3 and AS 43.4 treated plants and by approximately 5-fold for OH 182.9 treated plants (Figure 2C). Repeated bioassays were also run using isolates of *F. graminearum* obtained from Ohio and Ontario, Canada. In general, antagonist treatments were similarly successful in increasing 100 kernel weights and reducing disease severity. Exceptions were that only two of the seven antagonist reduced FHB severity incited by the *F. graminearum* isolate from Ontario while four of the seven reduced FHB severity incited by the Ohio isolate of *F. graminearum* (Khan et al., 2001). Greenhouse tests using four of the antagonists were also conducted on two durum wheat cultivars. *Bacillus* strains 43.3 and 43.4 were again the most successful in reducing FHB severity (by up to 94% compared to the control) but *Cryptococcus* sp. strains OH 71.4 and OH 182.9 also reduced FHB severity on durum cultivar Renville as did OH 71.4 on cultivar Ben ($P \leq 0.05$, FPLSD; Schisler et al., 1999).

Table 1. Antagonist strain designation and identification of bacteria and yeasts that reduce the severity of FHB of wheat

Antagonist	NRRL accession no.[1]	Identification
AS 43.3	B-30210	*Bacillus subtilis/amyloliquefaciens*[2]
As 43.4	B-30211	*Bacillus subtilis/amyloliquefaciens*[2]
OH 71.4	Y-30213	*Cryptococcus* sp. (synonym *Torula aurea*)[3]
OH 72.4	Y-30214	ND[4]
OH 131.1	B-30212	*Bacillus subtilis*[5]
OH 181.1	Y-30215	*Cryptococcus* sp. nov. [3]
OH 182.9	Y-30216	*Cryptococcus nodaensis* [3]

[1]NRRL patent culture collection, National Center for Agricultural Utilization Research, Peoria, IL.
[2]Identification by DSMZ, Braunschweig, Germany, based on 16S rDNA sequence homologies and biochemical and physiological tests of taxonomic utility.
[3]Identification based on nucleotide sequence divergence in domain D1/D2 of large subunit 26S rDNA and on divergence in ITS 1/5.8/ITS2 rDNA. C.P. Kurtzman, personal communication.
[4]Yeast, not determined.
[5]Identification by MIDI Labs, Newark, DE, based on 16S rDNA sequence homologies and biochemical and physiological tests of taxonomic utility.

Determination of the Influence of Order and Timing of Inoculum Application on Fusarium Head Blight. Greenhouse assays were then conducted to determine the influence of the order and timing of antagonist and pathogen inoculum application to wheat heads on FHB development. A spray inoculation method was used to mimic the arrival of inoculum at the infection court in the field. An aqueous suspension containing 5×10^5 conidia/ml of *F. graminearum* isolate Z-3639 and a wetting agent was applied as a mist onto wheat heads, immediately before or after and 4 h before or after treating heads with 5 ml of an aqueous suspension containing antagonist cells. Antagonist cells were applied at 10% or 50% of fully colonized broth. Heads sprayed only with conidia of *F. graminearum* served as controls.

Interestingly, all four of the antagonists tested significantly reduced FHB severity, regardless of the sequence, timing and concentration of inoculum application ($P \leq 0.05$, Table 2), though some antagonists did not increase 100-kernel weight when applied 4 h after inoculum of *F. graminearum* (Khan et al., 2001). Though it can reasonably be expected that sufficient delay in the application of antagonist inoculum would ultimately result in a loss of antagonist effectiveness, these results support the contention that biological reduction of FHB in field applications would not require that antagonist inoculum arrive before pathogen inoculum.

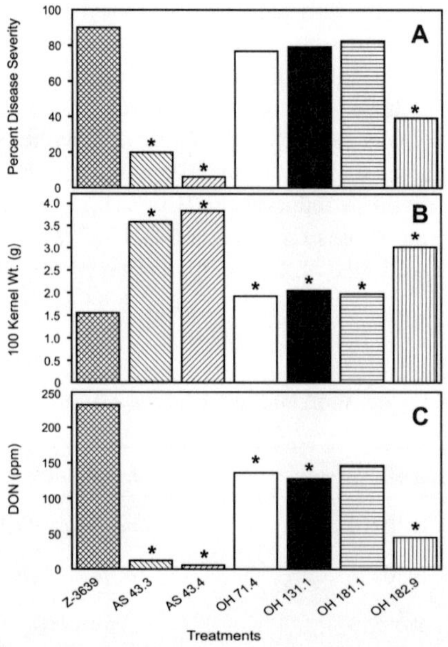

Figure 2$_{A-C}$. Influence of microbial antagonists on FHB disease development incited by *F. graminearum* Z-3639 in greenhouse studies. The pathogen only control of *F. graminearum* Z-3639 and pathogen plus antagonist are indicated on the x axis. Disease parameters of severity, 100 kernel weight and deoxynivalenol content of grain are shown on histograms A, B and C, respectively. Bars with an "*" are significantly different from the Z-3639 control ($P \leq 0.05$, FPLSD).

Field Trials of Antagonist Effectiveness in Reducing Fusarium Head Blight. All antagonist strains except OH 72.4 were tested on soft red winter wheat cultivars "Pioneer 2545" and "Freedom" in field experiments in Wooster, OH, and Peoria, IL, in 1999. These cultivars were selected because of their widespread use throughout the Midwestern United States. Additionally, since "Freedom" resists *F. graminearum* spread within heads, we were interested in evaluating the potential additive role biocontrol may provide in reducing FHB on this cultivar. Inoculum of *F. graminearum* was in the form of ascospores released from pathogen-infested corn kernels that were spread throughout test plots two weeks prior to wheat flowering. Antagonists were sprayed onto wheat at flowering at a concentration of 10% or 50% of fully colonized liquid culture broth. While no appreciable disease was observed on "Freedom" at Wooster, OH, 4 of the 6 antagonists tested at the 10% culture broth level decreased FHB severity by as much as 56% on Pioneer 2545 ($P \leq 0.05$, Figure 3A). Five of the six antagonists also increased 100 kernel weight (Figure 3B). Though 5 of the 6 antagonists showed a tendency towards decreasing DON content in grain, only 1 of the 6 significantly decreased kernel content of the mycotoxin ($P \leq 0.05$, Figure 3C). The efficacy of antagonists was not substantially changed when the dose was increased to the 50% culture broth level (Boehm et al., 1999). Though disease pressure was very light at the Peoria field site in 1999, 4 of 6 antagonists on "Pioneer" and 5 of 6 on "Freedom" reduced FHB severity at 1 of the 2 doses tested (Boehm et al., 1999). In the 2000 field season at Peoria, all 6 of the antagonists tested reduced FHB severity on Pioneer 2545 and 4 of 4 reduced disease severity by an average of 50% on durum wheat cultivar "Renville" (Schisler et al., unpublished). Thus, results at multiple sites and in multiple years demonstrate the considerable potential of these microbial antagonists to contribute to the management of FHB under field conditions. Further field trials are necessary to fully understand the role these antagonists might play in reducing the content of DON in grain and on additional cultivars of wheat.

Table 2. Percent FHB disease severity when varying the time and sequence of pathogen and antagonist inoculum application to wheat heads.

	Pathogen inoculum applied:			
Treatment	4 h before antagonist	Immediately before antagonist	Immediately after antagonist	4 h after antagonist
F. graminearum (Z-3639)	59	86	81	85
AS 43.3 (10%)[1]	13*[2]	2*	3*	21*
AS 43.3 (50%)	5*	1*	0*	15*
AS 43.4 (10%)	42*	3*	3*	30*
AS 43.4 (50%)	19*	33*	0*	18*
OH 71.4 (10%)	26*	24*	18*	49*
OH 71.4 (50%)	28*	51*	37*	63*
OH 182.9 (10%)	43*	64*	60*	45*
OH 182.9 (50%)	43*	49*	60*	58*

[1]Antagonists were applied as a mist at concentrations of 10% or 50 % of a fully colonized, semi-defined complete liquid medium.
[2]Within a column, means followed by an asterisk are significantly different from the *F. graminearum* Z-3639 control (Fisher's protected LSD, P≤0.05).

Other Research Efforts on the Biocontrol of Fusarium Head Blight

Research on biologically controlling FHB was initiated with *in vitro* studies in Brazil in by Luz (1988). Perondi et al. (1996) reported excellent reduction in FHB severity in a field trial in Brazil where a *Pseudomonas fluorescens* strain and a *Sporobolomyces roseus* strain reduced FHB by greater than 75% versus the control. Several microorganisms from the continuation of this work as well as recently selected strains of antagonists are being tested in the United States by Stockwell et al. (1999a). Antagonist strains such as *Paenibacillus macerans* have shown promising results in greenhouse trials when applied to wheat heads. Selection and testing of antagonists that could colonize crop debris and reduce the production of ascospore primary inoculum of the pathogen is also anticipated (Stockwell et al., 1999a).

Fusarium head blight is a problem of considerable concern in large portions of the wheat producing land in the People's Republic of China (Bai and Shaner, 1994). Work on biologically controlling FHB has been underway since at least the early 1990's with Fuming et al. (1994) reporting that several *Bacillus* and coccoid bacterial strains reduced FHB by more than 50% at two different field sites.

Fusarium head blight is also a disease of importance in the southern wheat growing regions of Russia. Kolombet and co-workers (1999) have discovered an isolate of *Trichoderma viride* that is antagonistic towards a wide spectrum of phytopathogenic fungi, including *F. graminearum* and *F. culmorum in vitro.* In field trials, pretreatment of wheat and barley seeds with *T. viride* effectively inhibited the development of FHB at antagonist doses of 20-30 g/tone of seeds. The level of disease control obtained was equivalent to that achieved with 2 kg of a standard chemical preparation per tone of seeds. Additionally, an 8-10-fold decrease of the content of Fusarium toxins in grain was shown after treatment with *T. viride* compared to the chemical fungicide treatment (Kolombet et al., 1999; and Kolombet, personal communication).

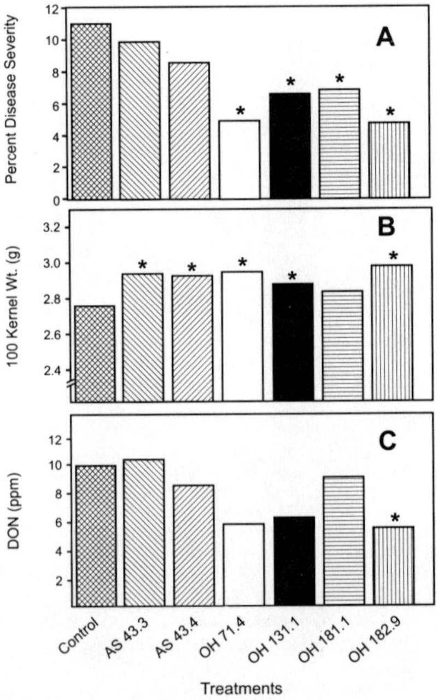

Figure $3_{A\text{-}C}$. 1999 Wooster, Ohio field experiment to determine the influence of microbial antagonists on FHB disease development incited by *F. graminearum* in field studies on the soft red winter wheat "Pioneer 2545". Antagonists were applied at a rate of 10% of fully colonized liquid culture broth. The control of buffer and wetting agent and antagonist strain treatments are indicated on the "x" axis. Disease parameters of severity, 100 kernel weight and deoxynivalenol content of grain are shown on histograms A, B and C, respectively. Bars with an "*" are significantly different than the control ($P \leq 0.05$, FPLSD).

Future Research Priorities in Developing Strategies and Microorganisms to Biocontrol Fusarium Head Blight

In order to increase the likelihood of bringing one or a mixture of microbial antagonists to the agricultural market place as an economically viable product for use in reducing FHB, several areas of additional research effort are crucial. Certainly, determining the mode of action of key biological control agents and focusing research on discovering additional biocontrol agents that utilize a variety of compatible mechanisms in reducing FHB should be a research priority for all laboratories involved in biologically controlling FHB. The discovery of a microorganism or suite of strains that could be utilized to preemptively colonize crop residues, thereby depriving the pathogen of nutrients and physical space normally used to overwinter and produce primary inoculum during wheat flowering, would benefit all other efforts to biologically protect wheat heads during flowering. The consensus of several papers indicating that conventional tillage reduces FHB (Teich and Hamilton, 1985; Miller et al., 1998; Dill-Macky and Jones, 2000), provides strong circumstantial support of likely success in removing crop debris from pathogen use by employing aggressive saprophytic colonists rather than by physical removal of debris via plowing. Indeed, Fernandez et al. (1992), demonstrated the potential of this approach using multiple applications of *Trichoderma harzianum* on wheat straw in the field under favorable moisture conditions. Additionally, mycoparasitic bacteria and fungi are known to be effective in reducing plant pathogen activity in other pathosystems (Huang, 1991). It is virtually certain that mycoparasites are active against *F. graminearum* in natural environments during the long period of host absence and saprophytic existence in the late summer, winter and early spring. Preliminary indications from scanning electron microscopy studies by Khan and Schisler (unpublished) indicate that *Cryptococcus nodaensis* OH182.9 adheres or is closely aligned with germ tubes of *F. graminearum* in wheat florets when co-inoculated with the pathogen and that pathogen germ tubes lined with cells of the yeast tend to lyse. Assays to determine if this yeast could reduce the amount of *F. graminearum* biomass that overwinters in crop debris have not yet been conducted.

Nutrient competition and antibiosis are mechanisms of biocontrol (Loper, 1993) that likely also play a major role in the biocontrol exhibited by strains in our, and other, studies. In the case of our antagonist AS 43.3, circumstantial evidence of antibiosis explaining at least a part of the biocontrol activity of this yeast antagonist is seen in Table 3. Cells of AS 43.3 were produced in liquid culture as described earlier and applied at 25%, 2.5% and 0.25% of fully colonized culture broth to flowering wheat heads as was conidial inoculum of the pathogen. When cells from 25% fully colonized culture broth were washed to remove culture metabolites before applying to wheat heads, FHB severity was only reduced to 55% compared to 20% with unwashed cells. Additionally, supernatant alone (cells removed) from AS 43.3 colonized culture broth reduced FHB severity (Table 2) compared to the pathogen control, indicating that the supernatant contained a biologically active compound produced by the antagonist. Competition for nutrients on leaf and head surfaces also deserves further study as a method for biologically reducing FHB though complete utilization of nutrients on plant surfaces, and therefore totally depriving a pathogen of nutrients, does not appear to be a realistic goal (Mercier and Lindow, 2000).

Table 3. Effect of *Bacillus* AS 43.4 colonized broth concentration, washed cells, and supernatant on FHB severity (*Fusarium graminearum* Z-3639) in greenhouse experiments.

Treatment	% Disease Severity
Z-3639	97[1] a
AS 43.4 (25%)[2]	20 c
AS 43.4 (2.5%)[2]	46 b
AS 43.4 (0.25%)[2]	58 b
Washed AS 43.4 (25%)[3]	55 b
AS 43.4 supernatant (25%)[4]	47 b

[1]Values not followed by the same letter are significantly different ($P \leq 0.05$, FPLSD).
[2]Value within parentheses represents the percentage of fully colonized culture medium and cells used in the treatment.
[3]Twenty-five percent of the cells in fully colonized broth were washed in weak buffer prior to use.
[4]Twenty-five percent strength supernatant from fully colonized broth used in the treatment.

Induced disease resistance is defined as the process of active resistance dependent on the host plant's physical or chemical barriers, activated by biotic or abiotic agents (inducing agents) (Kloepper et al., 1992) and is another potential mechanism of biological control being utilized by antagonists in our studies as well as others. In the studies of Kolombet et al. (1999) mentioned earlier, it seems quite likely that the substantial decrease in FHB severity on heads that was obtained by coating seeds at planting with a minimal amount of antagonist inoculum is due to the antagonist inducing disease resistance in the plants that grew from antagonist-coated seeds. Induced disease resistance via seed treatment with antagonists has been documented in field grown cucumbers (Wei et al., 1996). Wheat defense genes normally induced by *F. graminearum* infection (Pritsch et al., 2000) may be activated by some biocontrol agents in advance of the arrival of infecting hyphae of *F. graminearum*. The fact that some wheat cultivars can break down the mycotoxin DON (Miller and Arnison, 1986) suggests the intriguing possibility that some biocontrol agents may be able to stimulate this plant defense system as well.

Industrial scale production of bacterial and yeast biocontrol agents will, in almost every instance, take place in large-scale, stirred tank fermentors charged with liquid media. Fermentation technology is an extremely complicated and sophisticated scientific discipline unto itself (Chisti and Moo-Young, 1999). The fermentation environment can influence the efficacy of biocontrol agents (Schisler et al., 2000a). Antagonists can differ so dramatically in their efficacy when produced on solid vs. liquid substrates and in their amenability to production in economically feasible liquid culture media that eliminating from consideration those antagonists that are substandard in these commercial production criteria is advisable (Slininger et al., 1994; Schisler and Slininger, 1997). Research directed towards optimizing fermentation media and protocols to produce a maximal amount of biomass that is maximally effective must be conducted on FHB antagonists being considered for commercial development if one reasonably expects to achieve the production of an economically feasible, efficacious biocontrol agent product.

A large investment in microbial formulation research also must be made to bring a product for biologically controlling FHB to the marketplace. Because dewatered or dried antagonist cells offer convenience, handling and economic advantages, research on cell preservation should be a priority. Freeze drying, spray drying and air drying of biomass all have unique challenges to their successful utilization. The use of osmoprotectants (Hounsa et al., 1998), absorbents, as well as applying knowledge of cell stress factors associated with different types of drying (Crowe et al., 1990) will speed this research. Formulation research is also needed in the area of selective enhancement of antagonist populations through the use of nutrient amendments. Carbon sources such as uridine and others are utilized very sparingly in liquid culture by *F. graminearum* but are utilized as a sole carbon source by the FHB antagonist AS 43.3 (Ierulli and Schisler, unpublished). Adding compounds that specifically aid a biocontrol agent without benefit to the pathogen could significantly increase the efficacy of the biocontrol agent, especially if the agent is one that exhibits density dependent production of antimicrobial compounds (Pierson et al., 1994; Beattie and Lindow, 1999). Research to develop formulations with UV protectants to provide shielding of antagonists from the deleterious effects of sunlight (Moody et al., 1999) and to increase the adhesion of applied cells (Buck and Andrews, 1999) would help to insure that the majority of the investment made in producing and applying the antagonist provides a return of reduced FHB. Finally, it is not reasonable to assume that biological control measures alone will provide a complete solution to FHB. Research on developing biological control strains, especially yeast, that are compatible with the more efficacious and safe fungicides available for use against FHB could result in an integrated pest management approach that is more effective than either component alone while permitting less fungicide to be used (Chand-Goyal and Spotts, 1996).

CONCLUSION

The challenge of bringing a biological control product to the marketplace does not end when one has obtained effective antagonists that possess good efficacy when produced in economically feasible liquid media, are readily dried while maintaining viability and efficacy, and can be formulated with carriers and amendments that further enhance their ability as biocontrol agents. Regulatory hurdles such as the requirement for providing toxicological data for registration of biocontrol agents with the EPA can be extremely costly to complete (Cook, 1996), and government and university research facilities generally require at least the monetary participation of an agricultural company to complete the final steps towards a realized product. Thus, a considerable body of work in terms of investing research hours and capital must be completed before a product for the biocontrol of FHB in the marketplace can be realistically expected. However, the fact remains that this disease is without an adequate method of control and continues to devastate wheat growers around the globe on a financial and personal basis. Strategies and microorganisms for biologically controlling this disease have reached an advanced stage of development and offer the promise of being a valuable tool for substantially reducing FHB. With the cooperation of an agricultural company in the private sector, we are confident that the development of an environmentally safe biocontrol product effective against FHB can be achieved.

ACKNOWLEDGMENTS

The authors are grateful for the expert technical assistance of Jennifer Ierulli and Todd Hicks. We are grateful to Patrick Hart and Susan McCormick for analysis of DON content in grain and to Pat Lipps for initiating and maintaining the field plot site in Wooster, Ohio. We thank R. Dill-Macky for supplying wheat seed. This work was made possible in part from funding provided by the U.S. Wheat and Barley Scab Initiative.

REFERENCES

Accerbi, M., Rinaldi, V.E.A., and Ng, P.K.W., 1999, Utilization of highly deoxynivalenol-contaminated wheat via extrusion processing, *J. Food Prot.* 62:1485-1487.

Anderson, T., and Ward, R., 1999, Preface, in:*1999 National Fusarium Head Blight Forum*, University Printing, Michigan State University, East Lansing, MI.

Andrés, L. J., Riera, F. A., and Alvarez, R., 1997, Recovery and concentration by electrodialysis of tartaric acid from fruit juice industries waste waters, *J. Chem. Technol. Biotechnol.* 70:247-252.

Aoki, T., and O'Donnell, K., 1999, Morphological and molecular characterization of *Fusarium pseudograminearum* sp. nov., formerly recognized as the Group 1 population of *F. graminearum*, *Mycologia* 91:597-609.

Atanasoff, D., 1920, Fusarium-blight (scab) of wheat and other cereals, *J. Agric. Res.* 20:1-32.

Bai, G., and Shaner, G., 1994, Scab of wheat: prospects for control, *Plant Dis.* 78:760-766.

Bai, G.-H., Shaner, G., and Ohm H., 2000, Inheritance of resistance to *Fusarium graminearum* in wheat. *Theor Appl. Genet*: 100:1-8.

Baker, K.F., and Cook, R.J., 1974, *Biological Control of Plant Pathogens*, W.H. Freeman, San Francisco.

Baker, K.F., 1987, Evolving concepts of biological control of plant pathogens, *Ann. Rev. Phytopathol.* 25:67-85.

Beardall, J.M., and Miller, J.D., 1994, Diseases in humans with mycotoxins as possible causes, in:*Mycotoxins in Grain: Compounds Other than Aflatoxin*, J.D. Miller and H.L. Trenholm, eds., Eagan Press, St.Paul, MN.

Beattie, G.A., and Lindow, S.E., 1999, Bacterial colonization of leaves:a spectrum of strategies, *Phytopathology* 89:353-359.

Bechtel, D.B., Kaleiku, L.A., Graines, R.L., and Seitz, L.M., 1985, The effects of *Fusarium graminearum* infection on wheat kernels, *Cereal Chem.* 62:191-197.

Benbow, J.M., and Sugar, D., 1999, Fruit surface colonization and biological control of postharvest diseases of pear by preharvest yeast applications, *Plant Dis.* 83:839-844.

Boehm, M.J., Khan, N.I., and Schisler, D.A., 1999, USDA-ARS, Ohio State University cooperative research on biologically controlling Fusarium head blight: 3. field testing of antagonists, in: Proceedings of the *1999 National Fusarium Head Blight Forum*, J.A. Wagester et al., compilers, University Printing, Michigan State University, East Lansing, MI.

Bottalico, A., Logrieco, A., and Visconti, A., 1989, Fusarium species and their mycotoxins in infected cereals in the field and in stored grains, in: *Fusarium Mycotoxins, Taxonomy, and Pathogenicity*, J. Chelkowski, ed., Elsevier, NY.

Buck, J.W., and Andrews, J.H., 1999, Role of adhesion in the colonization of barley leaves by the yeast *Rhodosporidium toruloides*, *Can. J. Microbiol.* 45:433-440.

Bushnell, W.R., Somers, D.A., Giroux, R.W., Szabo, L.J., and Zeyen, R.J., 1998, Genetic engineering of disease resistance in cereals, *Can. J. Plant Pathol.* 20:137-149.

Casale, W.L., and Hart, L.P., 1988, Inhibition of ^{3}H-leucine incorporation by trichothecene mycotoxins in maize and wheat tissue, *Phytopathology* 78:1673-1677.

Campbell, R., 1989, *Biological Control of Microbial Plant Pathogens*, Cambridge University Press, Cambridge, U.K.

Chand-Goyal, R., and Spotts, R.A., 1996, Postharvest biological control of blue mold of apple and brown rot of sweet cherry by natural saprophytic yeasts alone or in combination with low doses of fungicides, *Biol. Control* 6:253-259.

Chisti, Y., and Moo-Young, M., 1999, Fermentation technology, bioprocessing, scale-up and manufacture, in: *Biotechnology: The Science and the Business*, D.G. Springham, ed., Overseas Publishers Association, Newark, NJ.

Clear, R.M., and Patrick, S.K., 1990, Fusarium species isolated from wheat samples containing tombstone (scab) kernels from Ontario, Manitoba, and Saskatchewan, *Can. J. Plant Sci.* 70:1057-1069.

Cook, R.J., and Baker, K.F, 1983, *The Nature and Practice of Biological Control of Plant Pathogens*, The American Phytopathological Society, St. Paul, MN.

Cook, R.J., and Veseth, R.J., 1991, *Wheat Health Management*, APS Press, St. Paul, MN.

Cook, R.J., 1996, Assuring the safe use of microbial biocontrol agents: a need for policy based on read rather than perceived risks, *Can. J. Plant Pathol.* 18:439-445.

Cotty, P.J., 1994, Influence of field application of an atoxigenic strain of *Aspergillus flavus* on the populations of *A. flavus* infecting cotton boils and on the aflatoxin content of cottonseed, *Phytopathology* 84:1270-1277.

Crowe, J.H., Carpenter, J.F., Crowe, L.M., and Anchordoguy, T.J., 1990, Are freezing and dehydration similar stress vectors? A comparison of modes of interaction of stabilizing solutes with biomolecules, *Cryobiology* 27:219-231.

Desjardins, A.E., Christ-Harned, E.A., McCormick, S.P., and Secor, G.A., 1993, Population structure and genetic analysis of field resistance to thiabendazole in *Gibberella pulicaris* from potato tubers, *Phytopathology* 83:164-170.

Desjardins, A.E., Proctor, R.H., Bai, G., McCormick, S.P., Shaner, G., Buechley, G., and Hohn, T.M., 1996, Reduced virulence of trichothecene-nonproducing mutants of *Gibberella zeae* in wheat field tests, *MPMI* 9:775-781.

Desjardins, A.E., and Plattner, R.D., 2000, Fumonisin B_1-nonproducing strains of *Fusarium verticillioides* cause maize (*Zea mays*) ear infection and ear rot, *J. Agric. Food Chem.* 48:(in press).

Dill-Macky, R., and Jones, R.K., 2000, The effect of previous crop residues and tillage on Fusarium head blight of wheat, *Plant Dis.* 84:71-76.

Dorner, J.W., Cole, R.J., and Blankenship, P.D., 1998, Effect of inoculum rate of biological control agents on preharvest aflatoxin contamination of peanuts, *Biol. Control* 12:171-176.

Droby, S., Lischinski, S., Cohen, L., Weiss, B., Daus, A., Chand-Goyal, T., Eckert, J.W., and Manulis, S., 1999, Characterization of an epiphytic yeast population of grapefruit capable of suppression of green mold decay caused by *Penicillium digitatum*, *Biol. Control* 16:27-34.

Fernandez, M.R., 1992, The effect of *Trichoderma harzianum* on fungal pathogens infesting wheat and black oat straw, *Soil Biol. Biochem.* 24:1031-1034.

Fernando, W.G.D., Paulitz, T.C., Seaman, W.L., and Martin, R.A., 1997, Fusarium head blight susceptibility of wheat inoculated at different growth stages, *Phytopathology* 87:S30 (abstract).

Forsyth, O.M., Yoshizawa, T., Morooka, N., and Tuite, J., 1977, Emetic and refusal activity of deoxynivalenol to swine, *Appl. Environ. Microbiol.* 34:547-552.

Fuming, D., Shimin, Z., Beiqu, Z., and An, Z., 1994, Control effects of four bacterial strains on wheat scab (*Fusarium graminearum*), *Acta Agriculturae Shanghai* 10:59-63.

Gupta, M.C., 1986, Biological control of *Fusarium moniliforme* Sheldon and *Pythium butleri* Subramaniam by *Aphelenchus avenae* Bastian in chitin and cellulose-amended soils, *Soil Biol. Biochem.* 18:327-329.

Hammerschmidt, R., 1999, Induced disease resistance: how do induced plants stop pathogens?, *Physiolog. Mol. Plant Pathol.*, 55:77-84.

Hanson, L.E., and Loria, R., 1994, Thiabendazole resistance in *Fusarium* spp. causing dry rot of potato in the Northeastern United States, *Phytopathology* 84:1078.

Hebbar, K.P., Davey, A.G., Merrin, J., and Dart, P.J., 1992, Rhizobacteria of maize antagonistic to *Fusarium moniliforme*, a soil-borne fungal pathogen: colonization of rhizosphere and roots, *Soil Biol. Biochem.* 24:989-997.

Hesseltine, C.W., Rogers, R.F., and Shotwell, O., 1978, Fungi, especially *Gibberella zeae*, and zearalenone occurrence in wheat, *Mycologia* 70:14-18.

Homdork, S., Fehrmann, H., and Beck, R., 2000, Influence of different storage conditions on the mycotoxin production and quality of *Fusarium*-infected wheat grain, *J. Phytopathol.* 148:7-15.

Hounsa, C.-G., Brandt, E.V., Thevelein, J., Hohhmann, S., Prior, B.A., 1998, Role of trehalose in survival of *Saccharomyces cerevisiae* under osmotic stress, *Microbiology* 144:671-680.

Huang, H.C., 1991, Control of soilborne pathogens by mycoparasites:prospects and constraints, in: *The Biological Control of Plant Diseases*, Bay-Petersen, J., ed., Food and Fertilizer Technology Center for the Asian and Pacific Region, Taipei.

Johnston, H.W., 1994, Resistance in advanced winter wheat breeding lines to scab, 1993. *Biol. Cultural Tests* 9:119.

Jones, R.K., 1999, Seedling blight development and control in spring wheat damaged by *Fusarium graminearum* Group 2, *Plant Dis.* 83:1013-1018.

Kang, Z., and Buchenauer, H., 1999, Immunocytochemical localization of fusarium toxins in infected wheat spikes by *Fusarium culmorum*, *Physiol. Mol. Plant Pathol.* 55:275-288.

Karlovsky, P., 1999, Biological detoxification of fungal toxins and its use in plant breeding, feed and food production, *Nat. Toxins* 7:1-23.

Kasuga, F., Hara-Kudo, Y., Saito, N., Kumagai, S., and Sugita-Konishi, 1998, *In vitro* effect of deoxynivalenol on the differentiation of human colonic cell lines Caco-2 and T84, *Mycopathologia* 142:161-167.

Khan, N.I., Schisler, D.A., Boehm, M.J., Slininger, P.J., and Bothast, R.J., 2001, Selection ad evaluation of microorganisms for biocontrol of Fusarium head blight of wheat incited by *Gibberella zeae*, *Plant Dis.* (In press).

Kloepper, J.W., Tuzun, S., Kuf, J.A., 1992, Proposed definitions related to induced disease resistance, *Biocontrol Sci. Technol.* 2:349-351.

Knasmüller, S., Bresgen, N., Kassie, F., Mersch-Sundermann, V., Gelderblom, W., Zöhrer, E., and Eckl, P.M., 1997, Genotoxic effects of three *Fusarium* mycotoxins, fumonisin B_1, moniliformin and vomitoxin in bacteria and in primary cultures of rat hepatocytes, *Mutation Res.* 391:39-48.

Kolombet, L.V., Ezov, D.V., Zigletsova, S.K., Bistrova, E.V., and Kosareva, N.I., 1999, Strain N16 *Trichoderma viride* Pers ex S.F. Gray as bioagent for plant protection against plant diseases and formulation for plant protection. Russian Patent N 99100509/13 (000451) January 10, 1999.

Langseth, W., Bernhoft, A., Rundberget, T., Kosiak, B., and Gareis, M., 1999, Mycotoxin production and cytotoxicity of *Fusarium* strains isolated from Norwegian cereals, *Mycopathologia*, 144:103-113.

Li, F.Q., Luo, X.Y., Yoshizawa, T., 1999, Mycotoxins (trichothecenes, zearalenone and fumonisins) in cereals associated with human red-mold intoxications stored since 1989 and 1991 in China, *Nat. Toxins*, 7:93-97.

Loper, J., 1993, Roles of competition and antibiosis in suppression of plant diseases by bacterial biological control agents, in: *Pest Management: Biologically Based Technologies*, R.D. Lumsden and J.L. Vaughn, eds., American Chemical Society, Washington, DC.

Luo, X.Y., 1992, Food poisoning caused by *Fusarium* toxins in China, in: *International Life Science Institute (ILSI),Proceedings of the Second Asian Conference of Food Safety*, ILSI, Bangkok, Thailand.

Luz, W.C. da., 1988, Biocontrol of fungal pathogens of wheat with bacteria and yeasts, in: Proceedings of the *5th International Congress of Plant Pathology*, Kyoto, Japan.

Martin, R.A., and Johnston, H.W., 1982, Effects and control of *Fusarium* diseases of cereal grains in the Atlantic Provinces, *Can. J. Plant Pathol.* 4:210-216.

Marasas, W.F.O., 1991, Toxigenic Fusaria, in: *Mycotoxins and Animal Foods*, J.E. Smith and R.S. Henderson, eds., CRC Press, Inc., Boca Raton, FL.

McMullen, M., and Bergstrom, G., 1999, Chemical and biological control of Fusarium head blight: 1999 projects and progress, in: Proceedings of the *1999 National Fusarium Head Blight Forum*, J.A. Wagester et al., compilers, University Printing, Michigan State University, East Lansing, MI.

McMullen, M., Jones, R., and Gallenberg, D., 1997, Scab of wheat and barley: a re-emerging disease of devastating impact, *Plant Dis.* 81:1340-1348.

Mercier, J., and Lindow, S.E., 2000, Role of leaf surface sugars in colonization of plants by bacterial epiphytes, *Appl. Environ. Microbiol.* 66:369-374.

Miller, J.D., and Arnison, P.G., 1986, Degradation of deoxynivalenol by suspension cultures of the fusarium head blight resistant wheat cultivar Frontana, *Can. J. Plant Pathol.* 8:147-150.

Miller, J.D., and Ewen, M.A., 1997, Toxic effects of deoxynivalenol on ribosomes and tissues of the spring wheat cultivars Frontana and Casavant, *Nat. Toxins* 5:234-237.

Miller, J.D., Culley, J., Fraser, K., Hubbard, S., Meloche, F., Ouellet, T., Seaman, W.L., Seifert, K.A.,

Turkington, K., and Voldeng, H., 1998, Effect of tillage practice on Fusarium head blight of wheat, *Can. J. Plant Pathol.* 20:95-103.
Moody, S.A., Newsham, K.K., Ayres, P.G., and Paul, N.D., 1999, Variation in the responses of litter and phylloplane fungi to UV-B radiation (290-315 nm), *Mycol. Res.* 11:1469-1477.
Moschini, R.C., and Fortugno,C., 1996, Predicting wheat head blight incidence using models based on meteorological factors in Pergamino, Argentina, *Eur. J. Plant Pathol.* 102:211-218.
Müller, H-.M., Reimann, J., Schumacher, U., and Schwadorf, K., 1997, *Fusarium* toxins in wheat harvested during six years in an area of southwest Germany, *Nat. Toxins* 5:24-30.
Munkvold, G.P., and Desjardins, A.E., 1997, Fumonisins in maize: can we reduce their occurrence?, *Plant Dis.* 81:556-565.
Niemira, B.A., Hammerschmidt, R., and Safir, G.R., 1996, Postharvest suppression of potato dry rot (*Fusarium sambucinum*) in prenuclear minitubers by arbuscular mycorrhizal fungal inoculum, *Am. Potato J.* 73:509-515.
Parry, D.W., Jenkinson, P., and McLeod L., 1995, *Fusarium* ear blight (scab) in small grain cereals-a review, *Plant Pathol.* 44:207-238.
Paulitz, T.C., 1999, Fusarium head blight: a re-emerging disease, *Phytoprotection* 80:127-133.
Pavlova, T.V., and Izmalkova A.G., 1995, Ear fusariose of wheat in Krasnodart kray, *Plant Prot.* 11:22B23.
Perondi, N.L., Luz, W.C. da, and Thomas, R., 1996, Controle microbiológico da giberela do trigo, *Fitopatologia Brasiliera* 21:243-249.
Pestka, J.J., and Bondy, G.S., 1994, Immunotoxic effects of mycotoxins, in: *Mycotoxins in Grain: Compounds Other than Aflatoxin*, J.D. Miller and H.L. Trenholm, eds., Eagan Press, St.Paul, MN.
Pierson, L.S. III, Keppenne, V.D., and Wood, D.W., 1994, Phenazine antibiotic biosynthesis in *Pseudomonas aureofaciens* 30-84 is regulated by PhzR in response to cell density, *J. Bacteriol.* 176:3966-3974.
Pritsch, C., Muehlbauer, G.J., Bushnell, W.R., Somers, D.A., and Vance, C.P., 2000, Fungal development and induction of defense response genes during early infection of wheat spikes by *Fusarium graminearum*, *MPMI* 13:159-169.
Polley, R.W., Turner, J.A., Cockerell, V., Robb, J., Scudamore, K.A., Sanders, M.F., and Magan, N., 1991, Survey of *Fusarium* species infecting winter wheat in England, Wales and Scotland, 1989 & 1990, in: *Project Report No. 39*, Home-Grown Cereals Authority, London.
Proctor, R.H., Hohn, T.M., and McCormick, S.P., 1995, Reduced virulence of *Gibberella zeae* caused by disruption of a trichothecene toxin biosynthetic gene, *MPMI* 8:593-601.
Rizzo, I., Lori, G., Vedoya, G., Carranza, M., Haidukowski, M., Varsavsky, E., Frade, H., Chiale, C., and Alippi, H., 1997, Sanitary factors and mycotoxins contamination in the Argentinian wheat crop 1993/94, *Mycotoxin Res.* 13:67-72.
Schisler, D.A., and Slininger, P.J., 1994, Selection and performance of bacterial strains for biologically controlling Fusarium dry rot of potatoes incited by *Gibberella pulicaris*, *Plant Dis.* 78:251-255.
Schisler, D.A., Kurtzman, C.P., Bothast, R.J., and Slininger, P.J., 1995, Evaluation of yeasts for biological control of Fusarium dry rot of potatoes, *Am. Potato J.* 72:339-353.
Schisler, D.A., and Slininger, P.J., 1997, Microbial selection strategies that enhance the likelihood of developing commercial biological control products, *J. Ind. Microbiol. Biotech.* 19:172-179.
Schisler, D.A., Slininger, P.J. , and Bothast, R.J., 1997, Effects of antagonist cell concentration and two-strain mixtures of biological control of Fusarium dry rot of potatoes, *Phytopathology* 87:177-183.
Schisler, D.A., Burkhead, K.D., Slininger, P.J., and Bothast, R.J., 1998, Selection, characterization and use of microbial antagonists for the control of Fusarium dry rot of potatoes, in: *Plant-Microbe Interactions and Biological Control*, G.J. Boland and L.D. Kuykendall, eds., Marcel Dekker, Inc., NY.
Schisler, D.A., Khan, N.I., Boehm, M.J., Slininger, P.J., and Bothast, R.J., 1999, USDA-ARS, Ohio State University cooperative research on biologically controlling Fusarium head blight: 1. antagonist selection and testing on durum wheat, in: Proceedings of the *1999 National Fusarium Head Blight Forum*, J.A. Wagester et al., compilers, University Printing, Michigan State University, East Lansing, MI.
Schisler, D.A., Slininger, P.J., Hanson, L.E., and Loria, R., 2000a, Potato cultivar, pathogen isolate and antagonist cultivation medium influence the efficacy and ranking of bacterial antagonists of Fusarium dry rot, *Biocontrol Sci. Technol.* 10:267-279.
Schisler, D.A., Slininger, P.J., Kleinkopf, G., Bothast, R.J., and Ostrowski, R.C., 2000b, Biological control of Fusarium dry rot of potato tubers under commercial storage conditions, *Am. J. Potato Res.* 77:29-40.
Slininger, P.J., Schisler, D.A., and Bothast R.J., 1994, Two-dimensional liquid culture focusing: a method of selecting commercially promising microbial isolates with demonstrated biological control capability, in: *Improving Plant Productivity with Rhizosphere Bacteria, 3rd International Workshop on Plant Growth-Promoting Rhizobacteria*, M.H. Ryder, P.M. Stephens, and G.D. Bowen, eds., Graphic Services, CSIRO Division of Soils:Glen Osmond, Adelaide, Australia.
Slininger, P.J., Burkhead, K.D., Schisler, D.A., and Bothast, R.J., 2000, Biological control of sprouting in potatoes, *U.S. Patent*, 6,107,247, August 22, 2000.

Snijders, C.H.A., 1990, Fusarium head blight and mycotoxin contamination of wheat, a review, *Neth. J. Plant Pathol.* 96:187-198.

Sokolov M.S., Pikushova, E.A., and Levashova G.I., 1998, Traditional and new ways to control winter wheat from ear and cereal diseases, *Agrochemistry* 3:67B77.

Stack, R.W., and McMullen, M.P., 1995, A visual scale to estimate severity of Fusarium head blight in wheat, *North Dakota State University Extension Service Bulletin PP-1095.*

Stockwell, C.A., Bergstrom, G.C., and da Luz, W.C., 1999a, Selection of microbial antagonists for biological control of Fusarium head blight of wheat, in: Proceedings of the *1999 National Fusarium Head Blight Forum*, J.A. Wagester et al., compilers, University Printing, Michigan State University, East Lansing, MI.

Stockwell, V.O., McLaughlin, R.J., Henkels, M.D., Loper, J.E., Sugar, D., and Roberts, R.G., 1999b, Epiphytic colonization of pear stigmas and hypanthia by bacteria during primary bloom, *Phytopathology* 89:1162-1168.

Strange, R. N., and Smith, H., 1971, A fungal growth stimulant in anthers which predisposes wheat to attack by *Fusarium graminearum*, *Physiol. Pl. Path.* 1:141-150.

Strange, R. N., and Smith, H., 1978, Specificity of choline and betaine as stimulants of *Fusarium graminearum*, *Trans. Br. Mycol. Soc.* 70:187-192.

Sutton, J.C., 1982, Epidemiology of wheat head blight and maize ear rot caused by *Fusarium graminearum*, *Can. J. Plant Pathol.* 4:195-209.

Suty, A., and Mauler-Machnik, A., 1997, Fusarium ear blight on wheat-epidemiology and control of *Gibberella zeae*, the teleomorph of *Fusarium graminearum* with Folicur, in: *Diagnosis and Identification of Plant Pathogens,* Proceedings of the 4th International Symposium of the European Foundation for Plant Pathology, H.-W. Dehne et al, eds., Kluwer Academic Publishers, Dordrecht.

Teich, A.H., and Hamilton, J.R., 1985, Effect of cultural practices, soil phosphorus, potassium, and pH on the incidence of Fusarium head blight and deoxynivalenol levels in wheat, *Appl. Environ. Microbiol.* 49:1429-1431.

Vesonder, R.F., and Hesseltine, C.W., 1980, Vomitoxin:natural occurrence on cereal grains and significance as a refusal and emetic factor to swine, *Process Biochem.* 16:12-15.

Vestal, E.F., 1964, Barley scab in South Korea in 1963 and 1964, *Plant Dis. Rep.* 48:754-755.

Wei, G., Kloepper, J.W., and Tuzun, S., 1991, Induction of systemic resistance of cucumber to *Colletotrichum orbiculare* by select strains of plant growth-promoting rhizobacteria, *Phytopathology* 81:1508-1512.

Wicklow, D.T., Horn, B.W., Shotwell, O.L., Hesseltine, C.W., and Caldwell, R.W., 1988, Fungal interference with *Aspergillus flavus* infection and aflatoxin contamination of maize grown in a controlled environment, *Phytopathology* 78:68-74.

Wilcoxson, R.D., 1996, Fungicides for control of Fusarium head blight, *Int. J. Tropical Plant Dis.* 14:27-50.

Will, M.E., Wilson, D.M., and Wicklow, D.T., 1994, Evaluation of *Paecilomyces lilacinus*, chitin, and cellulose amendments in the biological control of *Aspergillus flavus* fungi, *Biol. Fertil. Soils* 17:281-284.

Wilson, C.L., 1997, Biological control and plant diseases-a new paradigm, *J. Ind. Microbiol. Biotechnol.* 19:158-159.

Wilson, D.M., 1988, Potential for biological control of *Aspergillus flavus* and aflatoxin contamination, in: *Biocontrol of Plant Diseases, Volume II*, K.G. Mukerji and K.L. Garg, eds., CRC Press, Inc., Boca Raton, FL.

ANALYTICAL ASPECTS OF MYCOTOXINS: INTRODUCTION

Mary W. Trucksess[1] and Samuel W. Page[2]

Food and Drug Administration
[1]Center for Food Safety and Applied Nutrition
[2]Joint Institution for Food Safety and Applied Nutrition
Washington, DC 20204

Mycotoxins have a wide range of toxic effects, including carcinogenicity, neurotoxicity, immunotoxicity, and reproductive and developmental toxicity. Regulatory limits have been established for some of the well-known mycotoxins such as the aflatoxins, deoxynivalenol, ochratoxin A, and patulin in order to minimize human and animal exposure. Regulatory limits must be developed using science-based risk assessments to avoid unnecessary economic loss. Exposure data from surveillance studies are needed for risk assessments. Quality of data generated in these surveys depends on the analytical methods used and the adherence to quality assurance principles by skillful analysts.

Mycotoxin analytical methods consist of sampling, sample preparation and analysis. Because of the heterogeneous distribution of mycotoxins in grains and oilseeds much effort has been focused on sampling. The aim of sampling is to take a sample representative of the lot at the appropriate point using a selected collecting method. Then the sample is prepared to obtain a representative laboratory sample, using proper subsampling, grinding and mixing procedures.

Most analytical procedures consist of three basic steps: extraction, purification/separation, and quantitation. Mycotoxins are easily extracted into various mixtures of water and organic solvents including chloroform, methanol, acetonitrile, and acetone. Solid phase extraction (SPE), liquid-liquid partition, and precipitation are commonly used for purification. Traditional chromatographic techniques such as thin layer chromatography (TLC), gas chromatography (GC), and liquid chromatography (LC) are widely employed for separation and quantitation. The recently developed immunochemical methods also known as immunoaffinity-based methods (IA) utilize mycotoxin specific polyclonal or monoclonal antibodies to purify the toxin from sample extract and to directly extract and purify the toxin from liquid matrices. There are three types of IA for mycotoxins. The enzyme-linked immunosorbent assay (ELISA) and immunoaffinity column (IAC) assays are most commonly used. The use of biosensors for detection/quantitation of mycotoxins has recently been reported.

This section presents current methods of analysis by recognized experts in the various disciplines. The format of each paper is to review the methods for the particular mycotoxin; present one method in the theoretical terms; apply the method to the analysis;

Mycotoxins and Food Safety, edited by Trucksess *et al.*
Kluwer Academic/Plenum Publishers, 2002

and discuss the results of method performance. The first paper deals with sampling wheat for deoxynivalenol. The sampling variation seems less of a source of error than that found for other mycotoxin test procedures. The second paper discusses the current methods for aflatoxins as well as the novel approach of using biosensors. The third paper emphasizes the use of immunoaffinity-based cleanup and mass spectrometric (MS) analysis for fumonisins. It also describes the isolation of fumonisins from ELISA microtiter wells prior to LC/electrospray MS for confirmation of chemical identity. The fourth and fifth papers are concerned with immunoaffinity column cleanup of cyclopiazonic acid in peanuts or corn and ochratoxin A in grains, wine and beer before separation and quantification of the analytes by LC. Finally, the last two papers utilize liquid partition or solid phase extraction purification and GC/MS for quantitation or confirmation of patulin in juice and trichothecenes in foods with and without derivatization of the toxins.

SAMPLING WHEAT FOR DEOXYNIVALENOL

Thomas B. Whitaker,[1] Winston M. Hagler, Jr.,[2] Francis G. Giesbrecht,[3] Anders S. Johansson[1]

[1]U.S. Department of Agriculture, Agricultural Research Service, Box 7625, North Carolina State University, Raleigh, NC 27695-7625
[2]North Carolina State University, Department of Poultry Science, Mycotoxin Laboratory, Box 7636, Raleigh, NC 27695-7636
[3]North Carolina State University, Department of Statistics, Box 8203, Raleigh, NC 27695-8203

ABSTRACT

The variability associated with testing wheat for deoxynivalenol (DON) was measured using a 0.454 kg sample, a Romer mill, 25 g of comminuted subsample and the Romer Fluoroquant analytical method. The total variability was partitioned into sampling, sample preparation, and analytical variability components. Each variance component was found to be a function of the DON concentration and equations were developed to predict each variance component using regression techniques. The effects of sample size, subsample size, and number of aliquots on reducing the variability of the DON test procedure were also determined. Using the test procedure described above, the coefficient of variation (CV) associated with testing wheat at 5 ppm DON was found to be 13.4%. The CVs associated with sampling, sample preparation, and analysis were 6.3, 10.0, and 6.3%, respectively. The sample variations associated with testing wheat are relatively small when compared to CVs associated with testing other commodities for other mycotoxins such as aflatoxin in peanuts. Even with the use of a small sample size (0.454 kg), the sampling variation was not the largest source of error as found in other mycotoxin test procedures.

INTRODUCTION

Deoxynivalenol, sometimes called DON or vomitoxin, is a naturally occurring mycotoxin produced by the several fungal species in the genus *Fusarium* (Mirocha et al., 1976). *Fusarium* infects wheat and other small grains such as barley and corn. DON toxicosis is associated with feed refusal, vomiting, and reproductive problems in animals (Vesconder and Hesseltine, 1981). DON is most often associated with cool and wet environmental conditions where *Fusarium* tends to thrive. The U.S. Food and Drug Administration (FDA) currently has the following DON advisory limits: (a) 1 part per

Mycotoxins and Food Safety, edited by Trucksess *et al.*
Kluwer Academic/Plenum Publishers, 2002

million (ppm) on consumer-ready wheat products such as flour and bran, (b) 10 ppm on grains destined for ruminating beef and feedlot cattle over four months of age, swine, and chickens, and (c) 5 ppm on grain and grain products for all other animals (Food and Agriculture Organization, 1995). Advisory limits were established by FDA as guidelines for industry to use as a standards in the management and reduction of DON contamination in grain and grain products. However, the FDA will not automatically take regulatory action if limits are exceeded. Because of the FDA advisory limit, grain producers and processors routinely test grains for DON to determine if levels exceed the FDA advisory limits. The Federal Grain Inspection Packers and Stockyard Administration (GIPSA) of USDA provides official DON testing programs for grain producers and processors.

The test procedure used to measure DON in small grains is similar to that used to measure other mycotoxin in other grains. The test procedure consists of three steps. First is the sampling step where a random sample (test sample) is taken from the lot. Second is the sample preparation step where the entire test sample is comminuted in a mill or grinder and a subsample is removed from the comminuted test sample. Grinding and subsampling are collectively called the sample preparation step. Third is the analytical step where DON in the subsample is solvent extracted, purified, and quantified.

The variability associated with each of the three steps of the test procedure contributes to the total variability associated with the DON test procedure and makes it difficult to estimate the true DON concentration of a bulk lot with a high degree of confidence. As a result, it is difficult to accurately classify lots into categories required by management strategies and regulatory activities. If the variability of the DON test procedure can be reduced, the lot concentration can be estimated with more confidence, which means that producers and processors will suffer smaller economic losses due to fewer lots being misclassified by the testing program. It is important to be able to evaluate the performance of a DON sampling plan and to design sampling plans that reduce misclassification of producer's lots. To develop methods to design and evaluate DON sampling plans, the variability and distributional characteristics associated with a DON test procedure need to be determined.

Previous studies demonstrated that the variability among aflatoxin test results for corn, peanuts, and cottonseed is very large (Whitaker et al., 1976; Whitaker et al., 1979; Whitaker et al., 1994). However, a recent DON study with wheat (Hart and Schabenberger, 1998) indicated that the variability among probe sample concentrations may be much less than among aflatoxin test results for peanuts and other commodities. Less variability among samples with the same number of kernels suggest that the distribution among wheat kernels contaminated with DON may be different from the distribution among peanut kernels contaminated with aflatoxin (Whitaker et al., 1972).

It is important to design a DON test procedure that will have the lowest variability that resources will allow. Therefore, the objectives of this study were: (a) to measure the variability of the sampling, sample preparation, and analytical steps of a test procedure used to measure DON in wheat; and (b) to show how to decrease the variability of the test procedure and achieve more precise results.

EXPERIMENTAL

Variability Estimates

The sources of error associated with the DON test procedure are shown in Figure 1. The total variability is a function of the sampling, sample preparation, and analytical variability. Assuming independence among the random errors for each step of the DON procedure, the total variance (σ^2_t) is assumed to be equal to the sum of the sampling variance (σ^2_s), sample preparation variance (σ^2_{sp}), and analytical variance (σ^2_a).

$$\sigma^2_t = \sigma^2_s + \sigma^2_{sp} + \sigma^2_a \quad (1)$$

The variance model shown in Equation 1 has been described in detail by Whitaker et al. (1974). It is also assumed that sample selection and sample preparation methods are random in nature and no biases are introduced in the application of the DON test procedure. Experimental estimates of the true variance, σ^2 are denoted by S^2.

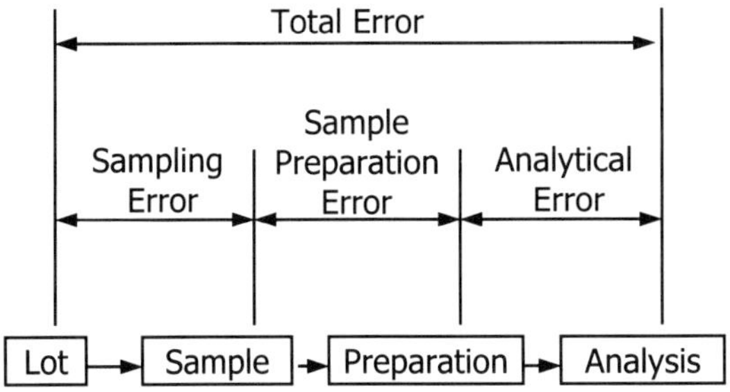

Figure 1. Partitioning the total variance associated with a mycotoxin test procedure into sampling, sample preparation, and analytical variance components.

Because analytical procedures are required to measure DON in the comminuted subsample, the sampling and sample preparation variances cannot be measured directly, but can be estimated indirectly. Experiments were designed to measure the total variance (S^2_t), combined sample preparation and analytical variance (S^2_{spa}), and the analytical variance (S^2_a). The combined sample preparation and analytical variances (S^2_{spa}) is defined in equation 2.

$$S^2_{spa} = S^2_{sp} + S^2_a \quad (2)$$

Knowing the total variance, combined sample preparation and analytical variance, and analytical variance, the sampling variance and sample preparation variance can be determined by subtraction using equations 1 and 2.
As a result,

$$S^2_{sp} = S^2_{spa} - S^2_a \quad (3)$$

and

$$S^2_s = S^2_t - (S^2_{sp} + S^2_a) \quad (4)$$

or

$$S^2_s = S^2_t - S^2_{spa} \quad (5)$$

EXPERIMENTAL DESIGN

Total Variance

A 20 kg bulk sample was taken from each of 24 commercial lots of wheat suspected of DON contamination and destined for further processing. Each bulk sample was riffle divided into thirty-two, 0.454 kg (one pound) samples. Each 0.454 kg sample was ground with a Romer mill (Romer Labs, Union, MO) with the grinding plates set to give the finest degree of grind. The adjustable gate on the Romer mill was set to automatically give about a 25 g subsample from the 0.454 kg sample during comminution. The DON in the 25 g subsample was extracted and quantified using the Romer DON FluoroQuant™ (Romer Labs, Union, MO) analytical method with a fluorometer (Malone et al., 1998). The total variance (S^2t) associated with testing DON in each of the 24 lots was estimated from the 32 DON measurements per lot. A total of 24 x 32 or 768 samples were analyzed for DON to measure total variability.

Sample Preparation Plus Analytical Variance

Twenty comminuted samples were selected from the 768 comminuted wheat samples described above to provide samples with a wide range in DON concentration. Eight, 25 g subsamples were taken from each of the 20 comminuted samples using a riffle divider. The Romer Fluoroquant method was used to measure DON in four aliquots taken from each subsample extract. The combined sample preparation and analytical variance (S^2_{spa}) and the analytical (S^2_a) variance were estimated from the 20 x 8 x 4, or 640 DON measurements in the nested design using SAS procedures (Statistical Analysis System Institute, 1997). Using equations 3 and 5, the sampling and sample preparation variances were estimated.

The sampling variance, S^2_s, is specific for a 0.454 kg sample; the sample preparation variance, S^2_{sp}, is specific to grinding the sample with a Romer mill and using a 25 g comminuted subsample for extraction; and the analytical variance, S^2_a, is specific to the Romer Fluoroquant method (Whitaker et al., 1996). DON measurements are reported in parts per million (ppm).

RESULTS

Total Variance

The 32 DON measurements used to measure total variance are shown in Table 1 for each of the 24 lots. Within each lot, DON test results are ranked from low to high. The lot concentrations were estimated by averaging the 32 DON measurements for each lot. The estimated lot concentrations ranged from 0.02 to 14.38 ppm. The 24 lots are also ranked by lot concentration from low to high. Table 1 allows examination of the range of DON test results for each lot. For example, sample test results for lot 21 ranged from 0.00 to 0.65 ppm and the 32 DON sample values averaged 0.27 ppm.

The estimated total variance, coefficient of variation, and DON concentration are shown in Table 2 for each of the 24 lots. The lots are ranked from low to high concentration. A full-log plot of total variance (S^2_t) versus DON concentration (C) is shown in Figure 2. Consistent with other mycotoxin studies, the variability among DON test results appears to increase with DON concentration. Because the plot is approximately linear in a full-log plot, a regression equation of the form

$$S^2_t = a\,C^b \qquad (6)$$

Table 1. Deoxynivalenol (DON) test results measured in 32 samples of wheat from each of 24 lots; DON test procedure reflects 0.454 kg sample, Romer mill, 25 g subsample and Romer Fluoroquant analytical method

Lot	Sample																													
	1	2	3	4	5	6	7	8	9	10	11	12	13	14	15	16	17	18	19	20	21	22	23	24	25	26	27	28	29	30	31
12	0.00	0.00	0.00	0.00	0.00	0.00	0.00	0.00	0.00	0.00	0.00	0.00	0.00	0.00	0.00	0.00	0.00	0.00	0.00	0.00	0.00	0.00	0.00	0.00	0.00	0.00	0.03	0.04	0.07	0.08	0.17
13	0.00	0.00	0.00	0.00	0.00	0.00	0.00	0.00	0.00	0.00	0.00	0.00	0.00	0.00	0.00	0.00	0.00	0.00	0.00	0.00	0.00	0.00	0.00	0.00	0.01	0.07	0.08	0.16	0.22	0.24	0.25
17	0.00	0.00	0.00	0.00	0.00	0.00	0.00	0.00	0.00	0.00	0.00	0.00	0.00	0.00	0.00	0.00	0.00	0.00	0.00	0.01	0.02	0.03	0.05	0.06	0.06	0.12	0.13	0.16	0.16	0.17	0.20
11	0.00	0.00	0.00	0.00	0.00	0.00	0.00	0.00	0.00	0.00	0.00	0.00	0.00	0.00	0.00	0.00	0.00	0.00	0.00	0.00	0.00	0.00	0.00	0.00	0.02	0.06	0.07	0.10	0.28	0.30	0.38
9	0.00	0.00	0.00	0.00	0.00	0.00	0.00	0.00	0.00	0.00	0.00	0.00	0.00	0.03	0.05	0.05	0.05	0.08	0.08	0.08	0.08	0.10	0.12	0.12	0.14	0.15	0.16	0.17	0.20	0.30	0.35
16	0.00	0.00	0.00	0.00	0.00	0.00	0.00	0.00	0.00	0.00	0.00	0.00	0.00	0.00	0.00	0.00	0.00	0.00	0.00	0.01	0.01	0.07	0.07	0.13	0.16	0.19	0.21	0.27	0.34	0.41	0.49
15	0.00	0.00	0.00	0.00	0.00	0.00	0.00	0.00	0.00	0.00	0.00	0.00	0.00	0.00	0.00	0.00	0.02	0.06	0.08	0.08	0.12	0.15	0.15	0.16	0.17	0.18	0.19	0.26	0.38	0.54	0.72
18	0.00	0.00	0.00	0.00	0.00	0.00	0.00	0.00	0.00	0.00	0.00	0.01	0.01	0.01	0.02	0.04	0.05	0.05	0.06	0.06	0.09	0.12	0.12	0.13	0.19	0.19	0.20	0.22	0.44	0.59	0.71
22	0.00	0.00	0.00	0.00	0.00	0.00	0.00	0.00	0.00	0.04	0.04	0.05	0.06	0.09	0.12	0.13	0.14	0.15	0.17	0.20	0.20	0.22	0.23	0.24	0.25	0.31	0.32	0.33	0.38	0.40	0.52
21	0.00	0.00	0.00	0.00	0.06	0.08	0.09	0.13	0.14	0.15	0.17	0.18	0.19	0.20	0.24	0.25	0.27	0.28	0.29	0.29	0.29	0.29	0.29	0.31	0.36	0.46	0.54	0.56	0.58	0.58	0.64
24	0.89	1.30	1.30	1.50	1.50	1.60	1.60	1.70	1.70	1.70	1.80	1.80	1.80	1.80	1.80	1.80	1.80	1.90	2.00	2.00	2.00	2.00	2.10	2.10	2.20	2.20	2.20	2.20	2.30	2.30	2.70
2	1.40	1.50	1.70	1.70	1.80	1.80	2.00	2.00	2.10	2.10	2.10	2.20	2.20	2.20	2.30	2.30	2.30	2.30	2.40	2.40	2.40	2.50	2.50	2.60	2.60	2.70	2.70	2.70	2.80	2.90	3.10
4	2.20	2.20	2.30	2.50	2.50	2.60	2.60	2.70	2.70	2.70	2.70	2.80	2.80	2.80	2.80	2.80	2.90	2.90	3.00	3.00	3.10	3.10	3.10	3.10	3.10	3.30	3.30	3.40	3.50	3.60	3.70
23	2.20	2.60	2.70	2.80	2.90	2.90	2.90	2.90	2.90	2.90	2.90	2.90	3.00	3.00	3.00	3.00	3.10	3.10	3.10	3.20	3.30	3.40	3.40	3.50	3.60	3.70	3.70	3.80	3.80	3.80	3.90
3	2.70	2.80	2.90	3.00	3.30	3.30	3.40	3.40	3.50	3.50	3.50	3.50	3.60	3.60	3.60	3.60	3.60	3.70	3.70	3.70	3.70	3.80	3.90	3.90	4.00	4.10	4.20	4.30	4.30	4.40	4.40
20	3.30	3.60	3.70	3.70	3.70	3.70	3.80	3.80	3.80	3.90	3.90	4.10	4.10	4.20	4.20	4.20	4.20	4.30	4.30	4.30	4.30	4.40	4.50	4.50	4.60	4.60	4.60	4.80	5.10	5.20	5.30
14	4.10	4.60	4.70	4.90	4.90	5.20	5.20	5.20	5.30	5.30	5.40	5.40	5.40	5.40	5.50	5.60	5.60	5.80	5.80	5.80	5.90	6.00	6.10	6.20	6.20	6.30	6.30	6.60	7.20	7.20	7.30
7	4.50	5.20	5.60	5.60	5.60	5.70	5.70	5.80	5.90	5.90	6.00	6.00	6.00	6.00	6.10	6.10	6.10	6.20	6.40	6.50	6.60	6.70	6.70	6.80	6.80	6.80	7.00	7.00	7.00	7.30	7.50
8	7.10	7.10	7.20	7.20	7.40	7.50	7.60	7.70	7.80	7.80	7.90	7.90	8.00	8.00	8.10	8.10	8.20	8.20	8.20	8.20	8.30	8.40	8.40	8.70	9.10	9.10	9.30	9.70	9.90	10.00	10.00
6	6.40	7.40	7.50	7.70	7.70	7.80	7.90	8.00	8.00	8.00	8.00	8.10	8.10	8.30	8.30	8.50	8.60	8.80	8.80	8.80	8.80	9.00	9.00	9.00	9.10	9.10	9.20	9.40	9.40	9.60	9.80
5	7.30	7.50	7.60	7.70	7.70	7.70	7.80	7.90	8.00	8.00	8.10	8.10	8.10	8.10	8.20	8.30	8.50	8.60	8.60	8.70	8.90	8.90	9.00	9.20	9.20	9.20	9.60	9.70	9.70	9.90	10.00
1	7.10	7.70	7.70	7.70	7.80	8.10	8.20	8.20	8.20	8.30	8.40	8.40	8.50	8.50	8.50	8.80	8.80	9.00	9.00	9.00	9.00	9.10	9.20	9.30	9.50	9.60	9.60	9.80	9.90	10.00	10.00
10	7.90	8.00	8.20	8.40	8.50	8.60	8.80	8.90	8.90	8.90	9.10	9.10	9.10	9.10	9.30	9.40	9.40	9.50	9.50	9.50	9.90	9.90	10.00	10.00	10.00	10.00	10.00	10.00	11.00	11.00	11.00
19	12.00	12.00	13.00	13.00	13.00	14.00	14.00	14.00	14.00	14.00	14.00	14.00	14.00	14.00	14.00	14.00	14.00	14.00	14.00	15.00	15.00	15.00	15.00	15.00	15.00	15.00	15.00	16.00	16.00	16.00	16.00

Where a and b are constants determined from the regression analysis is used to relate variance, S^2t, to the DON concentration, C. From the regression analysis, equation 6 becomes

$$S^2_t = 0.117\, C^{0.817} \tag{7}$$

with a correlation coefficient of 0.91 in the log scale. The standard error associated with the exponent (0.817) is 0.035. Regression equation 7 is also shown in Figure 2 with the variance data.

Table 2. Total variance and coefficient of variation (CV) associated with measuring deoxynivalenol in wheat lots. The test procedure reflects a 0.454 kg sample, a Romer mill, a 25 g subsample and the Romer Fluoroquant analytical method.

Lot #	Lot Concentration (ppm)	Total Variance	Coefficient of Variation (%)
12	0.02	0.00	261.8
13	0.04	0.01	209.9
17	0.05	0.01	168.1
11	0.05	0.01	231.1
9	0.09	0.02	145.2
16	0.10	0.03	180.9
15	0.13	0.04	161.7
18	0.13	0.04	164.9
22	0.16	0.02	98.3
21	0.27	0.04	72.7
24	1.88	0.15	20.2
2	2.31	0.22	20.1
4	2.94	0.21	15.6
23	3.19	0.19	13.7
3	3.68	0.22	12.9
20	4.28	0.34	13.5
14	5.75	0.66	14.1
7	6.28	0.50	11.2
8	8.32	0.79	10.7
6	8.50	0.62	9.3
5	8.59	0.76	10.2
1	8.78	0.61	8.9
10	9.50	1.05	10.8
19	14.38	1.27	7.9

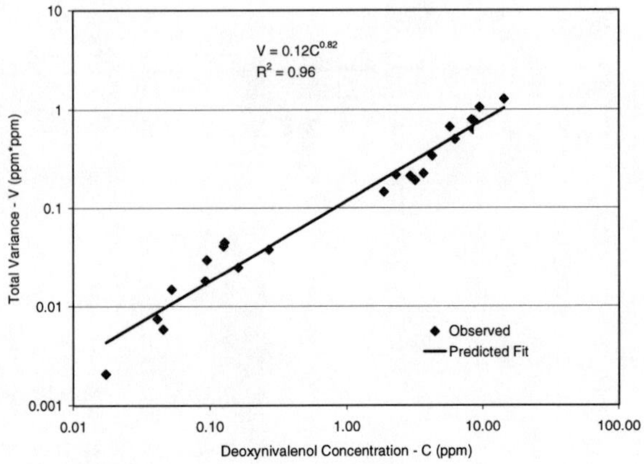

Figure 2. Full-log plot of total variance versus deoxynivalenol (DON) concentration when wheat is tested for DON, DON test reflects 0.454 kg sample, Romer mill, 25 g subsample and Romer Fluoroquant analytical method.

Sample Preparation Plus Analytical Variance

The data in the nested design were analyzed with a SAS procedure (that determines the combined sample preparation and analytical variance (S^2_{spa}) and the analytical variance (S^2_a). The combined sample preparation and analytical variance, analytical variance, and sample concentration are shown in Table 3 for each of the 20 samples. The samples are also ranked from low to high concentration. The combined sample preparation and analytical variance and the analytical variances in Table 3 appear to increase with DON concentration and are plotted versus DON concentration in full-log plots in Figures 3 and 4, respectively. From the regression analysis, the combined sample preparation and analytical variance and the analytical variance equations are shown in equations 8 and 9, respectively.

$$S^2_{spa} = 0.083\ C^{0.913} \tag{8}$$

$$S^2_a = 0.028\ C^{0.783} \tag{9}$$

The correlation coefficients associated with the regression in equations 8 and 9 are 0.97 and 0.88, respectively. The standard error associated with the exponents in equations 8 and 9 are 0.056 and 0.097, respectively. Regression equations 8 and 9 are also shown in Figures 3 and 4, respectively.

Table 3. Combined sample preparation and analytical variance and the analytical variance associated with testing wheat samples for deoxynivalenol. The test procedure reflects the use of a Romer mill, a 25 g subsample and the Romer Fluoroquant analytical method.

Lot #	Sample #	Sample Conc. (ppm)	Combined Sample Prep. & Analytical Variance	Analytical Variance
11	22	0.03	0.00	0.00
17	22	0.09	0.01	0.00
18	21	0.11	0.02	0.01
24	21	1.84	0.10	0.05
23	22	2.91	0.08	0.02
3	21	3.44	0.12	0.03
3	22	3.84	0.21	0.16
14	22	6.12	0.40	0.07
7	21	6.87	0.31	0.03
3	8	7.59	0.92	0.20
6	22	8.03	0.41	0.07
5	22	8.32	0.98	0.48
1	23	8.64	0.69	0.07
10	23	9.94	1.19	0.38
19	22	12.41	0.88	0.36
19	14	12.78	1.32	0.18
19	15	13.75	0.95	0.46
19	7	13.91	1.07	0.22
19	16	14.75	1.03	0.42
19	21	14.75	0.97	0.29

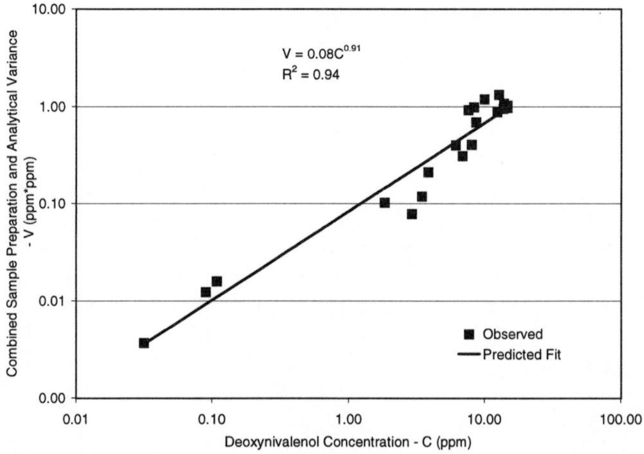

Figure 3. Full-log plot of combined sample preparation and analytical variance versus deoxynivalenol (DON) concentration when wheat is tested for DON. Variance reflects Romer mill, 25 g subsample, and Romer Fluoroquant analytical method.

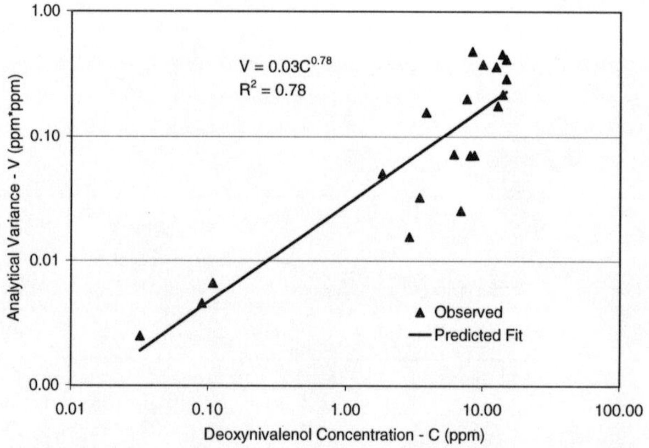

Figure 4. Full-log plot of analytical variance versus deoxynivalenol (DON) concentration when wheat is tested for DON. Variance reflects Romer Fluoroquant analytical method.

It was observed that the exponent values (b coefficient in equation 6) in equations 7, 8, and 9 were reasonably close in magnitude. It would be simpler mathematically when adding and subtracting equations 7, 8, and 9 to obtain sampling and the sample preparation variances, if all exponents in equation 7, 8, and 9 were the same value. Based upon the similarity of the estimated exponent values in equations 7, 8, and 9, and the standard error associated with each estimate, it was decided to fit the total, combined sample preparation and analytical, and analytical variance equations to the data in Tables 1 and 2 and force the b coefficient in equation 6 to be the same value for all three regression equations. The resulting regression equations for total, combined sample preparation and analytical, and analytical variances are shown in equations 10, 11, and 12, respectively.

$$S^2_t = 0.118\ C^{0.833} \tag{10}$$

$$S^2_{spa} = 0.092\ C^{0.833} \tag{11}$$

$$S^2_a = 0.026\ C^{0.833} \quad (12)$$

Using equations 3 and 4, the sample preparation variance and the sampling variance were estimated by subtraction and are given by equation 13 and 14, respectively.

$$S^2_{sp} = 0.066\ C^{0.833} \quad (13)$$

$$S^2_s = 0.026\ C^{0.833} \quad (14)$$

The variances in equations 12, 13, and 14, are specific for the DON test procedure used in this study (0.454 kg sample of wheat, Romer mill, 25 g comminuted subsample and Romer Fluoroquant analytical method).

Reducing Variability

The total variance associated with the DON test procedure can be reduced by increasing sample size or the number of 0.454 kg sampling units, increasing subsample size or the number of 25 g subsampling units taken from the Romer mill, and analyzing multiple aliquots for DON by the Fluoroquant method. Equations 12, 13, and 14 can be modified to predict the effects of sample size, subsample size and number of aliquots on reducing variability associated with each step of the DON testing procedure.

$$S^2_s = (0.454/ns)\ 0.026\ C^{0.833} \quad (15)$$

$$S^2_{sp} = (25/nss)\ 0.066\ C^{0.833} \quad (16)$$

$$S^2_a = (1/na)\ 0.026\ C^{0.833} \quad (17)$$

Where ns is the sample size in kg, nss is the subsample size in g when the sample is comminuted in a Romer mill, and na is the number of aliquots analyzed by the Romer Fluoroquant analytical method. Equation 15 was derived from the principal that if you double sample size (or double number of sample units), the sample variance is cut in half. This principal also applies to subsample size and number of aliquots quantified (equations 16 and 17). The total variance associated with the DON test procedure for a given sample size (ns), subsample size (nss), and number of aliquots (na) can be determined by adding equations 15, 16, and 17 together.

$$S^2_t = (0.454/ns)\ 0.026\ C^{0.833} + (25/nss)\ 0.066\ C^{0.833} + (1/na)\ 0.026\ C^{0.833} \quad (18)$$

For the DON test procedure used in this study, the variance and CV associated with a DON concentration of 5 ppm is shown in Table 4. For example, the CVs associated with sampling, sample preparation, and analysis are 6.3, 10.0 and 6.3 percent, respectively. The CV of 13.4 % shown in Table 4 for the DON test procedure is relatively low compared to testing other commodities for other mycotoxins such as testing peanuts for aflatoxin (CV=200%) (Whitaker et al., 1994). Variances in Table 4 associated with each step of the DON test procedure suggest that the best use of resources would be to increase the subsample size from 25 g to 50 g which would reduce the sample preparation variance from 0.252 to one-half the original value or to 0.126, which would make the sample preparation variance about the same magnitude as the sampling and analytical variability. The CV associated with a DON test procedure using a 0.454 kg sample, 50 g subsample comminuted in the Romer mill, and analyzing one aliquot by the Romer Fluoroquant method is shown over a range of DON concentrations in Figure 5. Because a different cost is associated with methods to reduce the variability of each step of the DON test procedure, it is important to consider cost as well as the expected reduction in variability when designing a DON test procedure.

Table 4. Variance and coefficients of variation (CV) associated with the deoxynivalenol (DON) test procedure using a single 0.454 kg sample, Romer mill, 25 g comminuted subsample, and a single aliquot quantitated by the Romer Fluoroquant method for a DON concentration of 5 ppm.

Variability Measure	Sampling	Sample Preparation	Analysis	Total
Variance	0.099	0.252	0.099	0.450
Coeff. Of Variation (%)	6.3	10.0	6.3	13.4
Variance Ratio (%)	22.0	56.0	22.0	100.0

Ratio of variances: sampling/total; sample preparation/total; analysis/total

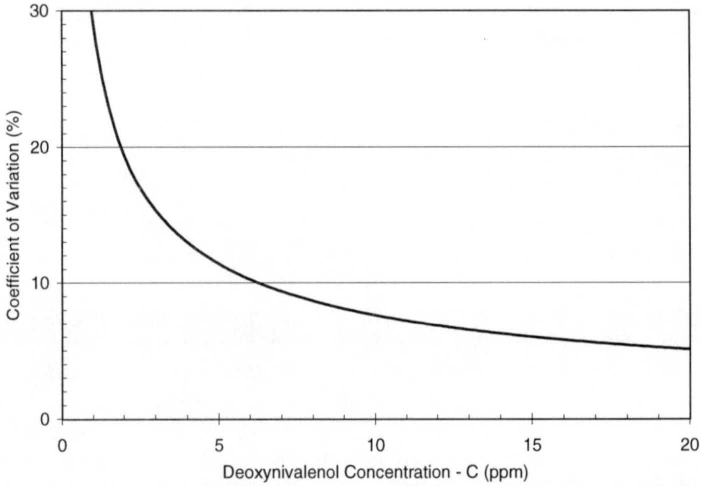

Figure 5. Coefficient of variation versus deoxynivalenol (DON) concentration when wheat is tested for DON. DON test reflects 0.454 kg sample, Romer mill, 50 g subsample, and Romer Fluoroquant analytical method.

The total variance, calculated from equation 18, can be used to estimate the range of DON test results expected when testing a contaminated wheat lot. The total variance associated with the DON test procedure described in Table 4 when testing a lot at 5 ppm is 0.450 or a CV of 13.4%. A variance of 0.450 suggests that replicate DON test results on a lot at 5 ppm will vary from 5-(1.96*0.671) to 5+(1.96*0.671) or 5 +/- 1.31 or from 3.69 ppm to 6.31 ppm 95 % of the time. This calculation assumes the distribution of DON test results are normally distributed which may or may not be the case. Studies with other mycotoxins suggest the distributions are positively skewed (Whitaker et al., 1972; Whitaker et al., 1996). DON test results in Table 1 suggest that the distribution among DON test results for a given lot is skewed. Table 1 indicates that for a given lot, more than 50 % of the DON test results are below the average of the 32 DON test results. This is particularly true for lots with low DON concentrations. Further studies are required to describe the distributional characteristics of DON test results for wheat.

SUMMARY AND DISCUSSION

For a 0.454 kg sample, Romer mill, 25 g subsample, and the Romer Fluoroquant analytical method, the coefficient of variation associated with testing wheat for deoxynivalenol was determined to be about 13.4 % at 5 ppm which is relatively low when compared to other mycotoxin test procedures and other commodities. The CV associated with taking a 0.454 kg sample was 6.3 % which is approximately the same magnitude as the CV associated with sample preparation and analysis. It is assumed that sampling variability is solely due to the distribution among contaminated kernels and that there are no biases associated with the sample selection process. The small variability associated with the sampling step (relative to other mycotoxins and other commodities) is due in part to the kernel count of wheat per unit mass (about 30 kernels per gram). The kernel count per unit mass for wheat is about 10 times larger than that for shelled corn and 30 times larger than that for shelled peanuts. The smaller variability also suggests that unlike aflatoxin, a larger percentage of kernels are contaminated and the distribution of contamination among wheat kernels may be less skewed than for aflatoxin. Studies are currently being conducted to find a suitable theoretical distribution to accurately simulate the observed distribution among DON test results taken from the same lot of wheat. The variability and distributional information will be combined and methods developed to evaluate the performance of DON sampling plans for wheat.

NOTES

The use of trade names in this publication does not imply endorsement by the USDA or the N.C. Agricultural Research Service of the products named nor criticism of similar ones not mentioned. Original manuscript was published in Journal of Association of Official Analytical Chemists, Int., 83:1285-1292.

REFERENCES

Food and Agriculture Organization, 1995, Worldwide Regulations for Mycotoxins, *FAO Food and Nutrition Paper 64*. Food and Agriculture Organization of the United Nations, Vialle delle Terme di Caracalla, 00100 Rome, Italy, pp. 43.

Hart, L.P. and Schabenberger, O., 1998, Variability of vomitoxin in truckloads of wheat in awheat scab epidemic year, *Plant Disease*. 82:625-630.

Malone, B.L., Humphrey, C.W., Romer, T.R., and Richard, J.L., 1998, One-step solid-phase cleanup and fluorometric analysis of deoxynivalenol in grains, *J. Assoc. Offic. Anal. Chem., Inc.* 82:448-452.

Mirocha, C.J., Pathre, S.V., Schauerhamer, B., and Christensen, C.M., 1976, Natural occurrence of fusarium toxins in feedstuffs, *Appl. Environ. Microbiol*. 32:553.

Statistical Analysis System Institute, Inc., 1997, *SAS/STAT Software: Changes and Enhancements through Release 6.12*, SAS Institute, Inc., Cary, NC 27511.

Vesonder, R.F. and Hesseltine, C.W., 1981, Vomitoxin: Natural occurrence on cereal grains and significance as a refusal and emetic factor to swine, *Process Biochem*. 16:12-15.

Whitaker, T.B., Dowell, F.E., Hagler, W.M. Jr., Giesbrecht, F.G. and Wu, J., 1994, Variability associated with sampling, sample preparation, and chemical testing for farmers' stock peanuts, *J. Assoc.Off. Anal. Chem., Int.* 77:107-116.

Whitaker, T.B., Giesbrecht, F.G., and Wu, J., 1996, Suitability of several statistical models to simulate observed distribution of sample test results in inspection of aflatoxin-contaminated peanut lots, *J. Assoc. Official Analytical Chem., Int.* 79:981-988.

Whitaker, T.B., Dickens, J.W. and Monroe, R.J., 1979, Variability associated with testing corn for aflatoxin, *J. Amer. Oil Chem. Soc*. 56:789-794.

Whitaker, T.B, Dickens, J.W. and Monroe, R.J., 1972, Comparison of the observed distribution of aflatoxin in shelled peanuts to the negative binomial distribution, *J. Amer. Oil Chem. Soc*. 49:590-593.

Whitaker, T.B., Dickens, J.W. and Monroe, R.J., 1974, Variability of aflatoxin test results, *J. Amer. Oil Chem. Soc*. 51:214-218 (1974).

Whitaker, T.B., Whitten, M.E. and Monroe, R.J., 1976, Variability associated with testing cottonseed for aflatoxin, *J. Amer. Oil Chem. Soc*. 53:502-505.

NOVEL ASSAYS AND SENSOR PLATFORMS FOR THE DETECTION OF AFLATOXINS

Chris M. Maragos

Mycotoxin Research Unit
USDA-ARS-NCAUR
Peoria, IL 61604

ABSTRACT

The importance of the aflatoxins from food safety and economic standpoints has continued to drive the development of new analytical methods for these mycotoxins. Currently the widely used methods for measurement of aflatoxins fall into two groups, the established chromatographic methods and traditional enzyme-linked immunosorbent assays (ELISAs). Recently substantial progress has been made in the application of new technologies to the monitoring of aflatoxins. In particular, several research groups have developed biosensors for detection of the toxins as well as presumptive tests for fungal infection. Biosensors have been developed in a variety of formats including surface plasmon resonance, fiber optic probes, and microbead-based assays. The sensitivity and selectivity of the biosensors and of the presumptive tests has reached the level the where the application of these techniques to the screening of foods warrants further investigation.

INTRODUCTION

Since their discovery in the 1960s the aflatoxins have received considerable attention from analytical chemists, and a substantial number of analytical methods have been developed for detection and quantitation of this important group of mycotoxins. The methods available range from those requiring technical skill to those which can be performed with minimal training. Because of the necessity to examine large numbers of samples in a short time the screening methods for aflatoxins generally require minimal technical expertise, while the methods used to confirm the presence of the toxins generally require more sophisticated instrumentation and technical skill. A third group of methods are those which are used primarily in research laboratories. Such methods may be more sensitive, or applicable to different matrices (such as tissues), but which have not been tested as extensively and may require detailed knowledge of the assay to perform. This review will be limited to examining the methods that use novel sensor technologies and which are currently used in research settings but which have the potential to find application in more routine settings. Readers interested in the widely used chromatographic methods for the aflatoxins are directed to excellent reviews on this subject by Sydenham and Shephard (1997), Holcomb et al. (1992), and the Official Methods of Analysis published by the AOAC International. Insights into the reliability of some of the commonly used methods for aflatoxins can be found in the review by Horwitz et al. (1993). Currently the commonly used screening methods for the aflatoxins include both the chromatographic techniques (TLC, HPLC) and the antibody-based techniques such as enzyme-linked immunosorbent assays (ELISAs). The ELISAs can be found in many different formats including microwell assays (microtiter plates, strips), dipstick assays, and immunofiltration assays. The ELISAs for the aflatoxins were first developed in the 1970s. The current state of mycotoxin ELISAs, including some of the commercial assays, have been reviewed by Chu (1996), Pestka (1994), and Trucksess and Wood (1997). More recently researchers have begun testing alternative analytical formats such as biosensors and machine vision devices in an effort to improve upon the speed, accuracy and cost of analysis.

BIOSENSORS

Biosensors are devices that use biological components to react or bind with a target molecule and transduce this event into a detectable signal. In the case where the biological component is an antibody or an antibody fragment the biosensor is termed an immunosensor. Immunosensors may offer certain advantages over more traditional antibody-based ELISA formats. The advantages may include more rapid assays, reusable sensor elements, and the capacity for continuous monitoring. For these reasons the number and type of immunosensors have expanded rapidly in recent years. Immunosensors that have been applied to the aflatoxins include a variety of instrument platforms such as those based on fiber optics, small particles (microbeads), and surface plasmon resonance.

The evolution of mycotoxin biosensors began with the modification of ELISAs. The modification consisted of using optical fibers to deliver light to, or collect light from, the site of the binding reaction. An example was an assay for T-2 and diacetoxyscirpenol that used optical fibers for detection of enzymatic products in the wells of microtiter plates (Scheper et al., 1994). A second such assay was recently described for the aflatoxins using both competitive and noncompetitive formats (Carter et al., 1997). In the noncompetitive format aflatoxin antibodies were attached to a permeable membrane, which was incubated with a sample for one hour. The membrane was then rinsed with buffer and then connected to a fluorometer for subsequent detection of the native fluorescence of the aflatoxins (excitation at 362 nm, emission at 425 nm). For this assay the signal was directly proportional to toxin content and 0.05 ng AFB_1/mL was clearly detected. A competitive version of the technique was also tested wherein an AFB_1-BSA conjugate was attached to the membrane which was then incubated with rabbit anti-aflatoxin antibodies. A second, enzyme-labeled anti-rabbit antibody, was then added. This was conceptually similar to an indirect ELISA, with a fluorogenic rather than a chromogenic substrate used for detection. Maize samples were extracted with methanol/water, filtered, and the filtrate shaken with aqueous sodium chloride/zinc acetate solution, filtered again and then extracted with benzene. The isolation procedure for peanut meal was simpler, however it involved an eight hour extraction step. Unfortunately the extensive sample cleanups that were required severely limit the utility of these methods because there would be no advantage over using an efficient chromatographic procedure.

Another fiber optic device that was developed for the aflatoxins used an integrated fiber optic refractometer to measure changes in refractive index at the surface of a planar ion-exchange waveguide via difference interferometry (Boiarski et al., 1996). The basis for the assay was the competition between free aflatoxin and aflatoxin labeled with peroxidase for aflatoxin specific antibodies attached to the surface of the sensor. Unlike with an ELISA, the effect of the enzyme upon the refractive index at the surface of the waveguide was the basis for the signal, not the generation of a colored or fluorescent product. Unfortunately the response of the device to free aflatoxin, the sensitivity of the device, and the effects of matrix components were not described.

Another group of immunosensors are those that utilize fiber optics in an evanescent wave-based format. Rather than using optical fibers as simple "light pipes" to deliver and collect light these sensors rely upon a physical effect that occurs when light travels through an optical fiber: the evanescent wave. When light is launched into an optical fiber, a small proportion escapes into the cladding or solution that surrounds the fiber. The intensity of this light, the evanescent wave (EW), decreases exponentially from the surface of the fiber. Antibodies attached to the surface of the fiber, at the fiber-solution interface, can be used to bind fluorescently tagged molecules within the EW. If light at the proper excitation wavelength is launched into the fiber, a small proportion of the emission from the tagged fluorophore can be coupled back into the fiber for detection. By trapping the analyte of interest near the interface, an element of selectivity is added that is not present with some of the other fiber-optic formats. EW-based immunosensors offer several advantages: high specificity, freedom from electromagnetic interference, ease of miniaturization, real-time monitoring, biocompatibility, and adaptability for remote sensing. Such sensors have been used to detect a variety of compounds including the fumonisins (Thompson and Maragos, 1996).

The fumonisin EW immunosensor was based upon the competition between unlabeled fumonisins and fumonisin B_1 (FB_1) labeled with fluorescein for antibody attached to the surface of optical fibers. The assay was capable of detecting high levels of FB_1 in maize, however, the effects of solvents upon the performance of the antibodies limited the sensitivity and utility of the device. Recently this EW immunosensor format was also investigated for the aflatoxins. Unlike the fumonisins the native fluorescence of the aflatoxins would allow a noncompetitive, rather than competitive, assay format. The aflatoxin sensor was capable of detecting 2 ng/mL aflatoxin B_1 with analysis times as short as 6 min and the sensor response was very reproducible. However, attempts to maximize the size of the evanescent wave resulted in a sensor that detected bulk solution fluorescence rather than fluorescence within the evanescent zone. The latter device was essentially acting as a fiber-optic fluorometer, not an immunosensor (Maragos and Thompson, 2000). While these two sensors were useful for demonstrating the assay principles they did not achieve the sensitivity or ease-of-use of the better ELISA methods. A portable version of the evanescent wave-based biosensor has been developed by the Naval Research Laboratory for detection of explosives (Anderson et al., 1996). The device is now commercially available, and could potentially be applied to analysis of mycotoxins in foods.

Fluorescence was also the basis behind liposome-based immunoassays that have been developed for several mycotoxins. Once such assay, for T-2 toxin used toxin-liposome conjugates filled with carboxyfluoresce in (Ligler et al., 1987). In a competitive format the labeled liposomes competed with free toxin for binding to toxin-specific antibody. Complement-mediated lysis of the liposomes over 30 minutes released the fluorophore and provided signal amplification. More recently tagged liposomes have been used in an assay for fumonisin B_1 (Ho and Durst, 2000). The method is a flow-injection liposome immunoanalysis (FILIA), wherein sulforhodamine B was encapsulated in fumonisin-tagged liposomes. The tagged liposomes competed with FB_1 for binding to antibody attached to a capillary column. After washing to remove non-specifically attached liposomes the bound liposomes were lysed with a detergent and the sulforhodamine fluorescence was detected. Unlike complement-mediated lysis the detergent-mediated lysis was very rapid. The range of the assay in buffer solutions was from 1 to 1000 ng FB_1/mL, with assays taking 11 min. While there are no reports in the literature of the application of this technique to the aflatoxins, the extension to this group of mycotoxins would appear plausible by substituting aflatoxin-tagged liposomes and aflatoxin-specific antibodies for the fumonisin-liposomes and fumonisin-antibodies respectively.

Other approaches to biosensors for aflatoxins include detection of the interaction of the toxin with lipid bilayers or DNA (Andreou and Nikolelis, 1998; Siontorou et al., 1998) and detection using microbead-based immunosensors (Strachan et al., 1997; Carlson et al., 2000). The first of these took advantage of the ability of aflatoxin M_1 to interact either with bilayer lipid membranes or to interfere with DNA hybridization (Siontorou et al., 1998). This is a novel approach to sensor development, however the interaction of aflatoxin M_1 with either transduction element was non-specific. This lack of specificity suggests that the extension of the technology from buffered solutions to actual samples of foods will be difficult, but may be possible as has been demonstrated with aflatoxin M_1 in milk (Andreou and Nikolelis, 1998).

The second approach, microbead immunoassay, is conceptually very similar to a technology that has been used extensively to isolate aflatoxins from foods, namely immunoaffinity columns (IAC). The use of IACs with mycotoxins has been reviewed by Scott and Trucksess (1997). Recently an immunosensor based upon the use of microbeads was described for detection of aflatoxin B_1 in nuts (Strachan et al., 1997). The beads were coated with aflatoxin B_1 conjugated to bovine serum albumin (AFB_1-BSA). AFB_1 in samples competed with the AFB_1-BSA for added rabbit anti-aflatoxin antibody. A second, fluorescently labeled goat anti-rabbit antibody, was then added to detect the amount of primary antibody bound. The signal obtained was inversely proportional to the toxin content. Nut samples were extracted with methanol/water/hexane. After removal of the hexane phase the aqueous phase was filtered and diluted in buffer for analysis. Following extraction each analysis took 8 min and measurement of 2.5 to 227 ng/g aflatoxin was demonstrated. Very recently a group at Johns Hopkins University (Carlson et al., 2000) tested a microbead-based immunosensor in a noncompetitive, rather than competitive format. The latter device is hand-held and contains a fluorometer that detects the native fluorescence of the aflatoxins, rather than the fluorescence of a labeled antibody. The microbeads were coated with anti-aflatoxin antibody to isolate the mycotoxin from buffered solution, essentially in a small scale IAC. After a washing step the toxin was eluted from the beads using a mixture of methanol and buffer (1+1). Fluorescence of the eluted aflatoxin was detected at 445 nm (excitation 365 nm). An advantage of the noncompetitive format was that the output from the device was directly, rather than inversely, proportional to the toxin content. The sensor was capable of detecting standard solutions of aflatoxin B_2 over the range of 0.1 to 50 ng/mL, with an assay time of 2 min. As with traditional IACs there is an upper limit or capacity for the device, and samples above the capacity could be diluted and re-analyzed as they are with IACs. While the device has yet to be demonstrated with aflatoxins in foods, the combination of a traditional analytical technique with novel instrumentation in a compact device makes the approach very promising.

An aflatoxin immunosensor has also been developed that relies upon the principle of surface plasmon resonance (SPR) (Van der Gaag, 1996, 1997, 1999). SPR methods measure the changes in mass that occur when molecules bind to the sensor surface. At a specific wavelength and angle of incidence, light will be absorbed by free electrons in the thin metal film of the sensor surface and the intensity of the reflected light will decrease. The specific angle of incidence where this occurs is known as the resonance angle. The resonance angle can be modified by the addition of mass onto the surface of the sensor, and this provides a mechanism for detecting molecular interactions. SPR immunosensors for mycotoxins use one of two approaches: a noncompetitive approach where antibodies are attached to the surface of the chip and the binding of toxin to the antibodies is detected (Mullet et al., 1998) or a competitive approach where the toxin is attached to the surface of the chip and the binding of antibodies to the toxin is detected (Van der Gaag, 1996, 1997, 1999). In the case of aflatoxins, the carboxymethyloxime of AFB_1 was produced and linked to a CM5 sensor chip (BIAcore

AB, Uppsala, Sweden). AFB_1 in samples then competed with the immobilized AFB_1 for anti-aflatoxin antibodies, and the sensor response was inversely proportional to the toxin content. The sensor had a limit of detection of 0.2 ng AFB_1/mL. Following an assay, the sensor surface could be regenerated by removing the antibody with 6 M Guanidine chloride, pH 4. The device has also been configured for zearalenone, ochratoxin A, and fumonisin B_1. Because the instrument has 4 channels it may be possible to analyze for all four mycotoxins simultaneously. Currently the instrument is expensive, which has limited the methods to research laboratories. However, the sensitivity and reusability of the technique combined with ongoing attempts to reduce the size and cost of the instrumentation suggest the SPR technology may find more widespread use in the future.

All of the aforementioned immunosensors function with either a toxin-conjugate attached to a solid surface (membrane, fiber, well, etc.) or with the toxin specific antibody attached to the solid surface. Attachment of antibodies to a surface can influence their ability to interact with the analyte of interest and solution-phase assays are a potential alternative. Among the solution phase immunosensors two types are those that are conducted in capillaries and those that use the physical property of fluorescence polarization. Recently attempts have been made to combine the specificity of immunoassay with the separation power of capillary electrophoresis (CE). This type of assay has been tested for the fumonisins (Maragos, 1997), but has not been applied to the aflatoxins. The miniaturization of CE instrumentation is an intriguing development that may eventually dramatically reduce the size and cost of the technology.

The second of the solution phase immunoassays depends upon the physical property of fluorescence polarization. The use of fluorescence polarization (FP) in immunoassay was first described about forty years ago and is currently undergoing a renaissance as a tool for modern analysis (Dandliker and Feigen, 1961; Haber and Bennett, 1962; Checovich et al., 1995). FP measurements are based upon the rate of rotation of a fluorescent tracer molecule in solution. Smaller molecules rotate faster and give lower polarization values than bigger molecules. Binding of antibody to the tracer forms an immuno-complex that rotates slower and gives rise to higher polarization values. Unlabeled toxin in the sample competes with the tracer and the polarization changes accordingly. An advantage of FP over other forms of immunoassays is that the polarization is independent of intensity and tracer concentration, and the assays can be conducted with colored or cloudy solutions. The technology has been used for a variety of applications, predominantly in the clinical area, where portable devices are available (Nasir and Jolley, 1999). No applications to mycotoxins have been reported, however the potential of this technology to yield rapid, easy to use assays suggests that research to produce such assays is warranted.

PRESUMPTIVE TESTS

Rapid tests for fungal contamination have been used extensively as presumptive tests for the presence of mycotoxins. By detecting fungal contamination rather than toxin content such tests can eliminate one of the major disadvantages of mycotoxin assays: the time consuming and labor intensive extraction step. The most widely used presumptive test for the aflatoxins has been the bright greenish-yellow fluorescence (BGYF) or "black light" test used to grade cereal grains. Under exposure to ultraviolet light (365 nm) maize infected with *Aspergillus flavus* or *A. parasiticus* may exhibit this fluorescence (Shotwell et al., 1972). The fluorescence derives from the enzymatic oxidation of kojic acid (Zeringue and Shih, 1998), a secondary metabolite of the fungus. BGYF may also be formed in the presence of air on matrices such as cellulose and silica gel (R. Vesonder, USDA-NCAUR, personal communication). A method for continuous screening of commodities based on BGYF and monitoring fluorescence at two wavelengths (420 and 490 nm) was developed by McClure and Farsaie (1980). The difficulty with presumptive tests such as BGYF has been correlating the fungal content with toxin content. The BGYF method, when used alone to screen for aflatoxin has tended to give many false positives but few false negatives. Although there is a clear association between BGYF and aflatoxin (Dickens and Whitaker, 1981), the correlation is insufficient to use the fluorescence as a predictor of toxin content (Shotwell and Hesseltine, 1981), particularly at lower levels of toxin content. This suggests that while the BGYF test may be a valuable aid in sorting commodities it should not be the sole analytical method used for estimating aflatoxin content.

Another approach to presumptive testing has been detection of the color changes that occur with fungal infection. Bichromatic sorters have been used for many years to remove discolored pistachios. More recently devices have been developed that combine rapid imaging of a commodity with recognition algorithms to detect discoloration. A machine vision system was developed that rapidly separated pistachios with small stains from unstained nuts (Pearson 1996). The device was also used to reduce the average aflatoxin level in pistachios through removal of stained nuts (Pearson and Schatzki, 1998). Multiple sorting by the same sorter did not improve aflatoxin levels significantly. Nevertheless, combining color sorting with the machine vision system in tandem may be an effective way to reduce aflatoxin levels in pistachios. Image analysis has also been applied to the detection of color and texture changes occurring from *Fusarium* infection (Ruan et al., 1998).

Rather than color or fluorescence, fungal contamination has also been detected using near infra red (NIR) spectroscopy. One advantage of a NIR technique is the potential to detect damage to nuts that may not be visibly moldy. The ratio of the transmittance at 700 and 1100 nm was used to detect *A. flavus* contamination in peanuts by Hirano et al. (1998). The difference in transmittance between infected and non-infected nuts may have resulted from the hydrolysis of peanut triglycerides by the fungus. Interestingly the same article also described the detection of internally moldy nuts using ^{1}H-NMR-CT, but the latter technology was not pursued further due to greater analysis times and higher operating costs. A sorter based upon the NIR device reduced the aflatoxin content to 4-18% of the levels before sorting, suggesting this is a very promising technology for removing contaminated nuts before processing. A peanut sorter based upon this technology has recently become commercially available (Anzai Manufacturing. Co., Ltd., Chiba, Japan).

Other variants of Fourier transform infrared spectrometry (FTIR) have also been used for assessment of fungal contamination. Two such variants of FTIR are diffuse reflectance spectroscopy (DRS) and photoacoustic spectroscopy (FTIR-PAS). Both techniques were applied to corn kernels infected with *Aspergillus flavus* (Greene et al., 1992). In preliminary tests with *A. flavus* infected corn FTIR-PAS was more sensitive than DRS (Greene et al., 1992). With FTIR-PAS an intact specimen was placed inside a closed cell containing a sensitive microphone. Infrared light was admitted through a window to irradiate the specimen. When the specimen absorbed the light, it became heated and warmed the surrounding layer of air or gas. The warm gas expanded, resulting in increased pressure within the cell. If the light was blocked off, the specimen cooled off and the gas pressure in the cell returned to its original level. Chopping or periodic modulation of the light produced alternating heating and cooling of the specimen at the modulation frequency, resulting in gas pressure waves, which were detected as sound by a microphone. If the specimen absorbs infrared radiation, some of the absorbed energy will be converted to heat and emitted as acoustic waves. Since the amplitude and frequency of the acoustic waves emitted depend on the amount and frequency of the infrared energy absorbed, the photoacoustic signal reflects the infrared spectrum of the specimen. Because of the inherently surface-profiling signal of FTIR-PAS it is a potentially sensitive detector of fungal infection on the surfaces of commodities. Gordon et al. (1997) applied the technique to corn kernels classified as either BGYF-positive or negative. The technology has promise as a presumptive test for aflatoxin contamination, however application of the method was limited because available photoacoustic detectors can only accommodate a single kernel at a time (Gordon et al., 1999). Another variant of IR spectroscopy, transient infrared spectroscopy (TIRS), has the potential to detect fungal infection on moving corn kernels and preliminary results have indicated TIRS may be useful for screening bulk quantities of corn (Gordon et al., 1999).

SUMMARY

In attempting to improve upon existing screening assays for mycotoxins a number of biosensors and imaging systems in a variety of instrument platforms have been developed. The biosensor platforms that have been examined range from hand-held devices to benchtop instruments and use technology as diverse as fiber optics, liposomes, small particles (beads), surface plasmon resonance, and microcapillaries. Several of these devices show substantial promise as rapid, sensitive, methods for measuring mycotoxins. The imaging systems that have been developed are presumptive tests for fungal, and by extension mycotoxin, contamination. These systems have used principles as varied as color, fluorescence, and infrared spectroscopy and have been successfully used to indicate fungal contamination. The utility of the techniques for indirectly reducing toxin levels, by removal of product contaminated with fungi, has also been demonstrated. Studies by multiple laboratories to validate the newer biosensor and imaging systems would contribute significantly to their more widespread acceptance, and will hopefully be conducted in the near future.

REFERENCES

Anderson, G.P., Breslin, K.A., and Ligler, F.S., 1996, Assay development for a portable fiberoptic biosensor, *ASAIO J.* 942.

Andreou, V., and Nikolelis, D., 1998, Flow injection monitoring of aflatoxin M_1 in milk and milk preparations using filter-supported bilayer lipid membranes, *Anal. Chem.* 70:2366.

Boiarski, A.A., Busch, J.R., Brody, R.S., Ridgway, R.W., Altman, W.P., and Golden, C., 1996, Integrated optic sensor for measuring aflatoxin-B_1 in corn, *SPIE* 2686:45.

Carlson, M.A., Bargeron, C.B., Benson, R.C., Fraser, A.B., Phillips, T.E., Velky, J.T., Groopman, J.D., Strickland, P.T., and Ko, H.W., 2000, An automated, handheld biosensor for aflatoxin, *Biosen. Bioelect.* 14:841.

Carter, R.M., Jacobs, M.B., Lubrano, G.J., and Guilbault, G.G., 1997, Rapid detection of aflatoxin B_1 with immunochemical optrodes, *Anal. Lett.* 30:1465.

Checovich, W.J., Bolger, R.E., and Burke, T., 1995, Fluorescence polarization-a new tool for cell and molecular biology, *Nature* 375:254.

Chu, F.S., 1996, Recent studies on immunoassays for mycotoxins, in: *Immunoassays for Residue Analysis*, R.C. Beier and L.H. Stanker, eds., American Chemical Society, Washington, DC.

Dandliker, W.B., and Feigen, G.A., 1961, Quantitation of the antigen-antibody reaction by the polarization of fluorescence, *Biochem. Biophy. Res. Commun.* 5:299.

Dickens, J.W., and Whitaker, T.B., 1981, Bright greenish-yellow fluorescence and aflatoxin in recently harvested yellow corn marketed in North Carolina, *J. Amer. Oil Chem. Soc.* 58:973A.

Gordon, S.H., Jones, R.W., McClelland, J.F., Wicklow, D.T., and Greene, R.V., 1999, Transient infrared spectroscopy for detection of toxigenic fungi in corn: potential for on-line evaluation, *J. Agric. Food Chem.* 47:5267.

Gordon, S.H., Schudy, R.B., Wheeler, B.C., Wicklow, D.T., and Greene, R.V., 1997, Identification of Fourier transform infrared photoacoustic spectral features for detection of *Aspergillus flavus* infection in corn, *Intl. J. Food Microbiol.* 35:179.

Greene, R.V., Gordon, S.H., Jackson, M.A., Bennett, G.A., McClelland, J.F., and Jones, R.W., 1992, Detection of fungal contamination in corn: potential of FTIR-PAS and -DRS, *J. Agric. Food Chem.* 40:1144.

Haber, E., and Bennett, J.C., 1962, Polarization of fluorescence as a measure of antigen-antibody interaction, *Proc. Natl. Acad. Sci. USA.* 48:1935.

Hirano, S., Okawara, N., and Narazki, S., 1998, Near infra red detection of internally moldy nuts, *Biosci. Biotechnol. Biochem.* 62:102.

Ho, J.A., and Durst, R.A., 2000, Development of a flow-injection liposome immunoanalysis system for fumonisin B_1, *Anal. Chim. Acta* 414:61.

Holcomb, M., Wilson, D.W., Trucksess, M.W., and Thompson, H.C., 1992, Determination of aflatoxins in food products by chromatography, *J. Chrom.* 624:341.

Horwitz, W., Albert, R., and Nesheim, S., 1993, Reliability of mycotoxin assays-an update. *J. AOAC Intl.* 76:461.

Ligler, F.S., Bredehorst, R., Talebian, A., Shriver, L.C., Hammer, C.F., Sheridan, J.P., Vogel, C.W., and Gaber, B.P., 1987, A homogeneous immunoassay for the mycotoxin T-2 utilizing liposomes, monoclonal antibodies, and complement, *Anal. Biochem.* 163:369.

Maragos C.M., 1997, Detection of the mycotoxin fumonisin B_1 by a combination of immunofluorescence and capillary electrophoresis, *Food Agric. Immunol.* 9:147.

Maragos, C.M., and Thompson, V.S., 2000, Fiber-optic immunosensor for mycotoxins, *Natural Toxins* 7:(in press).

McClure, W.F., and Farsaie, A., 1980, Dual-wavelength fiber optic photometer measures fluorescence of aflatoxin contaminated pistachio nuts, *Trans. ASAE* , 204.

Mullett, W., Lai, E.P.C., and Yeung, J.M., 1998, Immunoassay of fumonisins by a surface plasmon resonance biosensor, *Anal. Biochem.*, 258:161.

Nasir, M.S., and Jolley, M.E., 1999, Fluorescence polarization: an analytical tool for immunoassay and drug discovery, *Combinatorial Chemistry & High Throughput Screening*, 2:177.

Pearson, T.C., and Schatzki, T.F., 1998, Machine vision system for automated detection of aflatoxin-contaminated pistachios, *J. Agric. Food Chem.* 46:2248.

Pearson, T., 1996, Machine vision system for automated detection of stained pistachio nuts, *Lebensm. Technol.* 29:203.

Pestka, J.J., 1994, Application of immunology to the analysis and toxicity assessment of mycotoxins, *Food Agric. Immunol.* 6:219.

Ruan, R., Ning, S., Song, A., Ning, A., Jones, R., and Chen, P., 1998, Estimation of *Fusarium* scab in wheat using machine vision and a neural network, *Cereal Chem.* 75:455.

Scheper, T., Müller, C., Anders, K.D., Eberhardt, F., Plötz, F., Schelp, C., Thordsen, O., and Schügerl, K., 1994, Optical sensors for biotechnological applications, *Biosen. Bioelect.* 9: 73.

Scott, P.M., and Trucksess, M.W., 1997, Application of immunoaffinity columns to mycotoxin analysis, *J. AOAC Intl.* 80:941.

Shotwell, O.L., Goulden, M.L., and Hesseltine, C.W., 1972, Aflatoxin contamination: association with foreign material and characteristic fluorescence in damaged corn kernels, *Cereal Chem.* 49:458.

Shotwell, O.L. and Hesseltine, C.W., 1981, Use of bright greenish yellow fluorescence as a presumptive test for aflatoxin in corn, *Cereal Chem.* 58:124.

Siontorou, C.G., Nikolelis, D.P., Miernik, A., and Krull, U.J., 1998, Rapid methods for detection of aflatoxin M_1 based on electrochemical transduction by self-assembled metal-supported bilayer lipid membranes (s-BLMs) and on interferences with transduction of DNA hybridization, *Electrochimca Acta*, 43:3611.

Strachan, N.J.C., John, P.G., and Miller, I.G., 1997, Application of an automated particle-based immunosensor for the detection of aflatoxin B_1 in foods, *Food Agric. Immunol.* 9:177.

Sydenham, E.W., and Shephard, G.S., 1997,Chromatographic and allied methods of analysis for selected mycotoxins, in: *Progress in Food Contaminant Analysis*, J. Gilbert, ed., Chapman and Hall, New York.

Thompson, V.S., and Maragos, C.M., 1996, Fiber-optic immunosensor for the detection of fumonisin B_1, *J. Agric. Food Chem.*, 44:1041.

Trucksess, M.W., and Wood, G.E., 1997, Immunochemical methods for mycotoxins in foods, *Food Test. Anal.* 3:24.

Van der Gaag, B., Burggaaf, R.A., and Wahlström, L., 1996, Application and development of a BIAcore7 for the detection of mycotoxins in food and feed. Poster Presentation at the 110th AOAC International Annual Meeting and Exposition, Orlando, Fl. Sept. 8-12.

Van der Gaag, B., Stigter, E., and Pohl, S., 1999, Development of a biosensor for the detection of mycotoxins. Satellite workshop to the Gordon Conference on mycotoxins and phycotoxins, Salsbury Cove, ME, June 17-19.

Van der Gaag, B., and Stigter, E., 1997, Application development on the BIAcore for the detection of mycotoxins in food and feed. Poster Presentation at the 111th AOAC International Annual Meeting and Exposition, San Diego, CA. Sept 7-11.

Zeringue, H.J. Jr., and Shih, B.Y., 1998, Extraction and separation of the bright-greenish yellow fluorescent material from aflatoxigenic Aspergillus spp. Infected cotton lint by HPLC- UV/FL, *J. Agric. Food Chem.* 46:1071

ELECTROSPRAY MASS SPECTROMETRY FOR FUMONISIN DETECTION AND METHOD VALIDATION

Steven M. Musser, Robert M. Eppley, and Mary W. Trucksess

Center for Food Safety and Applied Nutrition
U.S. Food and Drug Administration
Washington, DC 20204

ABSTRACT

Fumonisins are a structurally related group of mycotoxins, characterized by a 19-20 carbon aminopolyhydroxy-alkyl chain which is diesterified with propane-1,2,3-tricarboxylic acid (tricarballylic acid). These mycotoxins are commonly found in corn and corn-based food products and have been linked to a variety of animal toxicities. The widespread prevalence of fumonisins and the toxicity associated with ingestion has resulted in a number of analytical methods for determining the amount of fumonisins present in foods. Among the most common of these methods are liquid chromatographic (LC) separation with fluorescence detection, enzyme-linked immunosorbent assay (ELISA) and LC /mass spectrometry. LC and ELISA give quantitative results while LC/MS provide quantitative analysis as well as confirmation of identity of the fumonisins.

INTRODUCTION

Most corn and corn-based food products are contaminated by a group of structurally related mycotoxins known as fumonisins. Fumonisins are produced by several species of *Fusarium*, however *F. verticilloides* (formally classified as *F. moniforme*), *F. proliferatum* and *F. nygami* are the principle fumonisin producing strains (Nelson et al., 1991;1992;1994). This group of mycotoxins is characterized by a 19-20 carbon aminopolyhydroxy-alkyl chain which is diesterified with two tricarballylic acid groups. The most abundant fumonisins found in naturally contaminated corn samples are fumonisins B_1, B_2 and B_3, although over 14 other fumonisins (Figure 1) are produced by various species of *Fusarium* (Musser et al., 1997). They include an N-acetyl "A" series (Cawood et al., 1991), a C1 des-methyl "C" series (Branham and Plattner, 1993; Seo et al. 1996), a hydroxypyridinium containing "P" series (Musser, 1996), a partially hydrolyzed

"B" series (Sydenham et al., 1995). The fully hydrolyzed forms of the B series (Thakur and Smith, 1996) are not naturally occurring, and have only been found in processed foods (Hopmans and Murphy, 1993).

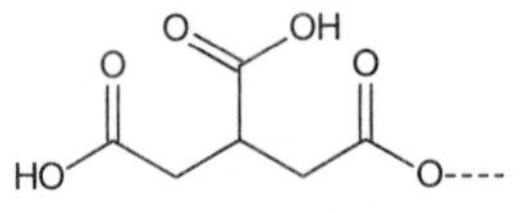

Tricarballylic Acid (TCA)

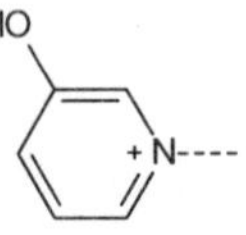

3-Hydroxypyridinium (3HP)

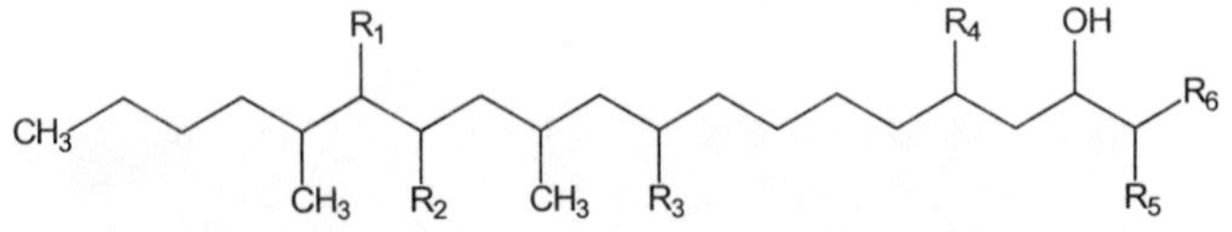

	R_1	R_2	R_3	R_4	R_5	R_6	M.W.
FA_1	TCA	TCA	OH	OH	$NHCOCH_3$	CH_3	763
FA_2	TCA	TCA	H	OH	$NHCOCH_3$	CH_3	747
FA_3	TCA	TCA	OH	H	$NHCOCH_3$	CH_3	747
FAK_1	=O	TCA	OH	OH	$NHCOCH_3$	CH_3	603
FB_1	TCA	TCA	OH	OH	NH_2	CH_3	721
FB_2	TCA	TCA	H	OH	NH_2	CH_3	705
FB_3	TCA	TCA	OH	H	NH_2	CH_3	705
FB_4	TCA	TCA	H	H	NH_2	CH_3	689
FC_1	TCA	TCA	OH	OH	NH_2	H	707
FC_2	TCA	TCA	H	OH	NH_2	H	691
FC_3	TCA	TCA	OH	H	NH_2	H	691
FP_1	TCA	TCA	OH	OH	3HP	CH_3	800
FP_2	TCA	TCA	H	OH	3HP	CH_3	784
FP_3	TCA	TCA	OH	H	3HP	CH_3	784
$PHFB_{1a}$	TCA	OH	OH	OH	NH_2	CH_3	563
$PHFB_{1b}$	OH	TCA	OH	OH	NH_2	CH_3	563
HFB_1	OH	OH	OH	OH	NH_2	CH_3	405

Figure 1. Chemical structures of known fumonisins.

Since their discovery and structural elucidation in 1988 (Bezuidenhout et al., 1988), a considerable amount of research has focused on the analysis and toxicity of fumonisins. A recent review of fumonisin toxicity in rodents, summarizes a variety of toxicological endpoints in rats and mice, which included cancer (Voss et al., 2001). In addition to rodent toxicoses, a number of other fumonisin associated animal toxicoses have been shown, including leukoencephalomalacia (ELEM) in horses (Kellerman et al., 1990; Ross et al., 1990) and pulmonary edema in swine (Harrison, et al., 1990). Epidemiology studies have demonstrated a strong correlation between human consumption of fumonisin-contaminated corn and esophageal cancer (Rheeder et al., 1992; Chu et al., 1994; Marasas et al., 1996), and the International Agency for Research on Cancer (IARC, 1993) has classified fumonisins as possible carcinogens.

A number of analytical methods have been developed which accurately measure fumonisin levels in a wide variety of matrices. The most common of these methods uses liquid chromatography (LC) along with precolumn fluorescent derivatization of fumonisins with o-phthaldialdehyde (OPA) (Shephard et al., 1990; Stack, 1998). This method has been collaboratively studied and provides a limit of determination of total fumonisins B_1, B_2 and B_3 at levels ≥ 0.8 µg/g (Sydenham et al., 1996). While generally yielding reliable results, the method has a number of limitations. First, the OPA derivative of fumonisins decomposes very rapidly, and can lead to under-reporting of levels. Second, only fumonisins containing a primary amine can be detected, leaving analogs such as the A series undetected. This is not a serious problem however, since most of the reported fumonisin toxicity has been linked to the B series, and the fumonisins that can not be derivitized with OPA are generally found at much lower levels than the B series. Finally, the derivatizing reagent reacts with all primary amines present in the sample extract and can lead to interferences that are difficult to separate by LC. This limitation can be largely overcome by using an immunoaffinity column to clean-up the sample extract prior to analysis by LC/florescence (Trucksess, et al., 1995, Maragos et al., 1997), although this procedure can introduce a bias into the assay due to cross-reactivity of the antibodies with similar analytes.

In addition to LC/fluorescence, enzyme-linked immunosorbent assay (ELISA) methods for fumonisins have become increasing popular as a rapid, inexpensive alternative to instrumental methods (Trucksess and Abouzied, 1996). ELISA methods based on both monoclonal and polyclonal antibodies provide good sensitivity and are comparable to instrument-based methods, with a limits of detection of 5-10 ng/g (Azcona-Olivera et al., 1992; Elissalde et al., 1995; Usleber et al., 1994; Yu et al., 1996). Although ELISA-based methods offer function as a cost effective high throughput screening technology, they are prone to both overestimation and underestimation of fumonisins. The reasons for this phenomenon have not been elucidated, but are likely a combination of several factors including: 1) antibody cross-reactivity with other fumonisins, 2) cross-reactivity with dissimilar analytes in the matrix, 3) the enzyme kinetics of the assay, and 4) the small linear portion of standard curve, which is often only one order of magnitude.

LC/MS methods have been used for the quantitative analysis of fumonisins in various food matrices (Lukacs et al., 1996; Musser, 1996; Newkirk et al., 1998; Hartl and Humpf, 1999), although the cost of the instrumentation generally precludes this method for routine use in laboratories which require high sample throughput. While LC/MS does have certain limitations, the technique is particularly useful for the identification of new fumonisins, and for the confirmation of analytical results obtained from other analytical techniques. LC/MS is a powerful analytical problem-solving tool, due to its capability of performing selected-ion monitoring and positive-ion chemical ionization for quantitation and confirmation of the analytes. Our discussion is focus on the use of LC/MS technique for validation of results obtained using LC or ELISA methods.

EXPERIMENTAL

Liquid Chromatography/Mass Spectrometry

Samples were prepared following a previously published method (Musser and Plattner, 1997). Chromatographic separations were carried out on a YMC Inc (Waters Corp. Millford, MA) J-sphere C18 column (2 x 250 mm) with low carbon loading (12%) and used a flow rate of 200 µl/min. Both binary gradients and isocratic conditions were used depending on the type of analysis. For rapid LC/MS/MS analysis of fumonisins B_1, B_2 and B_3, isocratic conditions consisting of 0.1 % formic acid in acetonitrile/water (35:65) were used. A binary acetonitrile/water gradient was used for the LC/MS analysis of culture material and the identification of new fumonisins (Musser and Plattner, 1997). Generally the gradient ran from 23% acetonitrile to 45% acetonitrile in 45 minutes, and contained

0.1% formic acid. A Finnigan (San Jose, CA) Model TSQ-7000 triple quadrupole mass spectrometer, equipped with an electrospray ion source was used for all MS experiments. The 200 µl/min flow rate from the LC was introduced directly into the ion source without splitting. A collision cell pressure of 2.5 torr and collision energy of 18 volts were used for MS/MS experiments.

OPA/HPLC Method

This method has been described in detail elsewhere (Sydenham et al., 1996), and follows the following procedure. A 25 g test sample of finely ground corn product is extracted with 100 mL of a methanol/water (75:25) solution for 3 minutes. The extract is centrifuged at 500 x g for 10 minutes at 4°C, the supernatant is removed and filtered through standard filter paper. The pH of the resulting solution is adjusted to 5.8-6.5 and 10 mL of the extract is applied to a strong anion exchange cartridge (SAX). The column is washed with a methanol/water (75:25) solution. The fumonisins are eluted with 10 mL of a methanol solution containing 1% acetic acid. The eluent is evaporated to dryness under a steady flow of nitrogen at 55°C. The dry sample is reconstituted in 200 µL of methanol. A 50 µL aliquot of the test extract is added to 200 µL of the OPA derivatizing solution, mixed, and incubated at room temperature for one minute. Then 10 µL of the resulting solution is injected onto the HPLC column. An HPLC column packed with C18 particles is used for the separation. Generally, a mobile phase consisting of 75-80% aqueous methanol and buffered to pH 3.35 with 0.1 M phosphate buffer will produce a good separation of the major fumonisins FB_1-FB_3.

Immunochemical Method

The method is based on a commercial ELISA product supplied by Neogen Corporation (Veratox Quantitative Fumonisin Test, Neogen Corporation, Lansing, MI). The details of the method can be obtained from the manufacturer or found a previous publication (Trucksess et al., 1996). Briefly, the sample is extracted with a methanol/water (70:30) solution. For quantitation, the sample extract is diluted 10x with binding buffer. A portion of the diluted test extract is mixed with an equal volume of antibody-enzyme conjugate solution. The resulting mixture is added to the well, then mixed for 20 minutes. The wells are emptied, then washed with water. A solution containing tetrametylbenzidine/peroxide is added and allowed to react for 10 minutes. Following this brief reaction period, sulfuric acid is added to terminate the reaction. The ELISA wells are quantitated based on absorbance measured at 450 nm.

For the purposes of confirmation by LC/MS, the methanol/water extract is diluted 10x with binding buffer in the absence of the conjugate. The diluted extract is then added to the well and mixed for 20 minutes. Methanol is added to the well to elute the fumonisins. The methanol solution is removed, evaporated to dryness and subject to LC/MS analysis.

RESULTS AND DISCUSSION

LC/MS with electrospray ionization has been proven as effective analytical method in our laboratory, as well as several other laboratories throughout the world. The methodology has been demonstrated on numerous corn containing products, as well as culture material and animal feeds (Musser and Plattner, 1997; Newkirk et al., 1998). The ability to select specific ions corresponding to the fumonisins of interest gives the methodology specificity not found in the other analytical methods, thus providing a high confidence level on analytical determinations. Compared with the LC-fluorescence method, detection limits are similar, and are generally in the 5-10 ppb range (Trucksess et al. 1995).

The routine analysis of fumonisins by LC/MS is made practical by the soft ionization technique, electrospray. Unless additional energy is used in the ion source or through collisional activation, the only ions observed for fumonisins in positive ion mode correspond to the protonated molecule $[M+H]^+$. Although fumonisins may also be observed in the negative ion mode, sensitivity is severely compromised. In addition, ions for hydrolyzed fumonisins are difficult to generate in the negative ion mode. Because little to no fragmentation is produced during the analysis and no derivatization or special sample preparation is required, it is possible to directly analyze culture materials and some food products by simply extracting the sample, filtering and injecting the sample on the LC column. Figure 2, demonstrates the advantages of good chromatography combined with LC/MS for fumonisin analysis. A detailed examination of the chromatogram shows the identity and relative abundance of all of the naturally occurring fumonisins obtained from a culture extract of *Fusarium moniliforme*. By taking advantage of the retention times and masses found for fumonisin standards, it is possible to directly analyze many corn-containing products. Figure 3 represents the direct analysis of corn screenings associated with an ELEM case.

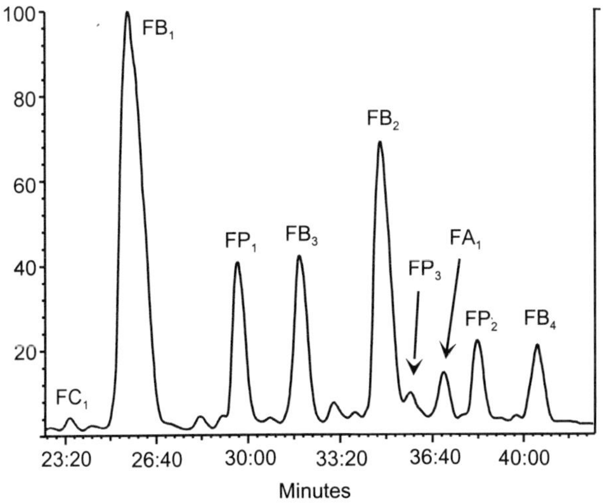

Figure 2. Total ion chromatogram for an LC/MS analysis of a culture extract from *Fusarium verticilloides*.

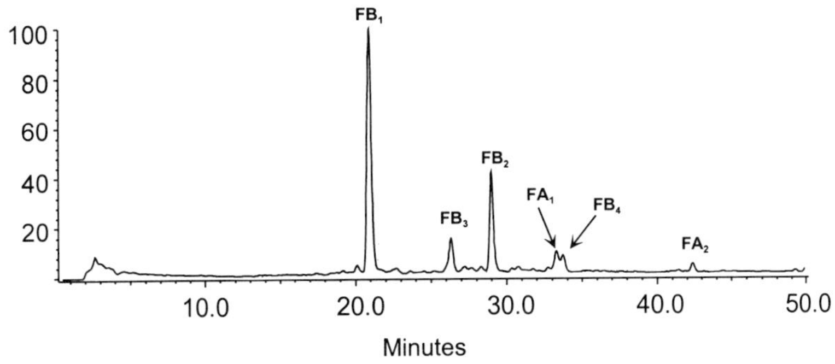

Figure 3. Summed ion chromatogram from an LC/MS analysis of corn screenings associated with an ELEM case.

Another powerful feature of some mass spectrometers is the ability to perform MS/MS experiments, where the ion of interest is selected in one experiment and then induced to fragment in another. MS/MS experiments are useful for two important reasons. First, since they allow for selection of only the ion of interest, structurally related analogs may be identified based on similar fragmentation patterns observed with known compounds. This experiment is illustrated in Figure 4, in which the P series of fumonisins can be identified based on the sequential loss of the TCA side-chain from the parent molecule. Since this loss is characteristic of all fumonisins, there is a high probability the unknown is also a fumonisin. The second, and most prevalent use of MS/MS experiments, is to remove background noise by increasing specificity. By selecting the ion corresponding to the protonated molecule, causing it to fragment and then monitoring only the fragments, signal to noise ratios can often be increased two orders of magnitude. This has the combined effect of increasing sensitivity and decreasing analysis time. Figure 5 is an MS/MS analysis of the same culture extract shown in Figure 2, in which only FB_1, FB_2 and FB_3 are selected for detection. The most noticeable difference between the two figures is the analysis time. Using the MS/MS method, the analysis is complete in 12 minutes versus 36 minutes for the full scan method. The most significant limitation of the MS/MS method is that other fumonisins present in the sample are not detected.

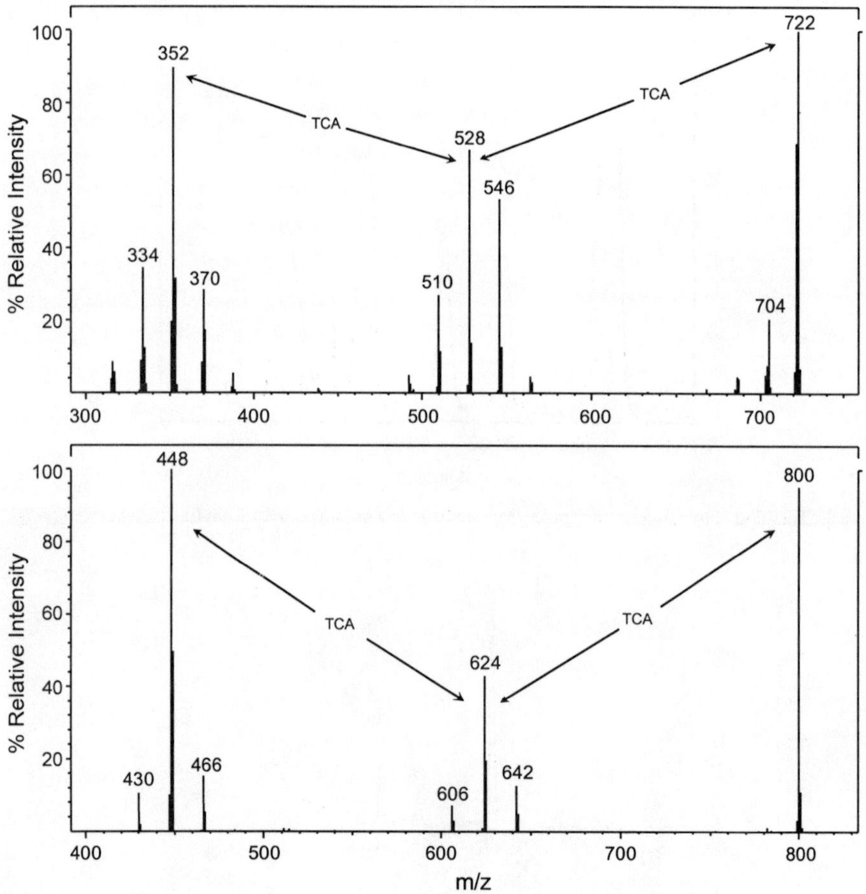

Figure 4. MS/MS spectra of fumonisin B_1 (top) and fumonisin P_1 (bottom) showing the constant mass loss of the TCA side chain common to all fumonisins.

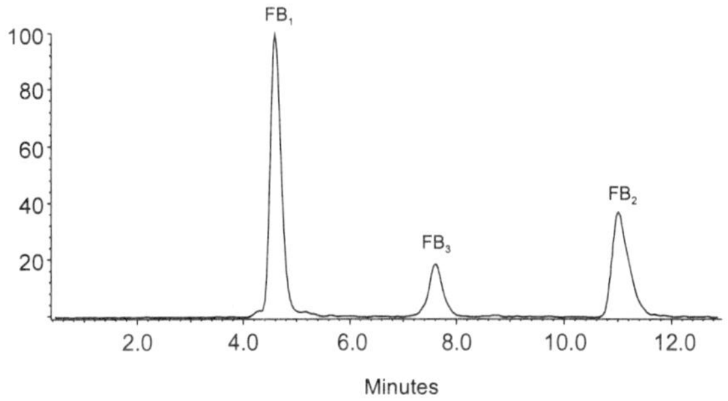

Figure 5. LC/MS/MS analysis of *Fusarium verticillioides* culture extract where ions corresponding to FB_1, FB_2 and FB_3 were the only ions being monitored.

The most widely used instrument-based method for quantitation of fumonisins is LC/fluorescence with precolumn derivatization. The extraction and derivatization process is easily automated (Dilken et al., 2001) and detection limits are in the in the range of 10-50 ppb. Good separations and quantitation of the commonly occurring fumonisins can be achieved on most C18-based HPLC columns (Figure 6). The method is fast and analyses are generally complete in 20 minutes or less. This method has been used to quantify fumonisins in a variety of food matrices including tortillas, sorghum syrup and breakfast cereals (Stack, 1998; Trucksess et al., 2000). A typical chromatogram for naturally contaminated food source, popcorn, is shown in Figure 7. Occasionally, when interferences or unexpectedly high levels of fumonisins are reported by the LC/fluorescence method, it is necessary to confirm the results with LC/MS. One such case involved tortilla flour in which samples were negative by ELISA assay, yet consistently showed the presence of hydrolyzed fumonisin B_1 by LC/fluorescence. Analysis by LC/MS (Figure 8) confirmed the ELISA results and identified a matrix interference with a molecular weight of 405 amu, which co-eluted with the OPA derivative of hydrolyzed FB_1. Recently, Vesonder et al. (2000) described a similar compound isolated from dairy cattle feed. The compound is described as an alpha amino acid, unrelated to fumonisin, but produced by several species of *Fusarium*.

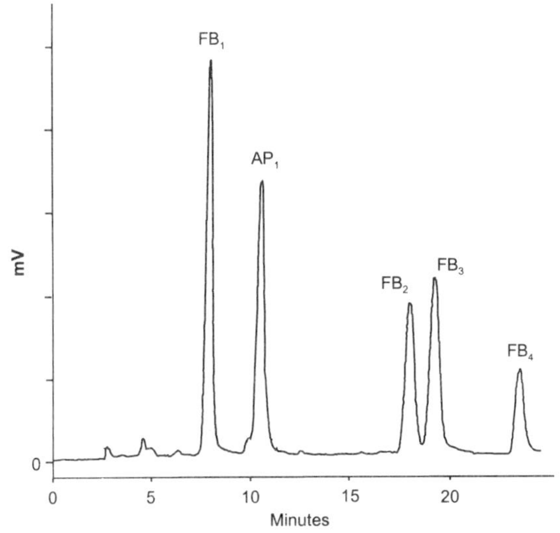

Figure 6. LC/fluorescence chromatogram for fumonisins standards.

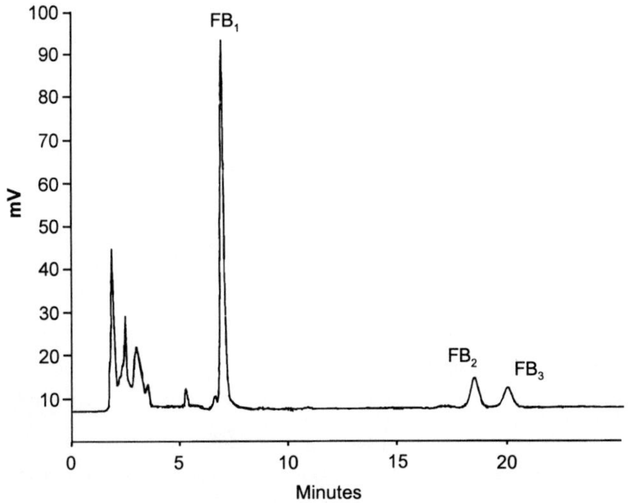

Figure 7. LC/fluorescence chromatogram of naturally contaminated popcorn extracts.

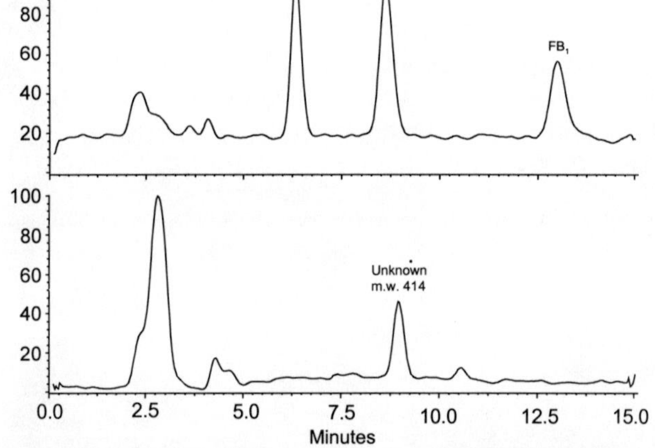

Figure 8. LC/MS/MS analysis of fumonisin standards (top) and tortilla flour extract which contained a matrix interference in the LC/florescence assay (bottom).

ELISA-based methods for fumonisin analyses are also used as a rapid means of screening food samples and estimating fumonisin contamination levels. Although ELISA-based methods provide a simple cost-effective method of analysis, their principal limitation for the analysis of fumonisins in foods and feeds is the overestimation of fumonisin levels. ELISA results were confirmed directly from the analysis wells by an alternative analytical method. LC/MS analysis (Figure 9) of washings from ELISA wells treated with naturally contaminated sample of hog feed clearly demonstrated the ability of LC/MS to directly confirm ELISA results. In addition to this finding, we were able to show the assay does not bias results toward FB_2 and FB_3, as the observed ratios for the fumonisins are very close to those found by other analytical techniques (Figure 7). This observation means the fumonisins bound to the ELISA wells are effectively removed, and the antibody shows binding proportional to the total fumonisins present in the sample, allowing the ELISA method to accurately reflect total fumonisin levels present in the sample. Although this was not an exhaustive study is does demonstrate the usefulness of LC/MS methods for evaluating ELISA method performance.

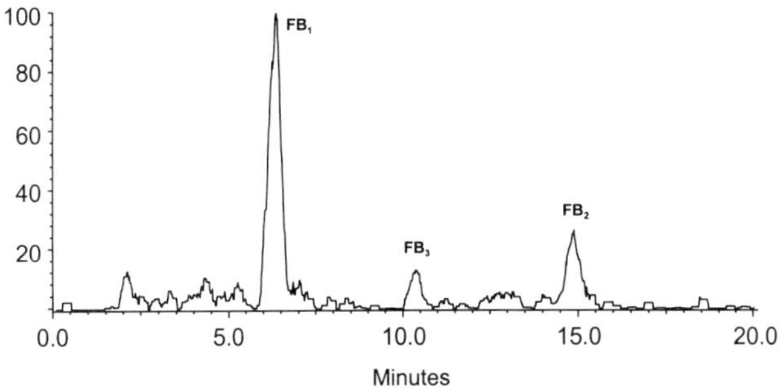

Figure 9. LC/MS/MS analysis of ELISA well washings, which had been treated with naturally contaminated hog feed extracts.

CONCLUSIONS

HPLC, immunochemical and LC/MS methods make valuable contributions to the analysis of fumonisins in foods, and the choice of the appropriate method depends on the desired specificity and precision of the results. LC/MS offers an excellent means of both confirming analytical results obtained from other analytical methods, and providing information on the reliability and performance characteristics of new methods.

REFERENCES

Azcona-Oliviera, J. I., Abouzied, M. M., Plattner, R. D., and Pestka, J. J. , 1992b, Production of monoclonal antibodies to the mycotoxins fumonisins B_1,B_2 and B_3. *J. Agric. Food Chem.* *40*:531-534.

Bezuidenhout, G.C., Gelderblom, W.C.A., Gorst-Allam, C.P., Horak, R.M., Marasas, W.F.O., Spiteller, G., and Vleggaar, R. 1988, Structure elucidation of the fumonisins, mycotoxins from *Fusarium moniliforme*. *J. Chem Soc. Chem Commun.* 743-745.

Branham, B.E. and Plattner, R.D., 1993, Isolation and characterization of a new fumonisin from liquid cultures of *Fusarium moniliforme*. *J. Nat. Prod.*, *56*:1630-1633.

Cawood, M.E., Gelderblom, W.C.A.,Vleggaar, R., Behrend, Y., Thiel, P.G., and Marasas, W.F.O., 1991, Isolation of fumonisin mycotoxins; a quantitative approach. *J. Agric. Food Chem.* *39*:1958-1962.

Chu, F. S., and Li, G. Y. 1994, Simultaneous occurrence of fumonisin B_1 and other mycotoxins in moldy corn collected from the People's Republic of China in regions with high incidences of esophageal cancer. *Appl. Environ. Microbiol.* *60*: 847-852.

Dilkin, P., Mallmann, C.A., de Almeida, C.A.A. and Correa, B., 2001, Robotic Automated clean-up for Detection of Fumonisins B_1 and B_2 in Corn and Corn-Based Feed by High-Performance Liquid Chromatography. *J. Chromatogr. A.* *925*: 151-157.

Elissalde, H. M., Kamps-Holtzapple, C., Beier, R. C., Plattner, R. D., Rowe, L. D., and Stanker, L. H., 1995, Development of an improved monoclonal antibody-based ELISA for fumonisin B1-3 and the use of molecular modeling to explain observed detection limits. *Food Agric Immunol.* 7:109-122.

Harrison, L. R., Colvin, B. M., Greene, J. T., Newman, L. E., and Cole, J. R., Jr. 1990, Pulmonary edema and hydrothorax in swine produced by fumonisin B_1, a toxic metabolite of *Fusarium moniliforme*. *J. Vet. Diagn. Invest.* 2:217-221.

Hartl, M., and Humpf, H.-U., 1999, Simultaneous determination of fumonisin B_1 and hydrolyzed fumonisin B_1 in corn products by liquid chromatography/electrospray ionization mass spectrometry. *J. Agric. Food Chem.*, 47:5078-5083.

Hopmans, E.C., and Murphy, P.A., 1993, Detection of fumonisins B_1, B_2, and B_3 and hydrolyzed fumonisin FB_1 in corn-containing foods. *J. Agric. Food Chem.* 41:1655-1658.

IARC (International Agency for Research on Cancer), 1993, *Some Naturally Occurring Substances: Food Items and Constituents, Heterocyclic Amines and Mycotoxins*; IARC Monographs on the Evaluation of Carcinogenic Risk to Humans; IARC: Lyon, Vol. 56.

Kellerman, T. S., Marasas, W. F. O., Thiel, P. G., Gelderblom, W. C. A., Cawood, M., and Coetzer, J. A. W., 1990, Leukoenzephalomalacia in two horses induced by oral dosing of fumonisin B1. *Onderstepoort J. Vet. Res.* 57:269-275.

Lukacs, Z., Schaper, S., Herderich, M., Schreier, P., and Humpf, H.-U., 1996, Identification and determination of fumonisin FB_1 and FB_2 in corn and corn products by high-performance liquid chromatography-electrospray-ionization tandem mass spectrometry (HPLC-ESI- MS-MS). *Chromatographia* 43:124-128.

Maragos, C. M., Bennett, G. A., and Richard, J. L., 1997, Affinity column clean up for the analysis of fumonisins and their hydrolysis products in corn. *Food Agric. Immunol.* 9:3-12.

Marasas, W. F. O., 1996, Fumonisins: History, worldwide occurrence and impact. *Adv. Exp. Med. Biol.* 392:1-17.

Musser, S.M., 1996, Quantitation and identification of fumonisins by liquid chromatography/mass spectrometry. In *Fumonisins in food*, Jackson, L; DeVries, J.W.; Bullerman, L.B., eds.; Plenum Publishing Corp., New York, pp. 65-74.

Musser, S.M., and Plattner, R.D., 1996, Identification of a new series of fumonisins containing 3-hydroxypyridine. *J. Nat. Prod.* 59:970-972.

Musser, S.M., and Plattner, R.D., 1997, Fumonisin composition in cultures of *Fusarium moniliforme, Fusarium proliferatum*, and *Fusarium nygami*. *J. Ag. Food Chem.* 45:1169-1173.

Nelson, P.E.., Plattner, R.D., Shackelford, D.D., and Desjardins, A.E., 1991, Production of fumonisins by *Fusarium moniliforme* strains from various substrates and geographical areas. *Appl. Environ. Microbiol.* 57:2410-2412.

Nelson, P.E.. Plattner, R.D., Shackelford, D.D., and Desjardins, A.E., 1992, Fumonisin B_1 production by *Fusarium* species other than *F. moniliforme* in section *Liseola* and by some related species. *Appl. Environ. Microbiol.* 58:984-989.

Nelson, P.E., Juba, J.H., Ross, P.F., and Rice, L.G., 1994, Fumonisin production by Fusarium species on solid substrates. *JAOAC*, 77:522-524.

Newkirk, D. K., Benson, R. W., Howard, P. C., Churchwell, M. I., Doerge, D. R., and Roberts, D. W., 1998, On-line immunoaffinity capture, coupled with HPLC and electrospray ionization mass spectrometry, for automated determination of fumonisins. *J. Agric. Food Chem.* 46:1677-1688.

Rheeder, J. P., Marasas, W. F. O., Thiel, P. G., Sydenham, E. W., Shephard, G. S., and Schalkwijk, D. J., 1992, *Fusarium moniliforme* and fumonisins in corn in relation to human esophageal cancer in Transkei. *Phytopathologia* 82:353-357.

Ross, P.F., Nelson, P.E., Richard, J.L., Osweiler, G.D., Rice, L.G., Plattner, R.D., and Wilson, T.M., 1990, Production of fumonisins by *Fusarium moniliforme* and *Fusarium proliferatum* isolates associated with equine leukoencephalomalacia and a pulmonary edema syndrome in swine. *Appl. Environ. Microbiol.*, 56:3225-3226.

Seo, J-A., Kim, J-C., and Lee, Y-W., 1996, Isolation and Characterization of Two New Type C Fumonisins Produced by *Fusarium oxysporum*, *J. Nat. Prod.*, 59:1003-1005.

Shephard, G.S., Sydenham, E.W., Thiel, P.G., and Marasas, W.F.O., 1990, Quantitative determination of fumonisins $B_1$and B_2 by high performance liquid chromatography with fluorescence detection. *J. Liq. Chromatogr.* 13:2077-2087.

Stack, M. E., 1998, Analysis of fumonisin B_1 and its hydrolysis product in tortillas. *J. AOAC Int.* 81:737-740.

Sydenham, E.W.,Thiel, P.G., Shephard, G.S., Koch, K.R., and Hutton, T., 1995, Preparation and isolation of the partially hydrolyzed moiety of fumonisin B_1. *J. Agric. Food Chem.* 43:2400-2405.

Sydenham, E.W., Shepard, G.S., Thiel, P.G., Stockenstrom, S., Snijman, P.W., and Van Schalkwyk, D.J., 1996, Liquid chromatographic determination of fumonisin B_1, B_2, and B_3 in corn: IUPAC/AOAC interlaboratory collaborative study. *J. AOAC Int.* 79:688-695.

Thakur, R. A., and Smith, J. S., 1996, Determination of fumonisins B_1 and B_2 and their major hydrolysis products in corn, feed, and meat, using HPLC. *J. Agric. Food Chem.* 44:1047-1052.

Trucksess, M.W., Stack, M.E., Allen, S., and Barrion, N., 1995, Immunoaffinity column coupled with liquid chromatography for determination of fumonisin B_1 in canned and frozen sweet corn. JAOAC Int. 78:705-710.

Trucksess, M.W., and Abouzied, M.M., 1996, Evaluation and application of immunochemical methods for fumonisin B_1 in corn. In *Immunoassays for Residue Analysis*, Beier, R.C. and Stanker, L.H. eds, American Chemical Society: Washington, DC, pp. 358-367.

Trucksess, M.W., Cho, T-H, and Ready, D.E., 2000, Liquid chromatographic method for fumonisin B_1 in sorghum syrup and corn-based breakfast cereals. *Food Add. Contam.*, 17:161-166.
Usleber, E., Straka, M., and Terplan, G., 1994, Enzyme immunoassay for fumonisin B_1 applied to corn-based food. *J. Agric. Food.Chem.* 42:1392-1396.
Vesonder, R.F., Wu, W., Weisleder, D., Gordon, S.H.,Krick, T., Xie, W., Abbass, H.K., and McAlpin, C.E., 2000, Toxigenic strains of *Fusarium moniliforme* and *Fusarium proliferatum* isolated from dairy cattle feed produce fumonisins, monliformin and a new C21H38N2O6 metabolite phytotoxic to *Lemna minor L. J. Nat. Toxins*, 9:103-112.
Voss, K.A., Riley, R.T., Norred, W.P., Bacon, C.W., Meredith, F.I., Howard, P.C., Plattner, R.D., Collins, T.F., Hansen, D.K., and Porter, J.K., 2001, Overview of rodent toxicities: liver and kidney effects of fumonisins and *Fusarium moniliforme. Environ Health Perspect.* 109:259-266.
Yu, F. Y., and Chu, F. S.,1996, Production and characterization of antibodies against fumonisin B_1. *J. Food Prot.* 59:992-997.

RECENT ADVANCES IN ANALYTICAL METHODOLOGY FOR CYCLOPIAZONIC ACID

Joe W. Dorner

USDA, ARS, National Peanut Research Laboratory
P. O. Box 509
1011 Forrester Dr., SE
Dawson, GA 31742

ABSTRACT

Cyclopiazonic acid (CPA) is a toxic indole tetramic acid that has been isolated from numerous species of *Aspergillus* and *Penicillium*. It has been found as a natural contaminant of cheese, corn, peanuts and various feedstuffs. Historically, thin-layer chromatography has been the most widely used method for quantitative determination of CPA in fungal cultures and agricultural commodities. Several liquid chromatographic (LC) and spectrophotometric methods have also been used, but these require extensive, time-consuming cleanup procedures to achieve accurate quantitation. More recently, enzyme-linked immunosorbent assays (ELISA) have been developed for quantification of CPA, and an immunoaffinity column (IAC) has been developed for cleanup of sample extracts prior to quantification by ELISA or LC. In applying the IAC to the cleanup of peanut extracts, recovery of CPA from spiked samples ranged from 83.7% to 90.8%, and the method was successfully applied to the analysis of peanuts that were naturally contaminated with CPA.

INTRODUCTION

Cyclopiazonic acid (CPA) (Figure 1) is a toxic, indole tetramic acid that was originally isolated from *Penicillium cyclopium* (Holzapfel, 1968) and subsequently reported to be produced by numerous species of *Penicillium* and *Aspergillus* (Dorner et al., 1985). The taxonomy of Penicillia producing CPA is not straightforward. The original isolate (*P. cyclopium*) was later referred to as *P. griseofulvum* (de Jesus et al., 1981) and *P. verrucosum* var. *cyclopium* (Malik et al., 1986). Although CPA also has been reported to be produced by many other species, including *P. patulum*, *P. viridicatum*, *P. puberulum*, and *P. crustosum* (Bryden, 1991), Pitt et al. (1986) maintained that the correct name for most saprophytic Penicillia that produce CPA is *P. commune* with *P. palitans* as a synonym. Based primarily on chemotaxonomical features coupled with conidial color on Czapek yeast autolysate agar, Lund

(1995) concluded that *P. palitans* was not synonymous with *P. commune*, but was actually a distinct species. Pitt et al. (1986) classified all CPA-producing molds used in the manufacture of white cheeses that produce CPA as *P. camembertii*, a domesticated species derived from *P. commune*. Species of *Aspergillus* that produce CPA include *A. versicolor* (Ohmomo et al., 1973), *A. flavus* (Luk et al., 1977), *A. oryzae* (Orth, 1977), and *A. tamarii* (Dorner, 1983). *A. flavus*, which is primarily known as a producer of the aflatoxins, is a frequent contaminant of corn, peanuts, and other commodities. The ubiquitous nature of these fungi and their propensity to invade commodities commonly used for food and feed indicates that the potential for the contamination of commodities with CPA is widespread (Bryden, 1991).

Natural occurrence of CPA has been reported in a variety of commodities including corn (Gallagher et al., 1978; Widiastuti et al., 1988; Urano et al., 1992a), peanuts (Lansden and Davidson, 1983; Urano et al., 1992a), cheese (Still et al., 1978; Le Bars, 1979; Le Bars, 1990), millet (Rao and Husain, 1985), sunflower (Ross et al., 1991), and various feeds and feedstuffs (Balachandran and Parthasarathy, 1996). The toxin has also been shown to accumulate in meat and eggs of chickens (Norred et al., 1988; Dorner et al., 1994) and the milk of sheep (Dorner et al., 1994) dosed with CPA.

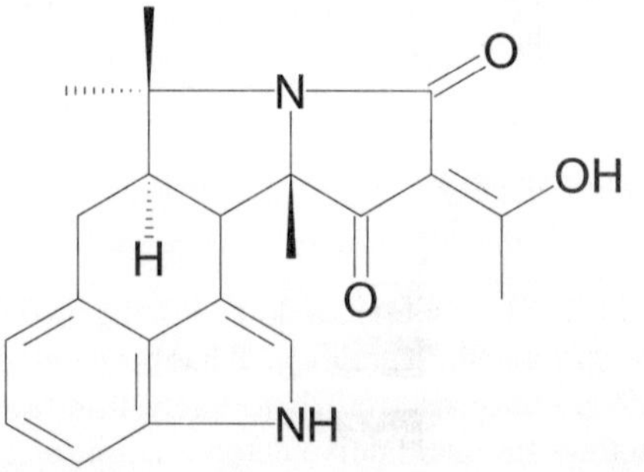

Figure 1. Chemical structure of cyclopiazonic acid (CPA).

Toxicosis resulting from consumption of CPA-contaminated food or feed has not been proven unequivocally, but CPA has been strongly implicated as the causative agent or one of the causative agents in several mycotoxicoses. Cole (1986) presented strong circumstantial evidence that CPA was probably involved along with aflatoxin in the outbreak termed turkey "X" disease (Sargeant et al., 1961), which resulted in the discovery of the aflatoxins. Support for this theory came when Bradburn et al. (1994) found CPA present at a concentration of 31 μg/kg in a sample of groundnut cake that had been saved from the original turkey "X" disease. CPA was strongly implicated in a human intoxication termed 'kodua poisoning' in which consumption of kodo millet produced symptoms of giddiness and nausea. The millet contained CPA and was heavily infected with CPA-producing strains of *A. flavus* and *A. tamarii* (Rao and Husain, 1985). CPA was also considered to be responsible for the death of quails that consumed feed containing 6000 μg/kg of the toxin (Stoltz et al., 1988). Several acute and many chronic toxicological studies have been conducted with CPA. A review of these studies has been published recently (Burdock and Flamm, 2000), and oral LD_{50} values for CPA ranged from 12 mg/kg in chickens (Wilson et al., 1986) to 64 mg/kg in mice (Purchase, 1971). The LD_{50} for intraperitoneal administration varied from 2.3 mg/kg in rats (Purchase, 1971) to 13 mg/kg in mice (Nishie et al., 1985).

A variety of analytical methods have been used to detect and quantify CPA in fungal cultures and agricultural commodities over the years. The purpose of this chapter is to review those methods and detail advances that have occurred more recently.

THIN-LAYER CHROMATOGRAPHY

Thin-layer chromatography (TLC) has been the method most widely used for the analysis of CPA. Steyn (1969) reported separation of CPA on silica gel plates impregnated with oxalic acid and developed with chloroform-methyl-isobutylketone (4:1, v/v). CPA had an R_F value of 0.65 in that system, but it did not move on plates without oxalic acid. CPA gave a violet color upon spraying with Erlich's reagent (*p*-dimethylaminobenzaldehyde) or became red-brown upon treatment with ferric chloride. When more polar systems, such as toluene-ethyl acetate-formic acid (5:4:1, v/v/v), were used with plates not treated with oxalic acid, there was good mobility of CPA, but with extensive tailing (Gallagher et al., 1978; Dorner et al., 1983). The use of oxalic acid-treated and non-treated plates with two different solvent systems provides good confirmation for the presence of CPA (Gallagher et al., 1978). Instead of using plates treated with oxalic acid, modification of solvent systems with ammonia or acetic acid has also been used to prevent tailing (Ohmomo et al., 1973; Gorst-Allman et al., 1979; Lansden, 1986).

Table 1. Characteristics of various analytical techniques used to quantify CPA.

Analytical technique	Analyte	Cleanup	Quantitation limit (ng/g)	Detection limit (ng/g)	Reference
TLC	Peanuts, Corn	SP	150	125	Lansden, 1986
TLC	Cheese	SP	N.D.	20	Le Bars, 1990
Spectro-photometry	Feed	SP, column chromatography	160	80	Chang-Yen and Bidasee,1990
HPLC	Corn	SPE	100	N.D.	Goto et al., 1987
HPLC	Meat	SP, minicolumn	16	16	Norred et al., 1987
HPLC	Corn, Peanuts	SP, SPE	50-100	N.D.	Urano et al., 1992b
ELISA	Corn	None	N.D.	100	Yu and Chu, 1998
ELISA	Corn	IAC	10	2	Yu et al., 1998
HPLC	Peanuts	IAC	10	2.5	Dorner et al., 2000

Abbreviations: SP, solvent partition; SPE, solid phase extraction; IAC, immunoaffinity column; N.D., not determined

TLC has been used to quantify CPA in cheese (Le Bars, 1990), corn (Gallagher et al., 1978; Widiastuti et al., 1988; Lee and Hagler, 1991), peanuts (Lansden and Davidson, 1983; Bradburn et al., 1994), sunflower (Ross et al., 1991), millet (Rao and Husain, 1985), and milk and eggs (Dorner et al., 1994). Typical quantitation limits for methods using TLC to quantify CPA in commodities have been around 150 ng/g (Lansden, 1986). Characteristics of TLC and other analytical techniques for CPA appear in Table 1.

COLORIMETRY/SPECTROPHOTOMETRY

Methods have been developed that take advantage of the color formed when CPA reacts with Erlich's reagent. Rathinavelu and Shanmugasundaram (1984) used TLC to purify extracts, eluted CPA from TLC plates, and added *p*-dimethylaminobenzaldehyde plus HCl to develop color which was measured with a colorimeter at 560 nm. Variations and improvements to this basic method have been reported by Rao and Husain (1987), Chang-Yen and Bidasee (1990), and Šimůnek et al. (1992). Chang-Yen and Bidasee (1990) used liquid-liquid partition and column chromatography to purify extracts before addition of Erlich's reagent and spectrophotometry at 580 nm. They reported recoveries of CPA from feed samples ranging from 79 to 105% with a detection limit in corn and poultry feed of 80 ng/g (Table 1).

HIGH PERFORMANCE LIQUID CHROMATOGRAPHY

High performance liquid chromatographic (HPLC) methods have been used to quantify CPA in various matrices. Lansden (1984) reported a reversed-phase system for quantifying CPA in peanuts that used a C_8 or C_{18} column and a mobile phase containing acetonitrile (40%), 2-propanol (30%), and water (20%) containing 1.0% ammonium acetate, 0.025% 4-dodecyldiethylenetriamine, and 0.001 M zinc acetate. CPA was detected by UV at 284 nm. The detection limit for pure CPA was 4 ng, and recoveries of CPA from peanuts (4 replications) spiked at 69, 210, and 955 ng/g were 85.9, 72.9, and 81.4%, respectively. Corresponding coefficients of variation (CV's) were 12.9, 6.4, and 0.4%.

Norred et al. (1987) modified the cleanup procedure of Lansden (1984), but used the same basic HPLC system to quantify CPA in poultry meat. After the meat was extracted with acidic chloroform-methanol (80:20, v/v), the extract was purified by solvent partition and minicolumn chromatography. Using sodium hydroxide in the partitioning step instead of sodium bicarbonate improved recovery of CPA added to extracts from 66% to 94%. Recovery of CPA added to meat samples at concentrations of 16 to 15,600 ng/g ranged from 67.2 to 89.3% with a minimum quantifiable limit of 16 ng/g.

Goto et al. (1987) used normal-phase HPLC with a silica gel column, a mobile phase consisting of ethyl acetate-2-propanol-25% aqueous ammonia (55:20:5, v/v/v), and a spectrophotometer operated at 284 nm to achieve a detection limit for pure CPA of 0.2 ng. Using an extraction solvent of chloroform-85% phosphoric acid and a silica cleanup cartridge, they reported an 82% recovery of CPA from maize with a minimum quantitation limit of 100 ng/g.

A reversed-phase HPLC method using a C_{18} column and a linear gradient of 0-4 mM $ZnSO_4$ in methanol-water (85:15, v/v) was developed to quantify CPA in corn and peanuts (Urano et al., 1992b). They reported quantitation limits of about 50 and 100 ng/g, respectively (Table 1), with CV's ranging from 3.5 to 7.4%. The method was used in a 1990 survey of corn and peanuts, which showed extensive contamination of both crops with CPA and aflatoxins (Urano et al., 1992a).

A simple method for analyzing extracts of fungal cultures for the presence of CPA was reported by Matsudo and Sasaki (1995). The system consisted of a C_{18} column with a mobile phase of 50 mM H_3PO_4 plus 1 mM $ZnSO_4$-acetonitrile (45:55, v/v) and UV detection at 284 nm. Indomethacin was added to sample extracts as an internal standard, and CPA concentrations were calculated on the basis of the ratio of the peak area of CPA to that of the internal standard. The detection limit for CPA was 0.3 ng.

Sobolev et al. (1998) reported a normal phase ion-pair partition HPLC system for detecting CPA simultaneously with other metabolites of various *Aspergillus* species. A silica gel column with a mobile phase of n-heptane-2-propanol-n-butanol-water-40% aqueous tetrabutylammonium hydroxide (2560:900:230:32:8, v/v/v/v/v) and a diode array detector allowed separation and detection of at least seven metabolites, including CPA, with a detection limit for CPA of 5 ng/injection.

ADVANCES IN METHODOLOGY

Analysis of CPA in foods and feeds has proven challenging and difficult. Much of that difficulty is associated with time-consuming, laborious cleanup procedures that result in relatively high quantitation limits and poor reproducibility. Typical quantitation limits for methods based on TLC, colorimetry/spectrophotometry, and HPLC are in the range of 50-150 ng/g. Many of these methods rely on cleanup accomplished by several liquid-liquid partition steps followed by column chromatography or solid phase extraction. The advances in analytical methodology for CPA that have been made in recent years are primarily associated with antibody-based systems, which have resulted in greater specificity and much lower quantitation limits. Monoclonal and polyclonal antibodies for CPA have been developed and utilized both in enzyme linked immunosorbent assays (ELISA) and in the production of immunoaffinity columns used in the cleanup of extracts prior to quantitation. Additionally, new methodology based on capillary electrophoresis has been reported, offering a new alternative for the determination of CPA.

Enzyme-Linked Immunosorbent Assay (ELISA)

The first ELISA for CPA was reported by Hahnau and Weiler (1991), and it utilized polyclonal antibodies to measure from 30 pg to 2 ng of CPA. The assay was used to detect CPA in fungi growing as agar surface cultures. However, limited availability of antisera prompted the authors to develop monoclonal antibodies that were used in ELISA analysis of fungal cultures and white-mold fermented cheeses (Hahnau and Weiler, 1993). The assay detected from 7 to 300 ng of CPA, and CPA was found in white-mold cheese extracts at concentrations as high as 7.9 μg/g, which was not considered a potential health hazard when compared with published oral toxicity data. The monoclonal antibodies were tested for cross-reactivity with a variety of structurally-related compounds, and cross-reactivity (20%) was found only with cyclopiazonic acid imine.

Huang and Chu (1993) also developed poly and monoclonal antibodies against CPA. The affinity of the polyclonal antibodies for CPA was at least an order of magnitude less than that of the monoclonals, which also had a slight cross-reactivity with cyclopiazonic acid imine. However, cross-reactivity with other related compounds was negative. An indirect competitive ELISA based on the monoclonal antibodies was used to analyze cultures of various species of *Aspergillus* (Huang et al., 1994). The detection limit for CPA in the liquid fungal cultures was about 0.01 ng/mL of media. The results showed a high degree of variability in the ability of various strains of *A. flavus* to produce CPA. Results from the ELISA analysis were in good agreement with TLC.

An improved direct competitive ELISA for the analysis of CPA in corn, peanuts, and mixed feed was reported by Yu and Chu (1998). Samples were extracted with 70% methanol in PBS, filtered, and diluted with PBS before assay. False positive results were obtained for corn, mixed feed, and peanuts at levels of 50, 200, and 500 ng/g, respectively. Therefore, limits of detection for the three commodities were estimated to be around 100, 300, and 600 ng/g, respectively. Mean recovery of CPA from all commodities was greater than 90% when false positive data were excluded.

Immunoaffinity Column Chromatography

Application of immunoaffinity columns to the cleanup of extracts prior to quantitation has been very successful, particularly in aflatoxin analysis. An AOAC International "official method" (method # 991.31) for aflatoxin (Scott, 1995) utilizes an immunoaffinity column for simple cleanup of a methanol-water extract, which can then be quantified by solution fluorometry or HPLC. Such methods are very popular because the selective affinity column provides an extremely clean extract that allows for highly sensitive measurements.

Yu et al. (1998) developed an immunoaffinity column for the cleanup of CPA extracts based on CPA-specific monoclonal antibodies (Yu and Chu, 1998). The antibodies were coupled to Sepharose gel, and 0.2 mL of CPA immunogel was packed in a clean filter tube and held in place between two frits. The column was filled with PBS containing 0.02% sodium azide and stored at 4°C until use. The column had an initial capacity of 4 μg of CPA. CPA was completely eluted from the column with 2 mL of 100% methanol at 0.5 mL/min. The column could be regenerated for additional use by immediate washing with 10 mL of PBS and storing in a cold room for > 14 h or overnight. The columns could be reused at least ten times at 60% of original capacity.

Immunoaffinity columns were used as a cleanup tool prior to ELISA analysis of corn, peanuts, and mixed feed for presence of CPA (Yu et al., 1998). Samples were extracted for 1 min in a Waring blender at high speed with methanol-50 mM Tris-buffered saline (70:30, v/v) at a ratio of 5 mL of solvent per g of sample. Extracts were filtered and a 5-mL aliquot was diluted with 5 mL deionized water prior to addition to the immunoaffinity column. Mean recoveries of CPA added to corn, peanuts, and mixed feed at 10-200 ng/g were 98.8, 94.8, and 102%, respectively, and mean CV's were all < 10%. Detection limits for corn, mixed feed, and peanuts were 2.0, 4.4, and 4.7 ng/g, respectively. That represents a tremendous improvement in sensitivity compared with ELISA analyses done without immunoaffinity column cleanup, which had detection limits in the range of 100-600 ng/g (Yu and Chu, 1998). Yu et al. (1998) also used the method to analyze naturally contaminated peanut samples, and results were compared with ELISA analyses that followed the solvent partition cleanup described by Lansden (1984). CPA concentrations in peanuts ranged from 2.7 to 384 ng/g, and the comparison of cleanup methods indicated that there was considerable loss of CPA using the partition cleanup method compared with the immunoaffinity column method.

The immunoaffinity column has been used for the cleanup of peanut extracts prior to quantitation by HPLC (Dorner et al., 2000). Peanuts were extracted in a Waring blender with methanol-1% sodium bicarbonate (70:30, v/v) at a ratio of 3 mL of solvent per g of peanuts. Saturated sodium bicarbonate was added to an aliquot of filtered extract, which was then placed in a freezer for 30 min. After filtration and partial evaporation, the extract was applied to the immunoaffinity column and washed with 15 mL of water followed by 2 mL of methanol-water (70:30, v/v). CPA was eluted in 2 mL of methanol, which was evaporated to dryness under nitrogen. The residue was dissolved in 0.5 mL of HPLC mobile phase, which consisted of hexane-reagent alcohol-10% TRIS (500:275:16, v/v/v), and injected into a system consisting of a silica gel column and a diode array detector at 282 nm. A flow rate of 1.2 mL/min resulted in elution of CPA in approximately 5 min with almost no interfering peaks. Typical chromatograms of standard CPA (20 ng injection) and naturally contaminated peanuts (187 ng/g) cleaned up with the immunoaffinity column are shown in Figures 2 and 3, respectively. Recovery of CPA from spiked peanut samples ranged from 83.7 to 90.8% with CV's of about 7% and a limit of detection of 2.5 ng/g. The method was used to analyze samples of peanuts grown under late-season drought conditions, and natural contamination with CPA ranged from 3.0 ng/g in highest quality seed to 8105.0 ng/g in the poorest quality, damaged seed.

Use of the immunoaffinity column as a cleanup tool prior to quantitation by ELISA or HPLC resulted in greatly reduced limits of detection for CPA with reasonable coefficients of variation. The immunoaffinity column addresses one of the major problems associated with

CPA analysis, which is the necessity of using laborious cleanup procedures that produce less than desirable results. However, the disadvantage of this methodology at this time is that immunoaffinity columns for CPA are not readily available. For laboratories that do not have access to the antibodies nor the expertise to generate the columns, it is not a viable option. Commercial manufacture and sale of immunoaffinity columns for CPA will probably not become a reality until a genuine need for such columns is apparent, as was the case for other mycotoxins, such as aflatoxin, fumonisin, and deoxynivalenol.

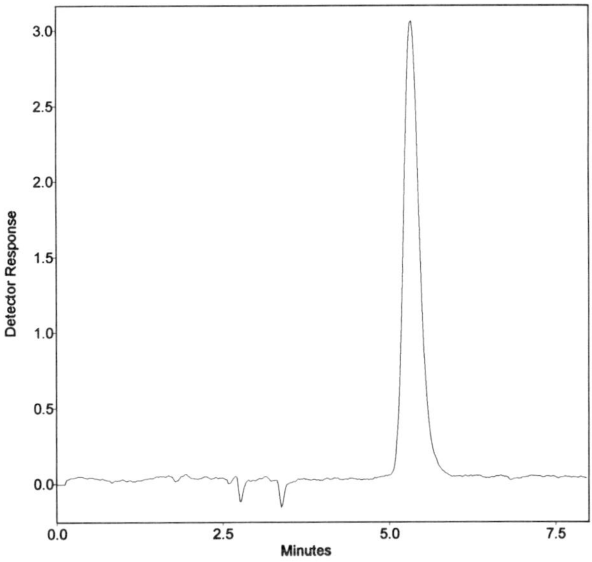

Figure 2. Chromatogram of a 20 ng injection of CPA standard.

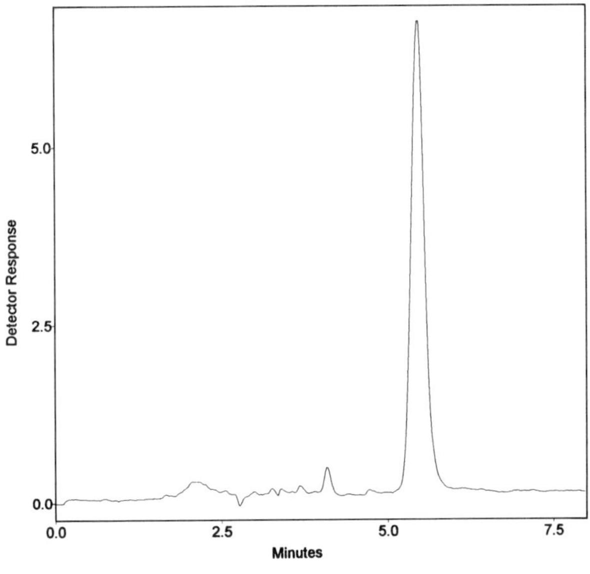

Figure 3. Chromatogram of a sample of peanuts naturally contaminated with 187 ng/g of CPA cleaned up with an immunoaffinity column.

Capillary Electrophoresis

Prasongsidh et al. (1998) developed a capillary electrophoresis (CE) method for quantifying CPA in milk. CPA was extracted from milk with methanol-2% sodium bicarbonate (70:30, v/v) and cleaned up by solvent partition and solid phase extraction. The CE system utilized a bare-fused-silica capillary-extended light path and an optimum mobile phase of 0.05 M sodium deoxycholate, 0.01 M disodium hydrogen phosphate, and 0.006 M disodium tetraborate at a pH of 9.3. For separation of CPA from aflatoxins and tenuazonic acid, 7% acetonitrile was added to the mobile phase, which produced a slightly longer migration time for CPA. CPA was detected with a diode array detector at 225 nm. The minimum quantifiable concentration of CPA in milk was 20 ng/mL, and recoveries of CPA from milk spiked with 20 to 500 ng/mL ranged from 77.9 to 81.0% with CV's ranging from 2.3 to 6.7%. The method was compared with the reversed-phase HPLC method reported by Urano et al. (1992b), and a two-fold increase in sensitivity was achieved with the CE method. The quantitation limit for the HPLC system used in this study was 50 ng/mL. However, recoveries and CV's for the two methods were similar for concentrations of 200-500 ng/mL. Whereas CE offers another alternative in methods for quantifying CPA, the reported method has the same need for improvement in cleanup as many of the other methods for TLC, HPLC, etc.

CONCLUSION

The production of CPA by numerous fungal species and its occurrence in a wide variety of commodities, foods, and feedstuffs points to its potential as a significant mycotoxin. However, its significance is poorly understood, partially because of difficulties associated with its analysis. There is no generally accepted "official" analytical method for CPA; therefore, many methods have been developed to address the specific needs of various researchers. Methods have utilized TLC, HPLC, colorimetry/spectrophotometry, ELISA, and CE for quantitation, but the major difficulty in analyzing for CPA involves cleanup procedures that lead to inconsistent recoveries, poor reproducibility, and relatively high detection limits. The major advances in CPA analysis have involved development of antibodies for CPA that have been used in ELISA analyses and in production of immunoaffinity columns for cleanup. Use of these has reduced limits of detection by 20-100 fold, but they are not readily available. The need still exists for the advances that have been made in analytical methodology to find their place in common use.

REFERENCES

Balachandran, C., and Parthasarathy, K.R., 1996, Occurrence of cyclopiazonic acid in feeds and feedstuffs in Tamil Nadu, India, *Mycopathologia* 133:159.

Bradburn, N., Coker, R.D., and Blunden, G., 1994, The aetiology of turkey 'X' disease, *Phytochemistry* 35:817.

Bryden, W.L., 1991, Occurrence and biological effects of cyclopiazonic acid, in: *Emerging Food Safety Problems Resulting from Microbial Contamination*, K. Mise, and J.L. Richard, ed., Toxic Microorganisms Panel of the UJNR, Tokyo.

Burdock, G.A., and Flamm, W.G., 2000, Review article: safety assessment of the mycotoxin cyclopiazonic acid, *Int. J. Toxicol.* 19:195.

Chang-Yen, I., and Bidasee, K., 1990, Improved spectrophotometric determination of cyclopiazonic acid in poultry feed and corn, *J. Assoc. Off. Anal. Chem.* 73:257.

Cole, R.J., 1986, Etiology of turkey "X" disease in retrospect: A case for the involvement of cyclopiazonic acid, *Mycotoxin Res.* 2:3.

de Jesus, A.E., Steyn, P.S., Vleggaar, R., Kirby, G.W., Varley, M.J., and Ferreira, N.P., 1981, Biosynthesis of α-cyclopiazonic acid. Steric course of proton removal during the cyclisation of β-cyclopiazonic acid in *Penicillium griseofulvum*, *J. Chem. Soc. Perkin Trans. I* 3292.

Dorner, J.W., 1983, Production of cyclopiazonic acid by *Aspergillus tamarii* Kita, *Appl. Environ. Microbiol.* 46:1435.
Dorner, J.W., Cole, R.J., Erlington, D.J., Suksupath, S., McDowell, G.H., and Bryden, W.L., 1994, Cyclopiazonic acid residues in milk and eggs, *J. Agric. Food Chem.* 42:1516.
Dorner, J.W., Cole, R.J., and Lomax, L.G., 1985, The toxicity of cyclopiazonic acid, in: *Trichothecenes and Other Mycotoxins*, J. Lacey, ed., John Wiley & Sons, Ltd., Chichester.
Dorner, J.W., Cole, R.J., Lomax, L.G., Gosser, H.S., and Diener, U.L., 1983, Cyclopiazonic acid production by *Aspergillus flavus* and its effects on broiler chickens, *Appl. Environ. Microbiol.* 46:698.
Dorner, J.W., Sobolev, V.S., Yu, W., and Chu, F.S., 2000, Immunochemical method for cyclopiazonic acid, in: *Mycotoxin Protocols*, M.W. Trucksess, and A.E. Pohland, ed., Humana Press, Totowa, New Jersey.
Gallagher, R.T., Richard, J.L., Stahr, H.M., and Cole, R.J., 1978, Cyclopiazonic acid production by aflatoxigenic and non-aflatoxigenic strains of *Aspergillus flavus*, *Mycopathologia* 66:31.
Gorst-Allman, C.P., and Steyn, P.S., 1979, Screening methods for the detection of thirteen common mycotoxins, *J. Chromatogr.* 175:325.
Goto, T., Shinshi, E., Tanaka, K., and Manabe, M., 1987, Analysis of cyclopiazonic acid by normal phase high-performance liquid chromatography, *Agric. Biol. Chem.* 51:2581.
Hahnau, S., and Weiler, E.W., 1991, Determination of the mycotoxin cyclopiazonic acid by enzyme immunoassay, *J. Agric. Food Chem.* 39:1887.
Hahnau, S., and Weiler, E.W., 1993, Monoclonal antibodies for the enzyme immunoassay of the mycotoxin cyclopiazonic acid, *J. Agric. Food Chem.* 41:1076.
Holzapfel, C.W., 1968, The isolation and structure of cyclopiazonic acid, a toxic metabolite of *Penicillium cyclopium* Westling, *Tetrahedron* 24:2101.
Huang, X., and Chu, F.S., 1993, Production and characterization of monoclonal and polyclonal antibodies against the mycotoxin cyclopiazonic acid, *J. Agric. Food Chem.* 41:329.
Huang, X., Dorner, J.W., and Chu, F.S., 1994, Production of aflatoxin and cyclopiazonic acid by various Aspergilli: an ELISA analysis, *Mycotoxin Res.* 10:101.
Lansden, J.A., 1984, Liquid chromatographic analysis system for cyclopiazonic acid in peanuts, *J. Assoc. Off. Anal. Chem.* 67:728.
Lansden, J.A., 1986, Determination of cyclopiazonic acid in peanuts and corn by thin layer chromatography, *J. Assoc. Off. Anal. Chem.* 69:965.
Lansden, J.A., and Davidson, J.I., 1983, Occurrence of cyclopiazonic acid in peanuts, *Appl. Environ. Microbiol.* 45:766.
Le Bars, J., 1979, Cyclopiazonic acid production by *Penicillium camemberti* Thom and natural occurrence of this mycotoxin in cheese, *Appl. Environ. Microbiol.* 38:1052.
Le Bars, J., 1990, Detection and occurrence of cyclopiazonic acid in cheeses, *J. Environ. Pathol., Toxicol., Oncol.* 10:136.
Lee, Y.J., and Hagler, W.M., Jr., 1991, Aflatoxin and cyclopiazonic acid production by *Aspergillus flavus* isolated from contaminated maize, *J. Food Sci.* 56:871.
Luk, K.C., Kobbe, B., and Townsend, J.M., 1977, Production of cyclopiazonic acid by *Aspergillus flavus* Link, *Appl. Environ. Microbiol.* 33:211.
Lund, F., 1995, Diagnostic characteriazation of *Penicillium palitans*, *P. commune* and *P. solitum*, *Lett. Appl. Microbiol.* 21:60.
Malik, R.K., Engel, G., and Teuber, M., 1986, Effect of some nutrients on the production of cyclopiazonic acid by *Penicillium verrucosum* var. *cyclopium*, *Appl. Microbiol. Biotechnol.* 24:71.
Matsudo, T., and Sasaki, M., 1995, Simple determination of cyclopiazonic acid, *Biosci. Biotech. Biochem.* 59:355.
Nishie, K., Cole, R.J., Dorner, J.W., 1985, Toxicity and neuropharmacology of cyclopiazonic acid, *Food Chem. Toxicol.* 23:831.
Norred, W.P., Cole, R.J., Dorner, J.W., and Lansden, J.A., 1987, Liquid chromatographic determination of cyclopiazonic acid in poultry meat, *J. Assoc. Off. Anal. Chem.* 70:121.
Norred, W.P., Porter, J.K., Dorner, J.W., and Cole, R.J., 1988, Occurrence of the mycotoxin cyclopiazonic acid in meat after oral administration to chickens, *J. Agric. Food Chem.* 36:113.
Ohmomo, S., Sugita, M., and Abe, M., 1973, Isolation of cyclopiazonic acid, cyclopiazonic acid imine and bissecodehydrocyclopiazonic acid from the cultures of *Aspergillus versicolor* (Vuill.) Tiraboschi, *J. Agr. Chem. Soc. Japan* 47:83.
Orth, R., 1977, Mycotoxins of *Aspergillus oryzae* strains for use in the food industry as starters and enzyme producing molds, *Ann. Nutr. Alim.* 31:617.
Pitt, J.I., Cruickshank, R.H., and Leistner, L., 1986, *Penicillium commune*, *P. camembertii*, the origin of white cheese moulds, and the production of cyclopiazonic acid, *Food Microbiol.* 3:363.
Prasongsidh, B.C., Kailasapathy, K., Skurray, G.R., and Bryden, W.L., 1998, Analysis of cyclopiazonic acid in milk by capillary electrophoresis, *Food Chem.* 61:515.

Purchase, I.F.H., 1971, The acute toxicity of the mycotoxin cyclopiazonic acid to rats, *Toxicol. Appl. Pharmacol.* 18:114.
Rao, L.B., and Husain, A., 1985, Presence of cyclopiazonic acid in kodo millet (*Paspalum scrobiculatum*) causing 'kodua poisoning' in man and its production by associated fungi, *Mycopathologia* 89:177.
Rao, B.L., and Husain, A., 1987, A simple colorimetric method for screening cyclopiazonic acid in agricultural commodities, *Mycopathologia* 97:89.
Rathinavelu, A., and Shanmugasundaram, E.R.B., 1984, Simple colorimetric estimation of cyclopiazonic acid in contaminated food and feeds, *J. Assoc. Off. Anal. Chem.* 67:38.
Ross, P.F., Rice, L.G., Casper, H., Crenshaw, J.D., and Richard, J.L., 1991, Novel occurrence of cyclopiazonic acid in sunflower seeds, *Vet. Hum. Toxicol.* 33:284.
Sargeant, K., Sheridan, A., O'Kelly, J., and Carnaghan, R.B.A., 1961, Toxicity associated with certain samples of groundnuts, *Nature* 192:1096.
Scott, P.M., 1995, Natural toxins, in: *Official Methods of Analysis of AOAC International*, P.A. Cunniff, ed., AOAC International, Arlington, VA.
Šimůnek, J., Březina, P., Matoušková, J., Bačová, L., Štětina, J., and Ježová, J., 1992, Determination of cyclopiazonic acid in dairy products, *Scripta Medica* 65:419.
Sobolev, V.S., Horn, B.W., Dorner, J.W., and Cole, R.J., 1998, Liquid chromatographic determination of major secondary metabolites produced by *Aspergillus* species from section *Flavi*, *J. AOAC Int.* 81:57.
Steyn, P.S., 1969, The separation and detection of several mycotoxins by thin-layer chromatography, *J. Chromatogr.* 45:473.
Still, P., Eckardt, C., and Leistner, L., 1978, Bildung von cyclopiazonsäure durch *Penicillium camemberti*-Isolate von Käse, *Fleischwirtschaft* 58:876.
Stoltz, D.R., Widiastuti, R., Maryam, R., Tri Akoso, B., Amang, and Unruh, D., 1988, Suspected cyclopiazonic acid mycotoxicosis of quail in Indonesia, *Toxicon* 26:39.
Urano, T., Trucksess, M.W., Beaver, R.W., Wilson, D.M., Dorner, J.W., and Dowell, F.E., 1992a, Co-occurrence of cyclopiazonic acid and aflatoxins in corn and peanuts, *J. AOAC Int.* 75:838.
Urano, T., Trucksess, M.W., Matusik, J., and Dorner, J.W., 1992b, Liquid chromatographic determination of cyclopiazonic acid in corn and peanuts, *J. AOAC Int.* 75:319.
Widiastuti, R., Maryam, R., Blaney, B.J., Stoltz, S., and Stoltz, D.R., 1988, Cyclopiazonic acid in combination with aflatoxins, zearalenone and ochratoxin A in Indonesian corn, *Mycopathologia* 104:153.
Wilson, M.E., Hagler, W.M., Jr., Ort, J.F., Brake, J.T., Cullen, J.M., Cole, R.J., 1986, Acute and subacute effects of cyclopiazonic acid in broiler chickens, *Poultry Sci.* 65 (Suppl. 1):145.
Yu, W., and Chu, F.S., 1998, Improved direct competitive enzyme-linked immunosorbent assay for cyclopiazonic acid in corn, peanuts, and mixed feed, *J. Agric. Food Chem.* 46:1012.
Yu, W., Dorner, J.W., and Chu, F.S., 1998, Immunoaffinity column as cleanup tool for a direct competitive enzyme-linked immunosorbent assay of cyclopiazonic acid in corn, peanuts, and mixed feed, *J. AOAC Int.* 81:1169.

METHODS OF ANALYSIS FOR OCHRATOXIN A

Peter M. Scott

Health Canada
Address Locator 2203D
Ottawa, Ontario K1A 0L2, Canada

ABSTRACT

The mycotoxin ochratoxin A (OTA) is produced by the fungi *Aspergillus alutaceus* and *Penicillium verrucosum* and has carcinogenic, nephrotoxic, teratogenic and immunosuppressive properties. The levels of OTA in foodstuffs are regulated in several countries, so reliable and sensitive methods are necessary for its determination. Procedures for extraction of OTA from ground foods generally use an organic solvent in the presence of acid or an extraction solvent containing aqueous sodium bicarbonate. Cleanup procedures include partition into aqueous sodium bicarbonate, solid phase extraction (SPE) columns and immunoaffinity chromatography. The latter technique allows detection of sub-ppb levels of OTA in a wide variety of foods and in plasma. The most widely used determinative procedure is reversed phase liquid chromatography (LC) with detection by fluorescence (excitation 330-340 nm, emission 460-470 nm) or, more recently, by tandem mass spectrometry. ELISA methods are also available. Certified reference materials containing OTA have been prepared.

INTRODUCTION

Ochratoxin A (OTA) was first reported in 1965 as a toxic metabolite of *Aspergillus ochraceus* (van der Merwe et al., 1965). This species and *Penicillium verrucosum,* as well as certain other *Aspergillus* spp.(e.g. *A. carbonarius)* and other *Penicillium* spp., are the principal producers of OTA (Kuiper-Goodman and Scott, 1989; Marquardt and Frohlich, 1992; Téren et al., 1996). The structure of OTA is shown in Figure 1. The non-toxic dechloro analogue, ochratoxin B (OTB), and the ethyl ester, ochratoxin C (OTC), are also fungal products. OTA is a potent kidney carcinogen and immunosuppressant in rodents, causes mycotoxicoses in farm animals, and is associated with human kidney disease (Bondy and Pestka, 2000; Creppy et al., 1993; Kuiper-Goodman and Scott, 1989; Stoev, 1998). There is abundant information on the natural occurrence of OTA in human foods and foodstuffs, including cereals, beans, coffee, cocoa beans, beer, dried vine fruit, wine, grape juice, spices, pig kidney and blood sausage, as well as in animal feed, human blood plasma and mother's milk (Miraglia et al., 1995; Pittet,

1998; Pohland et al., 1992; van Egmond and Speijers, 1994). Levels of OTA are regulated in several countries and a tolerance of 5 ppb in cereals in Europe has been proposed (Boutrif and Canet, 1998; Heilmann et al., 1999; Rosner, 1998; van Egmond, 1991a; van Egmond and Dekker, 1995), so sensitive, specific, accurate, precise and reliable methods are needed for its determination. In addition to meeting compliance needs, methods are required for monitoring and survey work and for research studies, covering a wide variety of matrices. A few methods have been adopted by international organizations, such as AOAC International (Nesheim, 1973b; Levi, 1975; Nesheim et al., 1992) and the Nordic Committee on Food Analysis (NMKL)(Larsson and Möller, 1996). However, the AOAC International methods published before 2000 have limits of detection that are too high for proposed regulations. There is considerable activity on collaborative study of more sensitive methods (see under **INTERLABORATORY METHOD PERFORMANCE**).

This review on methods of analysis for OTA is organized by procedure as was done by van Egmond (1991b) in an earlier review, covering sampling, extraction, cleanup, detection and determination, and confirmation, and includes discussion of method performance.

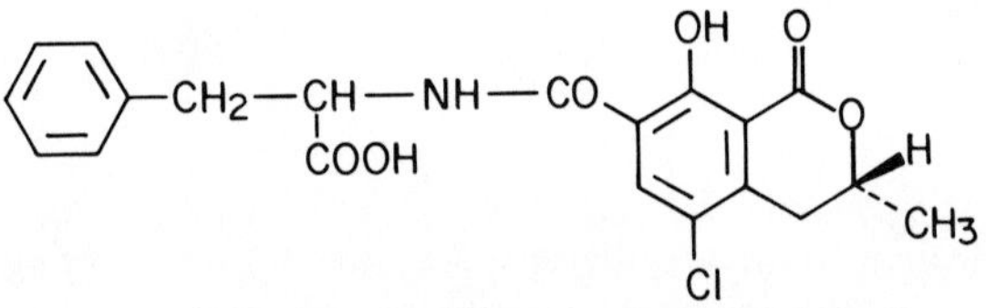

Figure 1. Chemical structure of OTA.

SAMPLING AND SUBSAMPLING

Except for green coffee beans, there are no published reports on schemes for sampling commodities containing OTA, a situation in contrast to the detailed sampling plans that have been devised for aflatoxins on a statistical basis (Park et al., 1998). For cereals, where sample size is mentioned, a 1-5 kg sample was typically taken (Czerwiecki, 1994; Eiwadh, 1992; Jørgensen et al., 1996; Jurjevic et al., 1999; Mühlemann et al., 1997; Shotwell et al., 1977; Soares and Rodriguez-Amaya, 1989; Trucksess et al., 1999; Veldman et al., 1992), but smaller samples have been used (Bacha et al., 1988). Attempts to take a representative sample of cereal from the container are infrequent (Bacha et al., 1988; Trucksess et al., 1999). For dried vine fruit 1 kg samples (usually a single bag) were analysed by MacDonald et al. (1999).

There has been considerable interest in the occurrence of OTA in coffee. Inhomogeneous distribution of low ppb levels of OTA in green coffee beans has been demonstrated (Studer-Rohr et al., 1995; Blanc et al., 1998; Heilmann et al., 1999). The distribution of OTA was characterized by a negative binomial function (Blanc et al., 1998). A sampling scheme tested by Heilmann et al. (1999) consisted of taking 60 samples of 125 g each during the discharge of bags into silo cells and producing three bulked samples of 2.25 kg from 30 of these samples. The homogeneity of the bulked samples (4.7-8.7 ppb OTA) contrasted with the heterogeneity of the other thirty 125 g samples (0.4-65 ppb OTA, mean 8.7 ppb). Further work is clearly needed to define sampling plans for OTA. It should be noted that some inhomogeneity may remain in ground, mixed samples of green coffee beans (Studer-Rohr et al., 1995); a 20% within-package coefficient of variation was found for roasted and ground coffee by Stegen et al. (1997).

Special sample preparation was used for dried vine fruits, which were blended for 20 minutes with water to form a slurry; this was stored frozen until analysis (MacDonald et al., 1999).

Sample preparation for animal kidney and liver was reviewed by Valenta (1998); details of sample size, which was up to 0.5 kg (Jørgensen, 1998; Valenta, 1998), are given in only a few papers.

EXTRACTION

Procedures for extraction of OTA from ground solid foods and feeds generally use an organic solvent in the presence of acid or an extraction solvent containing aqueous sodium bicarbonate. A shaker or high speed blender is used. The "classical" extraction solvent mixture is chloroform-0.1 *M o*-phosphoric acid (10+1) (Langseth et al., 1989; Nesheim et al., 1973ab; Nesheim et al., 1992), with or without addition of diatomaceous earth during the blending. Ethanol has been included as in the mixture dichloromethane-ethanol-0.1 *M o*-phosphoric acid (8+2+1) (Entwisle et al., 1997; Jørgensen and Vahl, 1999) and other acids such as citric acid (with dichloromethane) have been used (Barna-Vetró et al., 1996). To avoid use of chlorinated solvents there is a wide choice of alternative extraction solvent mixtures. Thus the chloroform can be replaced by ethyl acetate, as described by Clarke et al. (1994) for extraction of swine kidney. Toluene-acetic acid (99+1) and toluene-2 *M* hydrochloric acid-0.4 *M* magnesium chloride (8+8+4) were used by some laboratories in intercomparison studies on OTA in wheat (Hald et al., 1993; Wood et al., 1996). Aqueous methanol is a popular extraction solvent with several variations. Originally used as methanol-water (55+45) to extract OTA from cereals in the presence of *n*-hexane (Scott and Hand, 1967), it has been modified by changing the proportions and by the addition of salts, e.g. KCl (Soares et al., 1985), phosphate-buffered saline (PBS) at pH 7.4 (Bisson et al., 1994; Marley et al., 1995; Sharman et al., 1992), phosphoric acid (MacDonald et al., 1999), ascorbic acid (Hurst and Martin, 1998; Seidel et al., 1993), and alkali - for example, as methanol and 1 or 5% aqueous sodium bicarbonate to extract ground green and roasted coffee beans (Micco et al, 1989; Terada et al., 1986). Acetonitrile has also been used as the organic solvent under acidic, neutral or alkaline conditions (Akiyama et al., 1997; Bisson et al., 1994; Scudamore and MacDonald, 1998); acetonitrile-1% phosphoric acid gave the highest recovery of OTA from ground coffee beans and cereals compared to several other solvent mixtures (Akiyama et al., 1997). Biancardi and Riberzani (1996) favoured 3% acetic acid in acetonitrile because acetic acid has ready solubility in organic solvents and sufficient volatility to be evaporated before cleanup. It is possible to avoid organic solvents altogether. Thus Nakajima et al. (1990) extracted ground coffee beans with 1% aqueous sodium bicarbonate and ultrasonication and even water alone gave an average OTA recovery of 70% from coffee powder (Maierhofer et al., 1995).

The wide range of extraction solvents in use in European laboratories for extraction of OTA from solid foodstuffs is apparent from recent intercomparison studies on methods for determination of OTA in pig kidney (Entwisle et al., 1997) and wheat (Hald et al., 1993; Wood et al., 1996). Grouping results obtained by different laboratories according to extraction solvent showed that this factor did not greatly influence the results for OTA in wheat, except that dichloromethane-0.1 *M* phosphoric acid (10+1) was less efficient (Hald et al., 1993; Wood et al., 1996). A greater influence on results was the procedure used for spiking the blank wheat: toluene-acetic acid (99+1) as solvent for the spiking solution required 18 hours to evaporate from the ground wheat before analysis if low recoveries were to be avoided. The type of extraction solvent did not influence the results obtained by immunoaffinity column (IAC) cleanup or solid phase extraction column (SPE) cleanup (Wood et al., 1996). Similarly for pig kidney the choice of solvent, which for most laboratories included chloroform, and mode of extraction did not greatly influence the results obtained (Entwisle et al., 1997). For the actual certification exercise used in preparing two wheat reference materials a variety of extraction systems, most of which involved acidified solvents, were again used by the participating laboratories (Wood et al., 1997).

Liquid foods such as beer, wine, milk and coffee drink can be extracted with a water-immiscible organic solvent. For example, OTA was extracted from wine and grape juice with chloroform, following acidification with phosphoric acid and sodium chloride (Zimmerli and Dick, 1996a); beer, wine or grape juice was extracted with toluene after addition of 2 *M* hydrochloric acid and 0.4 *M* magnesium chloride (Degelmann et al., 1999; Majerus and Ottender, 1996; Ospital et al., 1998); milk to which ethanol and 1 *M* hydrochloric acid had been added could be extracted with chloroform (Breitholtz-Emanuelsson et al., 1993); and Terada et al. (1986) treated coffee drink with methanol and extracted the solution with chloroform and 20% hydrochloric acid. However, a more convenient procedure is direct addition of the liquid, which may be diluted, to a SPE column or an IAC. For example, Visconti et al. (1999) diluted wine with an equal volume of an aqueous solution containing 1% polyethylene glycol and 5% sodium bicarbonate and added the filtered solution to an IAC; Nakajima et al. (1999) added filtered, degassed beer directly to an IAC; while Scott and Kanhere (1995) mixed 5 ml degassed beer with 1 ml 2% sodium bicarbonate and 15% sodium chloride prior to immunoaffinity column (IAC) cleanup or extracted degassed beer on a C-18 SPE column.

The extraction of OTA from blood and animal tissues has been well reviewed by Valenta (1998). In most studies, chloroform was used with a solution of hydrochloric acid or phosphoric acid together with magnesium or sodium chloride (to increase the ionic strength) at pH 2.5 or less. The strongly acidic conditions are necessary to ensure complete extraction of OTA from proteinaceous matrices. Centrifugation is necessary to separate the phases. Problems with extraction of OTA from spiked freeze-dried pig kidney were reported by Entwisle et al. (1997). Blood plasma can also be extracted with methanol (Langseth et al., 1993; Scott et al., 1998). Hunt et al. (1979) added an enzyme, subtilisin A or papain, to methanol-water (1+1) at pH 9-10 and 50°C to extract pig kidney by dialysis and obtained twice the recovery of OTA from naturally contaminated samples compared to extraction without the enzyme. Dietrich et al. (1995) found that milk required enzyme treatment before adding to an IAC.

Supercritical fluid extraction (Anklam et al., 1998; Huopalahti and Jarvenpaa, 2000) has not been applied to OTA.

CLEANUP

Not all methods for OTA in foodstuffs incorporate a cleanup step. In particular, enzyme-linked immunosorbent assay (ELISA) methods may not require any cleanup. Cleanup may be as simple as adding a clarifying agent such as cupric sulfate and ammonium sulfate solutions (Soares and Rodriguez-Amaya, 1985,1989) or Carrez solution (potassium ferrocyanide, zinc acetate and acetic acid solution) (Czierwiecki, 1994) to the extract. Other procedures for cleanup of OTA after extraction include partition into aqueous sodium bicarbonate, liquid-liquid partition between two organic solvents, silica gel column chromatography, SPE columns and IACs. In European intercomparison studies on OTA in wheat (Hald et al., 1993; Wood et al., 1996) the most commonly used procedure was SPE on silica. In the second study, results obtained using an IAC were similar to those obtained by the method normally used by a given laboratory (Wood et al., 1996). In a European intercomparison study on OTA in pig kidney, SPE on C-18 was the most frequently used cleanup procedure and the number of laboratories using SPE on silica, sodium bicarbonate extraction or IAC cleanup was about the same (Entwisle et al., 1997).

The two earlier AOAC International official methods for OTA trapped the toxin on a column of diatomaceous earth impregnated with 5 or 1.25% sodium bicarbonate solution;

interferences were removed with hexane and chloroform and OTA was eluted with benzene-acetic acid (98+2) or chloroform-formic acid (99+1) (Levi, 1975; Nesheim et al., 1973ab). OTC required a separate column of diatomaceous earth impregnated with methanolic sodium bicarbonate solution (Nesheim et al., 1973ab). Similarly, Baumann and Zimmerli (1988) used an Extrelut column impregnated with 1% sodium bicarbonate solution to retain the OTA. A later AOAC method used liquid-liquid partitioning directly into 3% sodium bicarbonate solution, followed by adding an aliquot of this solution to a C-18 SPE column, which was acidified and the OTA eluted with ethyl acetate-methanol-acetic acid (90+5+0.5) (Nesheim et al., 1992). Other methods incorporating liquid-liquid extraction into sodium bicarbonate or sodium hydroxide solutions, followed by acidification and back extraction into dichloromethane or chloroform have also been described (Jørgensen et al., 1996, Jørgensen and Vahl, 1999; Lepom, 1986; Seidel et al., 1993; Steyn and van der Merwe, 1966).

Liquid-liquid partition between two organic solvents (Czerwiecki, 1994) and column chromatography on silica gel (Eppley, 1968; Ibe et al., 1984; Wilken et al., 1985) have largely been replaced as cleanup procedures by commercially available and disposable SPE columns or cartridges, of which those incorporating silica (Cohen and Lapointe, 1986; Howell and Taylor, 1981; Jiao et al., 1992; Langseth et al., 1989; Patel et al., 1997; Scott et al., 1991; Skaug, 1999; Seidel et al., 1993; Valenta and Goll, 1996), C-18 (Jornet et al., 2000; Hurst and Martin, 1998; Seidel et al., 1993; Terada et al., 1986), and immobilised antibodies for immunoaffinity chromatography (Scott and Trucksess, 1997) are the most widely used. Other types of SPE columns that have been used to clean up extracts of foodstuffs and feeds containing OTA are florisil (Seidel et al., 1993), phenyl (Dawlatana et al., 1996), cyano (Cohen and Lapointe, 1986), diethylaminopropyl weak anion exchange (Akiyama et al., 1997) and trimethylaminopropyl strong anion exchange (SAX) (Biancardi et al., 1996). All these columns offer a rapid and solvent efficient cleanup technique. However, the anion exchange columns offer more specificity than columns containing sorbents such as silica or C-18 since they exploit the carboxyl moiety of OTA; detection limits were 0.05-0.1 ppb by liquid chromatography (LC) in a range of foodstuffs, including green and roasted coffee beans and recoveries of OTA from spiked samples were 91-99%. Akiyama et al. (1997) were also able to determine OTB, which is not retained by an IAC, using diethylaminopropyl column cleanup. Further consideration to use of an anion exchange column for cleanup of OTA in food extracts should be given in view of the lower cost compared to that of IACs (*vide infra*). Other cleanup procedures reported in the literature include gel permeation (Dunne et al., 1993) and use of preparative reverse phase thin layer chromatography (TLC) (Frohlich et al., 1988). Valenta (1998) has discussed in detail cleanup procedures for OTA in animal and human tissues and biological fluids.

A cleanup technique now much in favour with analysts because of its specificity, resulting in very clean liquid chromatograms, is the use of antibody based IACs. However, there are some disadvantages: relatively high cost, limited shelf life of the columns even under refrigeration, and a maximum amount of toxin that can be applied to the IAC. In view of its importance, this cleanup procedure will be discussed in more detail. The sample extract or liquid matrix (e.g. milk, degassed beer), is loaded onto the IAC to bind the mycotoxin selectively (Scott and Trucksess, 1997). Impurities are removed by washing the column with water or buffer and the analyte is eluted with a small volume of a solvent such as methanol, acetonitrile or aqueous dimethylsulfoxide. Detection and quantitation following elution of the analyte is done by LC (see Table 1), fluorometer (Zabe et al., 1996), or capillary electrophoresis (CE) with laser-induced fluorescence (Corneli and Maragos, 1998). Three companies sell IACs specifically for cleanup of OTA in food extracts - Rhône-diagnostic technologies (Ochraprep® and Ochrascan®), Vicam (OchraTest™) and R-Biopharm (RIDA® ochratoxin A column). An example of the more specific cleanup of beer by an IAC compared to SPE on sequential C-18 and silica columns is shown in Figure 2. Considerable attention hasbeen given to development of IAC methods for determination of OTA in coffee beans, both green and roasted, and in coffee products. Performance of the IAC may be affected by the presence of caffeine in the extract and cleanup by SPE prior to IAC treatment is necessary. In addition to coffee, IACs have been applied to analysis of other foods, including cereals and

Table 1. Applications of IAC cleanup in the analysis of foods, feeds and biological fluids for OTA.

Matrix	Detection limit of method (using LC), ppb	Reference(s)
Green and roasted coffee beans	0.03-0.5	Jørgensen, 1998; Koch et al., 1996; Maierhofer et al., 1995; Nakajima et al., 1990,1997; Trucksess et al., 1999
Soluble coffee	0.1-0.5	Nakajima et al., 1990; Patel et al., 1997; Pittet et al., 1996
Coffee drink	0.025	Nakajima et al., 1990
Cereals and cereal products	0.02-0.25	Scudamore and MacDonald, 1998; Sharman et al., 1992; Solfrizzo et al., 1998; Trucksess et al., 1999
Beer	0.001-0.1	Jørgensen, 1998; Nakajima et al., 1999; Scott and Kanhere, 1995; Zimmerli and Dick,1995
Paprika powder	0.2	Bassen and Brun, 1999
Pulses	0.1	Jørgensen, 1998
Wine, grape juice	0.003-0.01	Visconti et al., 1999; Zimmerli and Dick,1995, 1996a
Meat, blood sausage	0.02-0.2	Jørgensen, 1998; Sharman et al., 1992
Dried vine fruits	0.2	MacDonald et al., 1999
Human serum/milk	0.005-0.01	Zimmerli and Dick, 1995
Cow's milk	0.005	Dietrich et al., 1995
Pet foods	0.3	Scudamore et al., 1997

cereal products, beer, wine, milk and meat, and also human body fluids for OTA. Detection limits of overall methods are as low as 0.001 ppb for liquids and 0.02 ppb for solid foodstuffs (Table 1). IACs for OTA may not be completely specific. OTC may also be retained and grapefruit juice contained an interference, even after IAC cleanup, which had the same LC retention time as OTA (Zimmerli and Dick, 1996a). IACs for both OTA and aflatoxins have been linked in series for isolation of these toxins from pet food extracts (Scudamore et al., 1997). Vicam have produced a single IAC, the AflaOchra HPLC™ column, that can be used to isolate both aflatoxins and OTA from food and feed extracts. This approach merits further consideration for multimycotoxin analysis. Immunoaffinity chromatography has been employed in an automated LC technique for OTA in cereal and animal products (Sharman et al., 1992). The affinity column clean-up was carried out either manually or with a commercial automated sample preparation system. A re-usable IAC is obviously of considerable value, not only to reduce costs but also for use in such an automated system. Flushing the column after use with phosphate buffered saline (PBS) containing sodium azide will regenerate the IAC; a detailed study indicated a regeneration time of 48 h in the refrigerator and the IAC of one company could be reused more than 30 times for isolation of OTA (Zimmerli and Dick, 1996b).

A recently introduced technique with potential wide applications to analytical chemistry is the use of molecularly imprinted polymers, i.e. polymers with cavities or imprints complementary in shape to an analyte of interest (Idziak, 2000). They can be regarded as artificial antibodies and their use in a cleanup column would have several advantages over immunoaffinity chromatography, including the saving in time needed to prepare antibodies and greater stability of columns. No mycotoxins have yet been imprinted onto a polymer but the technique offers exciting potential for their analysis.

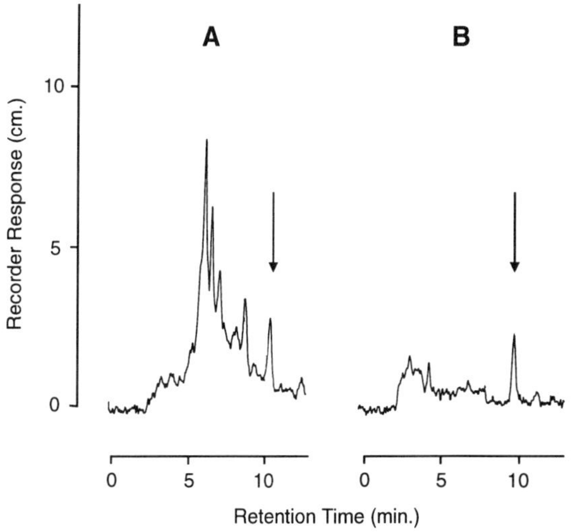

Figure 2. Liquid chromatograms of OTA in beer; A, after SPE cleanup; B, after IAC cleanup; detection by fluorescence. Estimated OTA concentrations (these injections) 0.17 ppb and 0.16 ppb, respectively (Scott and Kanhere, 1995). Beer equivalent injected, 50 µl; OTA indicated by arrows. Reproduced with permission from Taylor & Francis Ltd. (http://www.tandf.co.uk).

SEPARATION, DETECTION AND DETERMINATION

Thin-layer Chromatography (TLC)

TLC is featured in earlier AOAC methods (Levi, 1975; Nesheim et al., 1973ab) which use a silica gel adsorbent and an acidic solvent system. Visual detection of OTA is by greenish fluorescence under longwave ultraviolet light which changes to blue fluorescence on spraying the plate with methanolic sodium bicarbonate solution or exposing it to ammonia fumes; scanning densitometric analysis may also be carried out. The method detection limit of the order of 10 ppb for grains and other commodities (Nesheim et al., 1973; Soares et al., 1985) is inadequate for present day monitoring and compliance purposes, although detection limits below 10 ppb (2.4-4 ppb) were reported for rice and coconut by Asensio et al. (1982) and less than 1 ppb in some animal products by Kuiper-Goodman and Scott (1989), Mallmann et al. (1994) and Valenta (1998). Reverse phase TLC applied to analysis of cereals and animal feed (Biancardi and Riberzani, 1996) did not improve much on the higher detection limit although there was an increase in fluorescence intensity compared to normal phase TLC. High performance bi-directional TLC with scanning fluorodensitometry only gave a detection limit of 12 ppb in the analysis of rice (Dawlatana et al., 1996). However, an instrumental high performance TLC method used by one laboratory in an intercomparison study on OTA in wheat had a low detection limit of 0.7 ppb (Hald et al., 1993); it should be added that this was the only laboratory to use TLC out of 24 taking part in the study.

Minicolumns

Very little use has been made of minicolumns for detection of OTA and they are not very sensitive. The limit of detection in cereals and other commodities as a green fluorescent band at the lower end of a small column of silica gel or silica gel/alumina was 12-80 ppb (Hald and Krogh, 1975; Soares and Rodriguez-Amaya, 1985).

Liquid Chromatography (LC)

Reversed phase LC is the most widely used procedure for separation and determination of OTA in foods, feeds and biological fluids. Since OTA is a weak acid, the mobile phase must be acidic; usually mixtures of acetonitrile or methanol with aqueous *o*-phosphoric acid or acetic acid have been used. Detection of OTA is generally by fluorescence (excitation wavelength 330-340 nm and emission wavelength 420-470 nm) (Figure 2). Reinhard and Zimmerli (1999) studied the LC behavior of OTA (as well as citrinin) as a function of hydrophobicity and silanophilic activities of the stationary phase; composition, pH and type of acid in the mobile phase; and column temperature. They found that the brand of C-18 reversed phase material made no significant difference in the affinity of OTA. The capacity factor for OTA was not greatly affected by the composition of the mobile phase and showed little variation over the pH range 2.3-3.6; it decreased with increasing column temperature. There was a surprising increase in intensity of fluorescence for some stationary phases when maleic acid was used as the acid. Citrinin was more affected by the different variables studied. The type of lamp used in the fluorescence detector is, of course, also important for detection of OTA and an increased signal is obtained with a xenon lamp compared to a deuterium lamp (Breitholtz et al., 1991). Ion-pair LC has been used for determination of OTA in coffee products (Nakajima et al., 1990; Terada et al., 1986), human plasma (Breitholtz et al., 1991) and cheese (Vazquez et al., 1996). In this LC procedure, the mobile phase was made basic (pH 7.5-9) or acidified (pH 5.5) and low millimolar concentrations of cetyltrimethylammonium bromide, tetrabutylammonium bromide or tetrabutylammonium hydroxide were added. Greatly increased sensitivity was noted with the alkaline mobile phases. In the method for analysis of cheese, terbium chloride was included in the post-column reagent (which also contained trioctylphosphine oxide and triethylamine) for time-resolved luminescence, giving greater sensitivity than direct fluorescence (Vazquez et al., 1996). β-Cyclodextrin (0.1-0.2 m*M*) added to the mobile phase resulted in a slight increase (15%) in fluorescence intensity of OTA (Seidel et al., 1993) and better separation of OTA from interferences in cocoa bean extracts (Hurst and Martin, 1998). Another means of taking advantage of increased fluorescence of OTA under alkaline conditions is post-column introduction of 1.9-25% ammonia, which can increase the sensitivity up to ten times (Hunt et al., 1979; Langseth et al., 1993; Zimmerli and Dick, 1995) and a cleaner chromatogram was also obtained. The excitation wavelength was changed to 375-390 nm. Zimmerli and Dick (1995, 1996a) reported a detection limit for standard OTA of 3 pg (signal /noise = 3:1) by this technique. Miraglia et al. (1995) used post column treatment with 25% ammonia as a confirmation procedure in the analysis of human milk

Examples of matrixes analysed for OTA using an LC method include cereal grains, pet foods, coffee, beer, wine, grape juice, dried vine fruits, milk, kidney and other meat products, and human and animal biological fluids (see Table 1; Kuiper-Goodman and Scott, 1989).

Capillary Electrokinetic Techniques

Capillary zone electrophoresis (CZE), which achieves separation by differences in electrophoretic mobility of charged species, has been used for the separation and reliable quantitation of OTA in roasted coffee, corn and sorghum, with a detection limit (0.2 ppb) comparable to LC methods; fluorescence was induced by ultraviolet light from a He/Cd laser (Corneli and Maragos, 1998). Advantages of this procedure are the use of much smaller volumes of sample and organic solvents and use of less expensive capillary columns. Böhs et al. (1995) used reversed electroosmotic flow, diode array UV detection, and studied the effect of adding various cyclodextrins to the running buffer.

OTA was one of 10 mycotoxins, including neutral molecules, that were separated by micellar electrokinetic capillary chromatography (MECC) (Holland and Sepaniak, 1993).

ENZYME-LINKED IMMUNOSORBENT ASSAY (ELISA)

The use of ELISAs to test for OTA is important because of their ease of use and the large number of samples that can be processed per day. ELISA methods have been applied to analysis of barley and other grains, feed, animal tissues and serum as referenced in reviews by Chu (1992) and Wilson et al. (1998). Radioimmunoassay for OTA has been applied to surveys of cereals, cereal products, feedstuffs, and pig serum and tissues (Fukal, 1990, 1991; Rousseau et al., 1986) but does not appear to have been used in recent years. Examples of detection limits that can be obtained with commercially available ELISA kits and ELISAs developed in-house are shown in Table 2; in general these were lower where cleanup was used. De Saeger and Van Peteghem (1999) developed an enzyme immunoassay for OTA in wheat which uses a flow-through membrane, resulting in a rapid procedure that can be performed outside the laboratory; unfortunately, the procedure did not work for barley and maize because of interferences. An important consideration with ELISA is the specificity of the antibody. Cross reactivity with related compounds can vary widely. For example, Breitholtz-Emanuelsson et al. (1992) found 15-20% cross-reactivity of two antibodies toward OTC but only 0.08-0.2% toward OTB, and Jarkczyk (1999) reported 44% cross reaction with OTC and 14% with OTB for the RIDASCREEN® test kit. However, cross-reactivities for a hen egg yolk antibody as high as 400% and 100% were found for OTC and OTB, respectively, by Clarke et al. (1993); on the other hand cross reactions of only 0.01% and 1.4% for OTB and OTC, respectively, were observed with a polyclonal antibody used in an ELISA by Ruprich and Ostry (1991).

OTA has been determined by an enzyme immunosensor with an oxygen electrode (Aizawa, 1987).

SPECTROFLUORIMETRIC METHODS

Hult and Gatenbeck (1976) determined OTA by measuring loss in fluorescence when OTA is hydrolysed with carboxypeptidase A. They also applied the procedure to OTB, which hydrolyses faster and could be determined in the presence of OTA by using two incubations with enzyme at different temperatures and times (Hult et al., 1977). The method was applied to determination of OTA in pig blood with a detection limit of 2 ppb (Hult et al., 1980). The enzyme spectrofluorimetric method was compared with LC by Beker and Radić (1991).

Table 2. Selected applications of ELISA to detection and determination of OTA.

Test/reference	Food	Detection limit (ppb)	Cleanup
RIDASCREEN® - Weddeling et al., 1994	Grains Beer	0.4 0.1	+ +
RIDASCREEN® - Jarkzyk et al., 1999	Grains, feeds	0.4	+
EZ-SCREEN ®	Grains	10	
Veratox ®	Grains	10	
Agromed	Foods	0.5	
Kawamura et al., 1989	Chicken, wheat flour	1	
Lacey et al., 1991	Barley	1	+
Morgan et al., 1986	Barley	0.06	
Candlish et al., 1988	Barley	5	+
Lee and Chu, 1984	Wheat	1	+
Morgan et al., 1986	Kidney	0.05	+
Clarke et al., 1994	Kidney	7.8	
Valenta and Goll, 1996	Milk	0.01	+
De Saeger and Van Peteghem, 1999	Wheat	<4	
Barna-Vetró et al., 1996	Grains, feeds	0.50	+
Breitholtz-Emanuelsson et al., 1992	Plasma	0.1-0.2	

CONFIRMATION PROCEDURES

Chemical Derivatization

The simplest way to confirm positive findings of OTA is by chemical or enzymatic change. The reaction most often used is esterification of the carboxylic acid group of OTA using H_2SO_4, BF_3 or, more conveniently because of its volatility, HCl as catalysts for the reaction with methanol, ethanol or propanol (Breitholtz et al., 1991; Castegnaro et al., 1990; Hunt et al., 1980; Li et al., 1998; Nakajima et al., 1990; Scott and Kanhere, 1995; Takeda et al., 1991; Zimmerli and Dick, 1995). The resulting esters are determined using the same reversed phase LC system as for OTA and are more strongly retained. Takeda et al. (1991) partially formed the methyl ester at 15°C and used the peak height ratio as further confirmation. LC of the *O*-methyl methyl ester of OTA, prepared by reacting OTA with diazomethane, was reported by Phillips et al. (1983). Another means of confirmation is the addition of carboxypeptidase A to form the hydrolysis product ochratoxin α, which elutes before OTA. Application to confirmation of OTA in pig kidney extracts (Frohlich et al., 1997) produced higher amounts of ochratoxin α than expected from the measured OTA, suggesting the presence of other forms of OTA in the kidney. The procedure has also been applied to swine plasma samples (Hult and Gatenbeck, 1976; Ominski et al., 1996). One problem is that ochratoxin α elutes earlier in the chromatogram than OTA and matrix interferences can be expected.

LC-Mass Spectrometry (MS) and LC-Tandem MS (MS/MS)

LC-MS and LC-MS/MS procedures provide a more reliable and sensitive approach to confirmation of OTA. Abramson (1987) employed a direct liquid introduction (DLI) interface with a quadrupole mass spectrometer for analysis of barley extracts; the LC effluent, consisting of aqueous acetonitrile acidified with formic acid, was split so that only 2.5% of the mobile phase entered the DLI interface. He investigated both positive and negative chemical ionization (NICI). Using NICI and monitoring at *m/z* 403 (M^-), sub-ppb levels of OTA were detectable. Marquardt et al. (1988) and Ominski et al. (1996) confirmed OTA in swine serum by LC-DLI-NICI-MS. Rajakylä et al. (1987) used LC-thermospray MS for determination of OTA in wheat.

Recently, LC-electrospray (ESI) MS/MS has been applied for determination and confirmation of OTA in a variety of matrices (Table 3). The procedure can detect as little as 5-20 pg OTA injected, corresponding to detection limits of 0.5 ppb in plasma (Lau et al., 2000), 0.02 ppb in pig kidney and rye flour (Jørgensen and Vahl, 1999), and a quantitation limit of 0.05 ppb in beer (0.01 ppb was detectable) (Becker et al., 1998; Degelmann et al., 1999). Multiple reaction monitoring (MRM) mass chromatograms of a coffee extract are shown in Figure 3. For those samples that do not contain OTB, it can be used as an internal standard.

Table 3. Reversed phase LC-tandem MS of OTA

Matrix analysed	Ionization	Daughter ions monitored	Quantitation	Reference(s)
Plasma, coffee	ESI	$[M+H-HCOOH]^+$, $[M+H-Phe]^+$, $[M+Na-Phe]^+$	Standard addition, internal standard, external standard	Lau et al., 2000
Wheat, beer, coffee	ESI	$[M+H-H_2O-CO]^+$, $[M+H-Phe]^+$	External standard	Becker et al., 1998; Degelmann et al., 1999
Pig kidney, rye flour	ESI of methyl ester	$[M+H-MePhe]^+$	Internal Me(d_3) standard	Jørgensen and Vahl, 1999

Gas Chromatography (GC)-MS

Jiao et al. (1992) confirmed OTA present in food samples at sub-ppb levels by conversion into its *O*-methyl methyl ester with diazomethane, followed by capillary GC-MS using negative ion chemical ionization and multiple ion detection modes with *O*-methyl-d_3-OTA methyl-d_3 ester as internal standard.

INTERLABORATORY METHOD PERFORMANCE

Collaborative studies on TLC methods for OTA in barley and green coffee were carried out in the 1970's (Levi, 1975; Nesheim, 1973). Levels tested were 45 and 90 ppb for barley and 41-230 ppb for coffee, so these methods are now completely obsolete. However, it was shown that densitometry was more precise than visual estimation (Nesheim, 1973b). Recoveries of OTA from coffee were low (Levi, 1975).

The collaborative study of Nesheim et al. (1992) on an LC method for OTA in corn, barley and pig kidney showed between-laboratory relative standard deviations (RSD_R) of 20.7-31.7% at concentrations of 10-50 ppb OTA in corn and barley and 32.7-68.0% for kidney at concentrations of 5-10 ppb. Within-laboratory standard deviations (RSD_r) were 7.9-20.1% and mean recoveries were 72-82% for corn and barley but 53-97% for kidney. Because of these results, the method was not adopted as official, first action for pig kidney. The method, with some editorial modification, was tested by the Nordic Committee on Food Analysis (NMKL) for barley, wheat bran and rye at levels of 2-9 ppb (Larsson and Möller, 1996). RSD_R values found were 18-28%, RSD_r values were 12-21%, the Horwitz ratio was 0.50-0.77 and recoveries ranged from 64-72%, so this method was accepted by NMKL for determination of OTA in these cereal products at 2 ppb.

Recently, IAC methods for OTA in barley and roasted coffee were studied in an AOAC collaborative study at OTA concentrations of 0.1-4.5 ppb and 0.1-5.4 ppb, respectively (Entwisle et al., 2000a,b). Between-laboratory method performance at the lowest level was not acceptable for either commodity. For barley, at 1.3 ppb, RSD_R values found were 12-33%, RSD_r values were 4-24%, and mean recovery was 93%. For roasted coffee, at 1.2 ppb, RSD_R values were 13-26%, RSD_r values were 2-22% and mean recovery was 85%. The Horwitz ratio (HORRAT) was <2.0 in all cases. The methods have been adopted as AOAC International official, first action methods and other interlaboratory studies for sensitive methods applicable to green coffee, wine and beer, and baby food are planned (AOAC International, 2000).

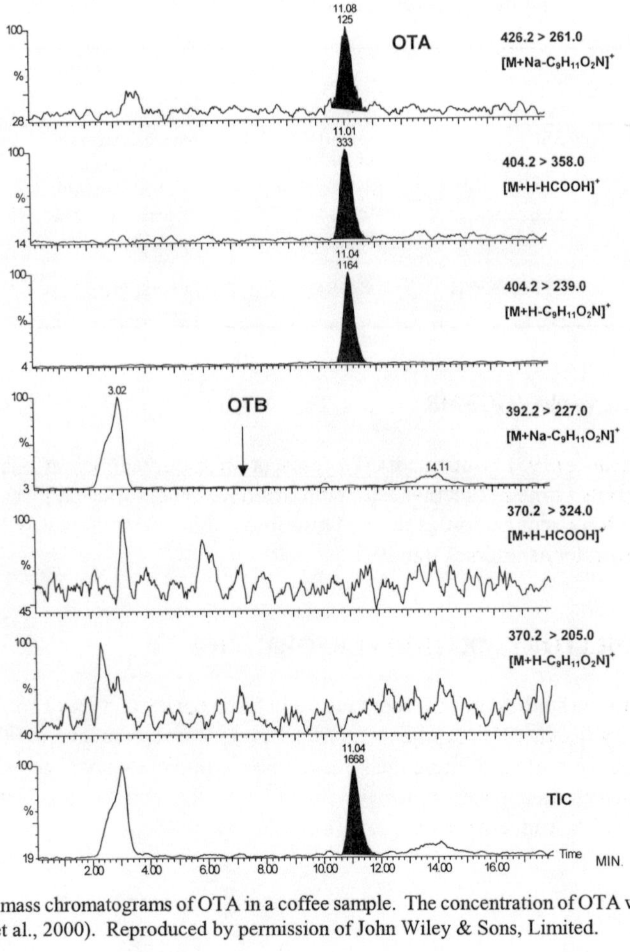

Figure 3. MRM mass chromatograms of OTA in a coffee sample. The concentration of OTA was determined to be 1.8 ppb (Lau et al., 2000). Reproduced by permission of John Wiley & Sons, Limited.

CERTIFIED REFERENCE MATERIALS AND PROFICIENCY TESTING

In an attempt to improve the accuracy of OTA determinations, certified reference materials (CRMs) for mycotoxins have been prepared and certified within the European Commission, Measurements and Testing Programme (Boenke, 1997). Two CRMs for OTA in wheat have been prepared and following two intercomparison studies and a certification exercise, the declared concentrations are <0.6 and 8.2 ppb (Boenke, 1997; Wood et al., 1997). A CRM for OTA in pig kidney is in course of preparation and one intercomparison study has been carried out (Boenke, 1997; Entwisle et al., 1997).

Proficiency testing is a means of assessing laboratory performance in analytical determinations. One important proficiency testing scheme called FAPAS (Food Analysis Performance Assessment Scheme) is organized from the United Kingdom and now includes wheat flour and ground coffee containing OTA as test materials (Gilbert et al., 2001).

REFERENCES

Abramson, D., 1987, Measurement of ochratoxin A in barley extracts by liquid chromatography-mass spectrometry, *J. Chromatogr.* 391:315.

Aizawa, M., 1987, Enzyme-linked immunosorbent assays using oxygen-sensing electrode, *Electrochem. Sens. Immunol. Anal.* p.269.

Akiyama, H., Chen, D., Miyahara, M., Goda, Y., and Toyoda, M., 1997, A rapid analysis of ochratoxin A in coffee beans and cereals, *J. Food Hyg. Soc. Japan* 38:406.

Anklam, E., Berg, H., Mathiasson, L., Sharman, M., and Ulberth, F., 1998, Supercritical fluid extraction (SFE) in food analysis: a review, *Food Addit. Contam.* 15:729.

AOAC International, 2000, Methods News, *Inside Laboratory Management* May issue:p.24.

Asensio, E., Sarmiento, I., and Dose, K., 1982, Quantitative determination of ochratoxin A in vegetable foods, *Fresenius Z. Anal. Chem.* 311:511.

Bacha, H., Hadidane, R., Creppy, E.E., Regnault, C., Ellouze, F., and Dirheimer, G., 1988, Monitoring and identification of fungal toxins in food products, animal feed and cereals in Tunisia, *J. Stored Prod. Res.* 24:199.

Barna-Vetró, I., Solti, L., Téren, J., Gyöngyösi, Á., Szabó, E., and Wölfling, A., 1996, Sensitive ELISA test for determination of ochratoxin A, *J. Agric. Food Chem.* 44:4071.

Bassen, B., and Brunn, W., 1999, Ochratoxin A in Paprikapulver, *Dtsche. Lebensm.-Rundsch.* 95.142.

Baumann, U., and Zimmerli, B., 1988, Einfache Ochratoxin-A-Bestimmung in Lebensmitteln (A simple determination of ochratoxin A in foods), *Mitt. Gebiete Lebensm. Hyg.* 79:151.

Becker, M., Degelmann, P., Herderich, M., Schreier, P., and Humpf, H.-U., 1998. Column liquid chromatography-electrospray ionisation-tandem mass spectrometry for the analysis of ochratoxin, *J. Chromatog. A* 818:260.

Beker, D., and Radić, B., 1991, Fast determination of ochratoxin A in serum by liquid chromatography: comparison with enzymic spectrofluorimetric method, *J. Chromatogr.* 570:441.

Biancardi, A., and Riberzani, A., 1996, Determination of ochratoxin A in cereals and feed by SAX-SPE clean up and LC fluorimetric detection, *J. Liq. Chromatogr. Rel. Technol.* 19:2395.

Bisson, E., Byass, L., Garner, A., and Garner, R.C., 1994, Analysis of wheat and kidney samples for ochratoxin A using immunoaffinity columns in conjunction with HPLC, *Food Agric. Immunol.* 6:311.

Blanc, M., Pittet, A., Muñoz-Box, R., and Viani, R., 1998, Behavior of ochratoxin A during green coffee roasting and soluble coffee manufacture, *J. Agric. Food Chem.* 46:673.

Böhs, B., Seidel, V., and Lindner, W., 1995, Analysis of selected mycotoxins by capillary electrophoresis, *Chromatographia* 41:631.

Boenke, A., 1997, The food and feed chains - a possible strategy for the production of Certified Reference Materials (CRMs) in the area of mycotoxins?, *Food Chem.* 60:255.

Bondy, G.S., and Pestka, J.J., 2000, Immunomodulation by fungal toxins, *J. Toxicol. Environ. Health B*, 3:109.

Boutrif, E., and Canet, C., 1998, Mycotoxin prevention and control: FAO Programmes, *Rev. Méd. Vét.*

Breitholtz, A., Olsen, M., Dahlback, Å., and Hult, K., 1991, Plasma ochratoxin A levels in three Swedish populations surveyed using an ion-pair HPLC technique, *Food Addit. Contam.* 8:183.

Breitholtz-Emanuelsson, A., Dalhammar, G., and Hult, K., 1992, Immunoassay of ochratoxin A, using antibodies developed against a new ochratoxin-albumin conjugate, *J. AOAC Int.* 75:824.

Breitholtz-Emanuelsson, A., Olsen, M., Oskarsson, A., Palminger, I., and Hult, K., 1993, Ochratoxin A in cow's milk and in human milk with corresponding human blood samples, *J. AOAC Int.* 76:842.

Candlish, A.A.G., Stimson, W.H., and Smith, J.E., 1988, Determination of ochratoxin A by monoclonal antibody-based enzyme immunoassay, *J. Assoc. Off. Anal. Chem.* 71:961.

Castegnaro, M., Maru,V., Maru G., and Ruiz-Lopez, M.-D., 1990, High-performance liquid chromatographic determination of ochratoxin A and its 4*R*-4-hydroxy metabolite in human urine, *Analyst* 115: 129.

Chu, F.S., 1992, Development and use of immunoassays in the detection of ecologically important mycotoxins, in: *Handbook of Applied Mycology. Volume 5: Mycotoxins in Ecological Systems*, D. Bhatnagar, E.B. Lillehoj and D.P. Arora, eds., Marcel Dekker, New York.

Clarke, J.R., Marquardt, R.R., Oosterveld, A., Frohlich, A.A., Madrid, F.J., and Dawood, M., 1993, Development of a quantitative and sensitive enzyme-linked immunosorbent assay for ochratoxin A using antibodies from the yolk of the laying hen, *J. Agric. Food Chem.* 41:1784.

Clarke, J.R., Marquardt, R.R., Frohlich, A.A., and Pitura, R.J., 1994, Quantification of ochratoxin A in swine kidneys by enzyme-linked immunosorbent assay using a simplified sample preparation procedure, *J. Food Prot.* 57:991.

Cohen, H., and Lapointe, M., 1986, Determination of ochratoxin A in animal feed and cereal grains by liquid chromatography with fluorescence detection, *J. Assoc. Off. Anal. Chem.* 69:957.

Corneli, S., and Maragos, C.M., 1998, Capillary electrophoresis with laser-induced fluorescence: method for the mycotoxin ochratoxin A, *J. Agric. Food Chem.* 46:3162.

Creppy, E.E., Castegnaro, M., and Dirheimer, G. (eds.), 1993, *Human Ochratoxicosis and its Pathologies*, John Libbey Eurotext, Montrouge, France.

Czerwiecki, L., 1994, Determination of ochratoxin A in infant and children cereal foods by reversed phase high performance liquid chromatography (RP-HPLC), *Pol. J. Food Nutr. Sci.* 3/44:67.

Dawlatana, M., Coker, R.D., Nagler, M., J., and Blunden, G., 1996, A normal phase HPTLC method for the quantitative determination of ochratoxin A in rice, *Chromatographia* 42:25.

De Saeger, S., and Van Peteghem, C., 1999, Flow-through membrane-based enzyme immunoassay for rapid detection of ochratoxin A in wheat, *J. Food Prot.* 62:65.

Degelmann, P., Becker, M., Herderich, M., and Humpf, H.-U., 1999, Determination of ochratoxin A in beer by high-performance liquid chromatography, *Chromatographia* 49:543.

Dietrich, R., Schneider, E., Usleber, E.,and Märtlbauer, E., 1995, Use of monoclonal antibodies for the analysis of mycotoxins, *Nat. Toxins* 3:288.

Dunne, C., Meaney, M., Smyth, M., and Tuinstra, L.G.M.T., 1993, Multimycotoxin detection and clean-up method for aflatoxins, ochratoxin and zearalenone in animal feed ingredients using high-performance liquid chromatography and gel permeation chromatography, *J. Chromatogr.* 629:229.

Entwisle, A.C., Jørgensen, K., Williams, A.C., Boenke, A., and Farnell, P.J., 1997, An intercomparison of methods for the determination of ochratoxin A in pig kidney, *Food Addit. Contam.* 14:223.

Entwisle, A.C., Williams, A.C., Mann, P.J., Russell, J., Slack, P.T., and Gilbert, J., 2000a, Combined phenyl silane and immunoaffinity column cleanup with liquid chromatography for the determination of ochratoxin A in roasted coffee: collaborative study, *J. AOAC Int.* 84:444.

Entwisle, A.C., Williams, A.C., Mann, P.J., Slack, P.T., and Gilbert, J., 2000b, Immunoaffinity column cleanup with liquid chromatography for the determination of ochratoxin A in barley: collaborative study, *J. AOAC Int.* 84:1377.

Eppley, R.M., 1968, Screening method for zearalenone, aflatoxin, and ochratoxin, *J. Assoc. Off. Anal. Chem.* 51:74.

Ewaidah, E.H., 1992, Ochratoxin A and aflatoxins in 1989 Saudi wheat, *Int. J. Food Sci. Technol.* 27:697.

Frohlich, A.A., Marquardt, R.R., and Bernatsky, A., 1988, Quantitation of ochratoxin A: use of reverse phase thin-layer chromatography for sample cleanup followed by liquid chromatography or direct fluorescence measurement, *J. Assoc. Off. Anal. Chem.* 71:949

Frohlich, A.A., Marquardt, R.R., and Clarke, J.R., 1997, Enzymatic and immunological approaches for the quantitation and confirmation of ochratoxin A in swine kidneys, *J. Food Prot.* 60:172.

Fukal, L., 1990, A survey of cereals, cereal products, feedstuffs and porcine kidneys for ochratoxin A by radioimmunoassay, *Food Addit. Contam.* 7:253.

Fukal, L., 1991, Spontaneous occurrence of ochratoxin A residues in Czechoslovak slaughter pigs determined by radioimmunoassay, *Dtsche. Lebensm.-Rundsch.* 87:316.

Gilbert, J., Mathieson, K., and Owen, L., 2001, Results from an international mycotoxin proficiency testing scheme, in: *Mycotoxins and Phycotoxins in Perspective at the Turn of the Millennium*, W. J. de Koe, R.A. Samson, H.P. Van Egmond, J. Gilbert and M. Sabino, eds., W.J. de Koe, Wageningen.

Hald, B., and Krogh, P., 1975, Detection of ochratoxin A in barley, using silica gel minicolumns, *J. Assoc. Off. Anal. Chem.* 58:156.

Hald, B., Wood, G.M., Boenke, A., Schurer, B., and Finglas, P., 1993, Ochratoxin A in wheat: an intercomparison of procedures, *Food Addit. Contam.* 10:185.

Heilmann, W., Rehfeldt, A.G., and Rotzoll, F., 1999, Behaviour and reduction of ochratoxin A in green coffee beans in response to various processing methods, *Eur. Food Res. Technol.* 209:297.

Holland, R.D., and Sepaniak, M.J., 1993, Qualitative analysis of mycotoxins using micellar electrokinetic capillary chromatography, *Anal. Chem.* 65:1140.

Howell, M.V. and Taylor, P.W., 1981, Determination of aflatoxins, ochratoxin A, and zearalenone in mixed feeds, with detection by thin layer chromatography or high performance liquid chromatography, *J. Assoc. Off. Anal. Chem.* 64:1356.

Hult, K., and Gatenbeck, S., 1976, A spectrophotometric procedure, using carboxypeptidase A, for the quantitative measurement of ochratoxin A, *J. Assoc. Off. Anal. Chem.* 59:128.

Hult, K., Hökby, E., and Gatenbeck, S., 1977, Analysis of ochratoxin B alone and in the presence of ochratoxin A, using carboxypeptidase A, *Appl. Environ. Microbiol.* 33:1275.

Hult, K., Hökby, E., Gatenbeck, S., and Rutqvist, L., 1980, Ochratoxin A in blood of slaughter pigs in Sweden: use in evaluation of toxin content of consumed feed, *Appl. Environ. Microbiol.* 39:828.

Hunt, D.C., Philp, L.A., and Crosby, N.T., 1979, Determination of ochratoxin A in pig's kidney using enzymic digestion, dialysis and high-performance liquid chromatography with post-column derivatisation, *Analyst* 104:1171.

Hunt, D.C., McConnie, B.R., and Crosby, N.T., 1980, Confirmation of ochratoxin A by chemical derivatisation and high-performance liquid chromatography, *Analyst* 105:89.

Huopalahti, R., and Jarvenpaa, E.P., 2000, Supercritical fluid extraction of mycotoxins from feeds, *Supercrit. Fluid Methods Protocols* 13:61.

Hurst, W.J., and Martin, A.M., Jr., 1998, High-performance liquid chromatographic determination of ochratoxin A in artificially contaminated cocoa beans using automated sample clean-up, *J. Chromatogr. A* 810:89.

Ibe, A., Nishijima, M., Yasuda, K., Saito, K., Kamimura, H., Nagayama, T., Ushiyama, H., Naoi, Y., and Nishima, T., 1984, High performance liquid chromatographic method for the determination of ochratoxins A and B in foods, *J. Food Hyg. Soc. Japan* 25:334.

Idziak, I., 2000, Molecular imprinting - a way to make smart polymers, *Can. Chem. News* June 2000, p.21.

Jarczyk, A., Jędrychowski, L., Wróbleweska, B., and Jędryczko, R., 1999, Relationship between ochratoxin A content in cereal grain and mixed meals determined by the ELISA and HPLC methods and an attempt to evaluate their usability for monitoring studies, *Pol. J. Food Nutr. Sci.* 8/49:53.

Jiao, Y., Blaas, W., Rühl, C., and Weber, R., 1992, Identification of ochratoxin A in food samples by chemical derivatization and gas chromatography-mass spectrometry, *J. Chromatogr.* 595:364.

Jornet, Busto, O., and Guasch, J., 2000, Solid-phase extraction applied to the determination of ochratoxin A in wines by reversed phase high-performance liquid chromatography, *J. Chromatogr. A* 882:29.

Jørgensen, K., 1998, Survey of pork, poultry, coffee, beer and pulses for ochratoxin A, *Food Addit. Contam.* 15:550.

Jørgensen, K., and Vahl, M., 1999, Analysis of ochratoxin A in pig kidney and rye flour using liquid chromatography tandem mass spectrometry (LC/MS/MS), *Food Addit. Contam.* 16:451.

Jørgensen, K., Rasmussen, G., and Thorup, I., 1996, Ochratoxin A in Danish cereals 1986-1992 and daily intake by the Danish population, *Food Addit. Contam.* 13:95.

Jurjevic, Z., Solfrizzo, M., Cvjetkovic, B., Avantaggiato, G., and Visconti, A., 1999, Ochratoxin A and fumonisins (B_1 and B_2) in maize from Balkan nephropathy endemic and non endemic areas of Croatia, *Mycotoxin Res.* 15:67.

Kawamura, O., Sato, S., Kajii, H., Nagayama, S., Ohtani, K., Chiba, J., and Ueno, Y., 1989, A sensitive enzyme-linked immunosorbent assay of ochratoxin A based on monoclonal antibodies, *Toxicon* 27:887.

Koch, M., Steinmeyer, S., Tiebach, R., Weber, R., and Weyerstahl, P., 1996, Bestimmung von Ochratoxin A in Röstkaffee, *Dtsche. Lebensm.-Rundsch.* 92:48.

Kuiper-Goodman, T., and Scott, P.M., 1989, Risk assessment of the mycotoxin ochratoxin A, *Biomed. Environ. Sci.* 2:179.

Lacey, J., Ramakrishna, N., Candlish, A.A.G., and Smith, J.E., 1991, Immunoassay of ochratoxin and other mycotoxins from a single extract of cereal grains utilizing monoclonal antibodies, in: *Mycotoxins, Endemic Nephropathy and Urinary Tract Tumours*, M. Castegnaro, R. Pleština, G.Dirheimer, I.N. Chernozemsky and H. Bartsch, eds., International Agency for Research on Cancer, Lyon.

Langseth, W., Ellingsen, Y., Nymoen, U., and Økland, E.M., 1989, High-performance liquid chromatographic determination of zearalenone and ochratoxin A in cereals and feed, *J. Chromatogr.* 478:269.

Langseth, W., Nymoen, U., and Bergsjø, B., 1993, Ochratoxin A in plasma of Norwegian swine determined by an HPLC column-switching method, *Nat. Toxins* 1:216.

Larsson, K., and Möller, T., 1996, Liquid chromatographic determination of ochratoxin A in barley, wheat bran, and rye by the AOAC/IUPAC/NMKL method: NMKL collaborative study, *J. AOAC Int.* 79:1102.

Lau, B.P.-Y., Scott, P.M., Lewis, D.A., and Kanhere, S.R., 2000, Quantitative determination of ochratoxin A by liquid chromatography/electrospray tandem mass spectrometry, *J. Mass Spectrom.* 35:23.

Lee, S.C., and Chu, F.S., 1984, Enzyme-linked immunosorbent assay of ochratoxin A in wheat, *J. Assoc. Off. Anal. Chem.* 67:45.

Lepom, P., 1986, Simultaneous determination of the mycotoxins citrinin and ochratoxin A in wheat and barley by high-performance liquid chromatography, *J. Chromatogr.* 355:335.

Levi, C.P., 1975, Collaborative study of a method for the determination of ochratoxin A in green coffee, *J. Assoc. Off. Anal. Chem.* 58:258.

Li, S., Marquardt, R.R., and Frohlich, A.A., 1998, Confirmation of ochratoxins in biological samples by conversion into methyl esters in acidified methanol, *J. Agric. Food Chem.* 46:4307.

MacDonald, S., Wilson, P., Barnes, K., Damant, A., Massey, R., Mortby, E., and Shephard, M.J., 1999, Ochratoxin in dried vine fruit: method development and survey, *Food Addit. Contam.* 16:253.

Maierhofer, P., Dietrich, R., and Märtlbauer, E., 1995, Nachweis von Ochratoxin A in grünem und geröstetem Kaffee (Detection of ochratoxin A in green and roasted coffee), in: *Proceedings 17. Mykotoxin-Workshop, Braunschweig-Völkenrode (FAL) 15.-17.mai 1995*, M. Goll, E. Oldenberg and H. Valenta, eds.

Majerus, P., and Otteneder, H., 1996, Nachweis und Vorkommen von Ochratoxin A in Wein und Traubensaft, *Dtsche. Lebensm.-Rundsch.* 92:388.

Mallmann, C.A., Santurio, J.M., Baldissera, M.A., and von Mickwitz, G., 1994, Determination of ochratoxin A in blood serum of pigs by using thin layer chromatography, *Rev. Microbiol., São Paulo* 25:107.

Marley, E.C., Nicol, W.C., and Candlish, A.A.G., 1995, Determination of ochratoxin A by immunoaffinity column clean-up and HPLC in wheat and pig liver, *Mycotoxin Res.* 11:111.

Marquardt, R.R., and Frohlich, A.A., 1992, A review of recent advances in understanding ochratoxicosis, *J. Anim. Sci.* 70:3968.

Marquardt, R.R., Frohlich, A.A., Sreemannarayana, O., Abramson, D., and Bernatsky, A., 1988, Ochratoxin A in blood of slaughter pigs in western Canada, *Can. J. Vet. Res.* 52:186.

Micco, C., Grossi, M., Miraglia, M., and Brera, C., 1989, A study of the contamination by ochratoxin A of green and roasted coffee beans, *Food Addit. Contam.* 6:333.

Miraglia, M., de Dominicis, A., Brera, C., Corneli, S., Cava, E., Menghetti, E., and Miraglia, E., 1995, Ochratoxin A levels in human milk and related food samples: an exposure assessment, *Nat.Toxins* 3:436.

Morgan, M.R.A., McNerney, R., and Chan, H.W.-S., 1983, Enzyme-linked immunosorbent assay of ochratoxin A in barley, *J. Assoc. Off. Anal. Chem.* 66:1481.

Morgan, M.R.A., McNerney, R., Chan, H.W.-S., and Anderson, P.H., 1986, Ochratoxin A in pig kidney determined by enzyme-linked immunosorbent assay (ELISA), *J. Sci. Food Agric.* 37:475.

Mühlemann, M., Lüthy, J., and Hübner, P., 1997, Mycotoxin contamination of food in Ecuador. B: ochratoxin A, deoxynivalenol, T-2 toxin and fumonisin, *Mitt. Gebiete Lebensm. Hyg.* 88:593.

Nakajima, M., Terada, H., Hisada, K., Tsubouchi, H., Yamamoto, K., Uda, T., Itoh, Y., Kawamura, O., and Ueno, Y., 1990, Determination of ochratoxin A in coffee beans and coffee products by monoclonal antibody affinity chromatography, *Food Agric. Immunol.* 2:189.

Nakajima, M., Tsubouchi, H., Miyabe, M., and Ueno, Y., 1997, Survey of aflatoxin B_1 and ochratoxin A in commercial green coffee beans by high-performance liquid chromatography linked with immunoaffinity chromatography, *Food Agric. Immunol.* 9:77.

Nakajima, M., Tsubouchi, H., and Miyabe, M., 1999, A survey of ochratoxin A and aflatoxins in domestic and imported beers in Japan by immunoaffinity and liquid chromatography, *J. AOAC Int.* 82:897.

Nesheim, S., Hardin, A., Francis, O.J., Jr., and Langham, W.S., 1973a, Analysis of ochratoxins A and B and their esters in barley, using partition and thin layer chromatography. I. Development of the method, *J. Assoc. Off. Anal. Chem.* 56:817.

Nesheim, S., 1973b, Analysis of ochratoxins A and B and their esters in barley, using partition and thin layer chromatography. II. Collaborative study, *J. Assoc. Off. Anal. Chem.* 56:822.

Nesheim, S., Stack, M.E., Trucksess, M.W., Eppley, R.W., and Krogh, P., 1992, Rapid solvent-efficient method for liquid chromatographic determination of ochratoxin A in corn, barley, and kidney: collaborative study, *J. AOAC Int.* 75:481.

Ominski, K.H., Frohlich, A.A., Marquardt, R.R., Crow, G.H., and Abramson, D., 1996, The incidence and distribution of ochratoxin A in western Canadian swine, *Food Addit. Contam.* 13:185.

Ospital, M., Cazabeil, J.-M., Betbeder, A.-M., Tricard, C., Creppy, E., and Medina, B., 1998, L'ochratoxine A dans les vins, *Rev. Fr. Oenol.* No. 169, 16.

Park, D.L., Njapau, H., and Coker, R.D., 1998, Sampling programs for mycotoxins: perspectives and recommendations, in: *Mycotoxins and Phycotoxins - Developments in Chemistry, Toxicology and Food Safety*, M. Miraglia, H.P. van Egmond, C. Brera and J. Gilbert, eds., Alaken, Fort Collins, CO.

Patel, S., Hazel, C.M., Winterton, A.G.M., and Gleadle, A.E., 1997, Survey of ochratoxin A in UK retail coffees, *Food Addit. Contam.* 14:217.

Phillips, T.D., Stein, A.F., Ivie, G.W., Kubena, LF., Hayes, A.W., and Heidelbaugh, N.D., 1983, High pressure liquid chromatographic determination of an *O*-methyl, methyl ester derivative of ochratoxin A, *J. Assoc. Off. Anal. Chem.* 66:570.

Pittet, A., 1998, Natural occurrence of mycotoxins in foods and feeds - an updated review, *Rev. Méd. Vét.* 149:479.

Pittet, A., Tornare, D., Huggett, A., and Viani, R., 1996, Liquid chromatographic determination of ochratoxin A in pure and adulterated soluble coffee using an immunoaffinity column cleanup procedure, *J. Agric. Food Chem.*44:3564.

Pohland, A.E., Nesheim, S., and Friedman, L., 1992, Ochratoxin A: a review, *Pure Appl. Chem.* 64:1029.

Rajakylä, E., Laasasenaho, K., and Sakkers, P.J.D., 1987, Determination of mycotoxins in grain by high-performance liquid chromatography and thermospray liquid chromatography-mass spectrometry, *J. Chromatogr.* 384:391.

Reinhard, H., and Zimmerli, B., 1999, Reversed-phase chromatographic behavior of the mycotoxins citrinin and ochratoxin A, *J. Chromatogr. A* 862:147.

Rosner, H., 1998, Mycotoxin regulations: an update, *Rev. Méd. Vét.* 149:679.

Rousseau, D.M., Siegers, G.A., and Van Peteghem, C.H., 1986, Solid-phase radioimmunassay of ochratoxin A in serum, *J. Agric. Food Chem.* 34:862.

Ruprich, J., and Ostry, V., 1991, Enzymo-immunological assays of the mycotoxin ochratoxin A (in Czech), *Vet. Med. (Praha)* 36:245.

Scott, P.M., and Hand, T.B., 1967, Method for the detection and estimation of ochratoxin A in some cereal products, *J. Assoc. Off. Anal. Chem.* 50:366.

Scott, P.M., and Kanhere, S.R., 1995, Determination of ochratoxin A in beer, *Food Addit. Contam.* 12:591.

Scott, P.M., and Trucksess, M.W., 1997, Application of immunoaffinity columns to mycotoxin analysis, *J. AOAC Int.* 80:941.

Scott, P.M., Kanhere, S.R., Canela, R., Lombaert, G.A., and Bacler, S., 1991, Determination of ochratoxin A in meat by liquid chromatography, *Prehrambeno-Tehnol. Biotehnol.Rev.* 29:61.

Scott, P.M., Kanhere, S.R., Lau, B.P.-Y., Lewis, D.A., Hayward, S., Ryan, J.J., and Kuiper-Goodman, T., 1998, Survey of Canadian human blood plasma for ochratoxin A, *Food Addit. Contam.* 15:555.

Scudamore, K.A., Hetmanski, M.T., Nawaz, S., Naylor, J., and Rainbird, S., 1997, Determination of mycotoxins in pet foods sold for domestic pets and wild birds using linked-column immunoassay clean-up and HPLC, *Food Addit. Contam.* 14:175.

Scudamore, K.A., and MacDonald, S.J., 1998, A collaborative study of an HPLC method for determination of ochratoxin A in wheat using immunoaffinity column clean-up, *Food Addit. Contam.* 15:401.

Seidel, V., Poglits, E., Schiller, K., and Lindner, W., 1993, Simultaneous determination of ochratoxin A and zearalenone in maize by reversed-phase high-performance liquid chromatography with fluorescence detection and β-cyclodextrin as mobile phase additive, *J. Chromatogr.* 635:227.

Sharman, M., MacDonald, S., and Gilbert, J., 1992, Automated liquid chromatographic determination of ochratoxin A in cereals and animal products using immunoaffinity column cleanup, *J. Chromatogr.* 603:285.

Shotwell, O.L., Goulden, M.L., Bennett, G.A., Plattner, R.D., and Hesseltine, C.W., 1977, Survey of 1975 wheat and soybeans for aflatoxin, zearalenone, and ochratoxin, *J. Assoc. Off. Anal. Chem.* 60:778.

Skaug, M.A., 1999, Analysis of Norwegian milk and infant formulas for ochratoxin A, *Food Addit. Contam.* 16:75.

Soares, L.M.V., and Rodriguez-Amaya, D.B.,1985, Screening and quantitation of ochratoxin A in corn, beans, rice, and cassava, *J. Assoc. Off. Anal. Chem.* 68:1128.

Soares, L.M.V., and Rodriguez-Amaya, D.B.,1989, Survey of aflatoxins, ochratoxin A, zearalenone, and sterigmatocystin in some Brazilian foods by using multi-toxin thin-layer chromatographic method, *J. Assoc. Off. Anal. Chem.* 72:22.

Solfrizzo, M., Avantaggiato, G., and Visconti, A., 1998, Use of various clean-up procedures for the analysis of ochratoxin A in cereals, *J. Chromatogr. A* 815:67.

Stegen, G.v.d., Jörissen, U., Pittet, A., Saccon, M., Steiner, W., Vicenzi, M., Winkler, M., Zapp, J., and Schlatter, C., 1997, Screening of European coffee final products for occurrence of ochratoxin A (OTA), *Food Addit. Contam.* 14:211.

Steyn, P.S., and v.d. Merwe, K.J., 1966, Detection and estimation of ochratoxin A, *Nature* 211:418.

Stoev, S.D., 1998, The role of ochratoxin A as a possible cause of Balkan endemic nephropathy and its risk evaluation, *Vet. Hum. Toxicol.* 40:352.

Studer-Rohr, I., Dietrich, D.R., Schlatter, J., and Schlatter, C., 1995, The occurrence of ochratoxin A in coffee, *Food Chem. Toxicol.* 33:341.

Takeda, T., Akiyama, Y., and Shibasaki, S., 1991, Solid-phase extraction and cleanup for liquid chromatographic analysis of ochratoxin A in pig serum, *Bull. Environ. Contamin. Toxicol.* 47:198.

Terada, H., Tsubouchi, H., Yamamoto, K., Hisada, K., and Sakabe, Y., 1986, Liquid chromatographic determination of ochratoxin A in coffee beans and coffee products, *J. Assoc. Off. Anal. Chem.* 69:960.

Téren, J., Varga, J., Hamari, Z., Rinyu, E., and Kevei, F., 1996, Immunochemical detection of ochratoxin A in black *Aspergillus* strains, *Mycopathologia* 134:171.

Trucksess, M.W., Giler, J., Young, K., White, K.D., and Page, S.W., 1999, Determination and survey of ochratoxin A in wheat, barley, and coffee B 1997, *J. AOAC Int.* 82:85.

Valenta, H., 1998, Chromatographic methods for the determination of ochratoxin A in animal and human tissues and fluids, *J. Chromatogr. A* 815:75.

Valenta, H., and Goll, M., 1996, Determination of ochratoxin A in regional samples of cow's milk from Germany, *Food Addit. Contam.* 13:669.

van der Merwe, K.J., Steyn, P.S., Fourie, L., Scott, De B., and Theron, J.J., 1965, Ochratoxin A, a toxic metabolite produced by *Aspergillus ochraceus* Wilh., *Nature* 205:1112.

van Egmond, H.P., 1991a, Worldwide regulations for ochratoxin A, in: *Mycotoxins, Endemic Nephropathy and Urinary Tract Tumours*, M. Castegnaro, R. Pleština, G.Dirheimer, I.N. Chernozemsky and H. Bartsch, eds., International Agency for Research on Cancer, Lyon.

van Egmond, H.P., 1991b, Methods for determining ochratoxin A and other nephrotoxic mycotoxins, in: *Mycotoxins, Endemic Nephropathy and Urinary Tract Tumours*, M. Castegnaro, R. Pleština, G.Dirheimer, I.N. Chernozemsky and H. Bartsch, eds., International Agency for Research on Cancer, Lyon.

van Egmond, H.P., and Speijers, G.J.A., 1994, Survey of data on the incidence and levels of ochratoxin A in food and animal feed worldwide, *J. Nat. Toxins* 3:125.

van Egmond, H.P., and Dekker, W.H., 1995, Worldwide regulations for mycotoxins in 1994, *Nat. Toxins* 3:332.

Vazquez, B.I., Fente, C., Franco, C., Cepeda, A., Prognon, P., and Mahuzier, G., 1996, Simultaneous high-performance liquid chromatographic determination of ochratoxin A and citrinin in cheese by time-resolved luminescence using terbium, *J. Chromatogr. A* 727:185.

Veldman, A., Borggreve, G.J., Mulders, E.J., and van de Lagemaat, D., 1992, Occurrence of the mycotoxins ochratoxin A, zearalenone and deoxynivalenol in feed components, *Food Addit. Contam.* 9:647.

Visconti, A., Pascale, M., and Centonze, G., 1999, Determination of ochratoxin A in wine by means of immunoaffinity column clean-up and high-performance liquid chromatography, *J. Chromatogr. A* 864:89.

Weddeling, K., Bäßler, H.M.S., Doerk, H., and Baron, G., 1994, Orientierende Versuche zur Anwendbarkeit enzymimmunologischer Verfahren zum Nachweis von Deoxynivalenol, Ochratoxin A und Zearalenon in Braugerste, Malz und Bier, *Monatssch. Brauwiss.* No.3:94.

Wilken, C., Baltes, W., Mehlitz, I., Tiebach, R., and Weber, R., 1985, Ochratoxin A in Schweinenieren - eine Methodenbeschreibung (Description of a method for the determination of ochratoxin A in pork kidneys), *Z. Lebensm. Unters. Forsch.* 180:496.

Wilson, D.M., Sydenham, E.W., Lombaert, G.A., Trucksess, M.W., Abramson, D., and Bennett, G.A., 1998, Mycotoxin analytical techniques, in: *Mycotoxins in Agriculture and Food Safety*, K.K. Sinha and D. Bhatnagar, eds., Marcel Dekker, New York.

Wood, G.M., Patel, S., Entwisle, A.C., and Boenke, A., 1996, Ochratoxin A in wheat: a second intercomparison of procedures, *Food Addit. Contam.*13:519.

Wood, G.M., Patel, S., Entwisle, A.C., Williams, A.C., Boenke, A., and Farnell, P.J., 1997, Ochratoxin A in wheat: certification of two reference materials, *Food Addit. Contam.* 14:237.

Zabe, N., Jackson, B., Hansen, T., and Prioli, R., 1996, Immunoaffinity column clean-up for HPLC to quantitate mycotoxins: ochratoxin A in wheat samples, Abstr. 110th AOAC Int. Ann. Mtg., September 8-12, Orlando, FL, p.58.

Zimmerli, B., and Dick, R., 1995, Determination of ochratoxin A at the ppt level in human blood, serum, milk and some foodstuffs by high-performance liquid chromatography with enhanced fluorescence detection and immunoaffinity column clean-up: methodology and Swiss data, *J. Chromatogr. B* 666:85.

Zimmerli, B., and Dick, R., 1996a, Ochratoxin A in table wine and grape-juice: occurrence and risk assessment, *Food Addit. Contam.* 13:655.

Zimmerli, B., and Dick, R., 1996b, Study to the repeated use of commercial immunoaffinity columns, *Mitt. Gebiete Lebensm. Hyg.* 87:732.

HPLC DETECTION OF PATULIN IN APPLE JUICE WITH GC/MS CONFIRMATION OF PATULIN IDENTITY

John A. G. Roach[1], Allan R. Brause[2], Thomas A. Eisele[3], and Heidi S. Rupp[4]

[1]U.S. Food and Drug Administration, CFSAN, 200 C St., SW, Washington, DC 20204, [2]Analytical Chemical Services of Columbia Inc., 9110 Red Branch Road, Suite K, Columbia, MD 21045, [3]Treetop Inc., 111 S. Railroad Avenue, Selah, WA 98942, [4]U.S. Food and Drug Administration, Pacific Regional Laboratory Northwest, Bothell, WA 98021

ABSTRACT

The official patulin LC procedure was further examined (AOAC 995.10). Juice or juice concentrate was extracted with ethyl acetate and cleaned up with sodium carbonate. Patulin in the dried extract was determined by reversed-phase LC with UV detection (280 nm) in 1% THF aqueous solution after evaporation of the ethyl acetate. An end-capped C18 column was required to separate patulin from hydroxymethylfurfural. Patulin was detected in approximately half of the >1000 extracts examined. Only ca 10% of the extracts contained patulin at levels greater than 50 μg/L (50 ppb). Some presumptive findings were confirmed by capillary gas chromatography/mass spectrometry as the trimethyl silyl derivative using electron ionization or as underivatized patulin using negative ion chemical ionization. Trifluoropropylmethyl polysiloxane capillary columns provided superior gas chromatography of underivatized patulin compared to phenyl/methyl polysiloxane and methyl polysiloxane columns.

HPLC DETECTION OF PATULIN IN APPLE JUICE

Chromatographic methods for detecting patulin in apple juice date back to a 1977 gas chromatography/mass spectrometry (GC/MS) assay for patulin as its acetate (Ralls et al., 1977). The article described storage rot in apple with subsequent formation of patulin. A subsequent article described an HPLC procedure that is similar to the current AOAC procedure (995.10) but with a more complicated clean up protocol (Stray, 1978). The method used a 90-10 methanol-ethyl acetate mobile phase. A "second peak" was noted 1-2 minutes after patulin. This peak was in all likelihood, hydroxymethylfurfural (HMF), which can occur in significant amounts in apple juice. This paper cited a Norwegian provisional upper limit of 50 μg/L (50 ppb) for patulin.

Mycotoxins and Food Safety, edited by Trucksess *et al.*
Kluwer Academic/Plenum Publishers, 2002

A collaborative study (Brause et al., 1996) evaluated several methods of analysis and found a modified method that worked best (Forbito and Babsky, 1985). The mobile phase was water with 0. 8% tetrahydrofuran (THF) and 0.02 % sodium azide. HMF eluted at ca. 18 min and patulin at 20 min by this procedure. A non-end capped 25cm 100 Å, 5μm C18 column (Reliasil C18, Column Engineering, Ontario, CA) was recommended.

The key to accurate HPLC determination is separating patulin from HMF. This is essential since HMF arises during processing and aging of juice. HMF can run as high as 20 mg/L (20 ppm) in processed apple juice. Eisele (private communication) recently found that a 25 cm end capped C18 column (Phenomenex, Torrance CA, Luna™) separated patulin and HMF by four min. Column Engineering suggested that a Reliasil C18 should do as well. This proved to be true (Figure 1A). Even better resolution is obtained with Reliasil BDX (Figure 1B) and Monitor™ (Figure 1C) columns, i.e. 5 and 6 minutes respectively.

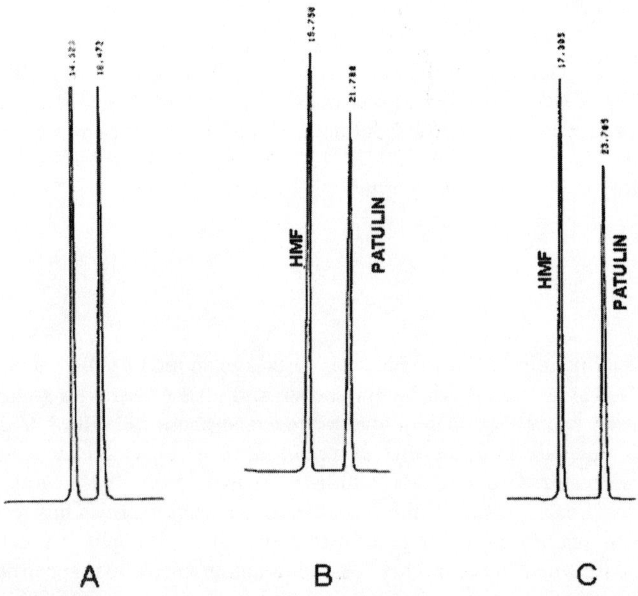

Figure 1. Patulin and hydroxymethylfurfural separation with (A) Reliasil, (B) Reliasil BDX, and (C) Monitor 250 x 4.6 mm LC columns.

In 1994, Brause provided patulin survey data to the U.S. Food and Drug Administration and the Technical Committee for Juice and Juice Products (TCJJP). The data were updated in 1998. These patulin survey results to mid 2000 are shown in Table 1.

Table 1. Results of a patulin survey conducted from 1994 – 2000

	<LOD[a]	**1-50**	**51-100**	**101-200**	**>200**	**TOTAL**
Collation mid 1994	154	184	7	0	0	**345**
Collation mid 1998	334	387	76	40	29	**866**
Collation mid 1998 to present	180	94	17	17	6	**314**
Summary totals	**668**	**665**	**100**	**57**	**35**	**1525**
Percentage of total samples	43.8	43.6	6.6	3.7	2.3	

[a] Less than limit of detection (<LOD). Patulin levels are in μg/L (ppb).

These results show that 87.4% of the samples contained acceptable levels of <50 μg/L (50 ppb) as per WHO guidance, EU guidance (Van Egmond, 1989) and recently proposed FDA guidelines (Federal Register 65, June 16, 2000, 37791-37792). The levels have remained fairly consistent from 1994 to the present.

HPLC quantitates patulin, but does not confirm it. Brause (unpublished data) developed a GC procedure that would supply complementary GC information in the event of a patulin LC response. Patulin eluted in ca 17 min from a 30 m x 0.25 mm DB-5 column when the GC oven was held at 190°C for 2 min and then heated at 2°/min to 240°C and 10°/min to 275°C with a 10 min hold at 275°C. The patulin response with a DB-5 column was linear down to 10 mg/L (10 ppm). A 500:1 concentration step from an apple juice extract gave a 20 μg/L (20 ppb) minimum detectable level (MDL).

GC/MS CONFIRMATION OF PATULIN

Recent noteworthy reports of the GC/MS confirmation of identity of patulin include high resolution single ion monitoring (Rychlik et al., 1999), low resolution multiple ion monitoring (Rupp et al., 2000), and low resolution limited mass scans (Roach et al., 2000). The high resolution assay quantitates patulin against ^{13}C-labelled patulin (Rychlik et al., 1998). The low resolution multiple ion monitoring assay and the high resolution single ion monitoring assay determine patulin as its trimethyl silyl derivative (TMS-patulin) with electron ionization (EI). The low resolution limited mass scan assay determines patulin directly with a variety of gas chromatographic columns and modes of ionization utilizing magnetic sector, quadrupole, and ion trap instrumentation.

The EI mass spectrum of TMS-patulin has abundant diagnostic fragment ions (Figure 2). The molecular ion, *m/z* 226, is relatively abundant. Rychlik and Schieberle (1999) report a limit of detection of 12 ng/L (12 parts per trillion) monitoring the molecular ions of incurred and labeled patulin at 5000 resolving power. Their report includes data obtained for apple juice, grape juice and moldy wheat bread.

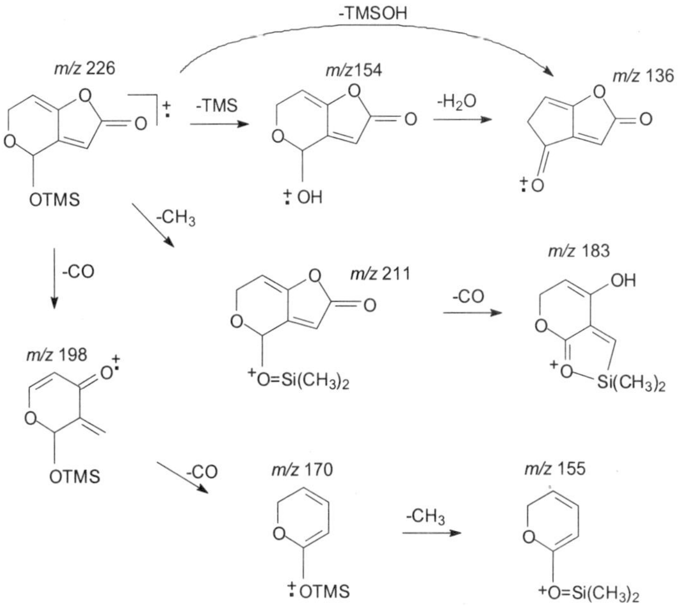

Figure 2. Suggested EI fragmentation of trimethyl silyl derivative of patulin.

Diagnostic TMS-patulin EI fragment ions are also relatively abundant. The low resolution multiple ion assay capitalizes on their abundance to detect patulin with excellent reproducibility, sensitivity and specificity. Rupp and Turnipseed (2000) report less than 7.1% interassay variation in relative abundances of the monitored ions (*m/z* 136, 154, 155, 170 183, 211 and 226) in analyses for spiked and incurred patulin in apple juice extracted by the official method (AOAC 995.10). The multiple ion assay identifies HMF as well as patulin. Patulin is quantitated by LC prior to preparation of TMS-patulin.

TMS-patulin chromatographs readily in methyl silicone or phenyl-methyl silicone phase GC columns. Patulin also chromatographs in these phases, but peak tailing is a problem with thin-film phase coatings frequently used for GC/MS. Mass spectra obtained by EI, positive ion chemical ionization (PCI) and negative ion chemical ionization (NCI) are all sufficient for the confirmation of patulin in food products (Figure 3).

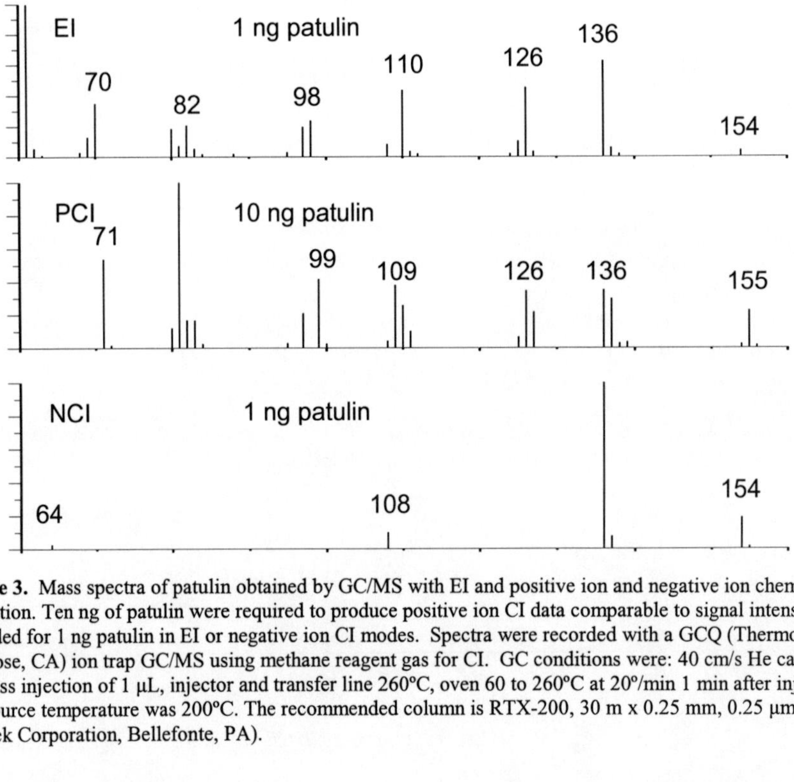

Figure 3. Mass spectra of patulin obtained by GC/MS with EI and positive ion and negative ion chemical ionization. Ten ng of patulin were required to produce positive ion CI data comparable to signal intensities recorded for 1 ng patulin in EI or negative ion CI modes. Spectra were recorded with a GCQ (Thermoquest, San Jose, CA) ion trap GC/MS using methane reagent gas for CI. GC conditions were: 40 cm/s He carrier, splitless injection of 1 µL, injector and transfer line 260°C, oven 60 to 260°C at 20°/min 1 min after injection. Ion source temperature was 200°C. The recommended column is RTX-200, 30 m x 0.25 mm, 0.25 µm film (Restek Corporation, Bellefonte, PA).

Limited mass scans of 60 to 160 daltons record the GC/MS response of patulin and document the absence of co-eluting interferences in the extracts. Rychlik and Schieberle (1998) contrasted the mass spectra of patulin and ^{13}C-labelled patulin to suggest plausible EI fragmentation pathways for patulin. Chemical ionization processes are significantly less exothermic. The observed PCI and NCI fragmentation arise in large part from the loss of small neutral molecules (Figure 4).

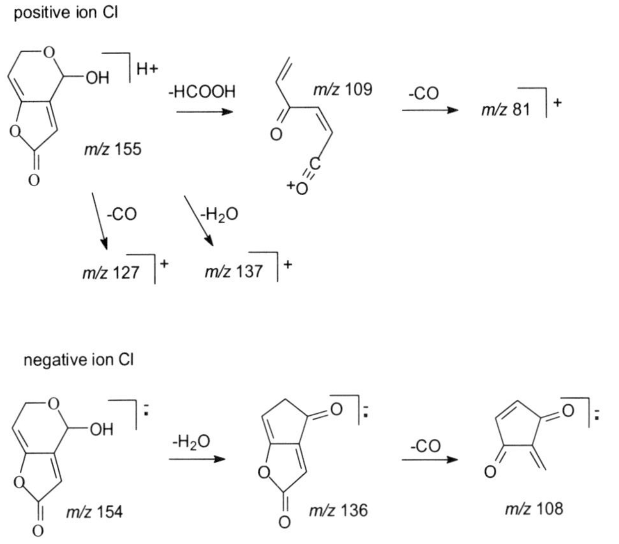

Figure 4. Suggested patulin chemical ionization fragmentation in positive ion and negative ion modes using methane reagent gas. Losses of small neutral molecules comprise the principal fragmentation pathways.

A trifluoropropylmethyl polysiloxane phase (RTX-200, Restek Corporation, Bellefonte, PA) GC column is superior to methyl and phenylmethyl silicone phases for the GC/MS analysis of patulin (Figure 5). Peak tailing is drastically reduced, concentrating the patulin into a narrow band as it enters the mass spectrometer. The improved chromatography significantly enhances the sensitivity of GC/MS for patulin.

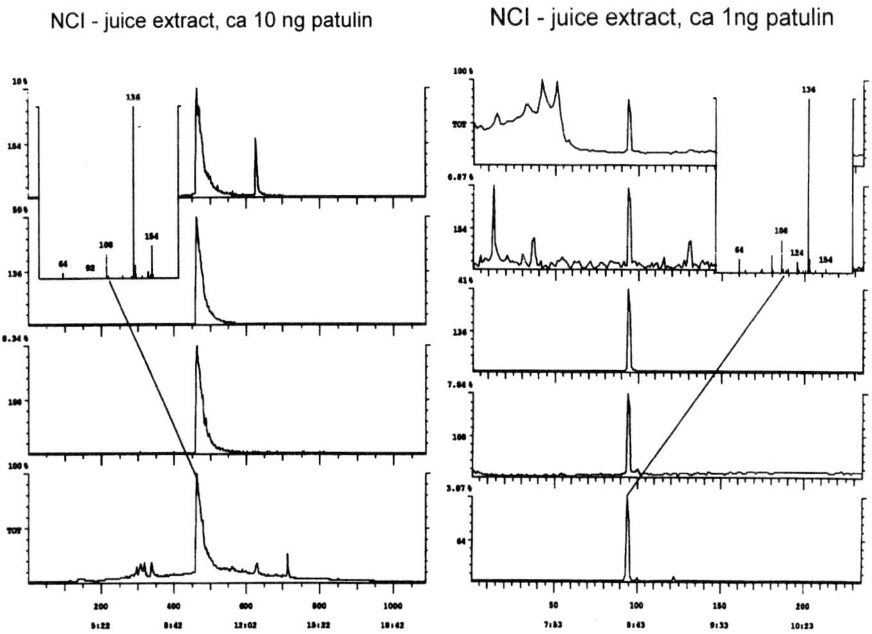

Figure 5. Negative ion GC/MS data for juice extracts analyzed with 5% phenylmethyl silicone (left) and trifluoropropylmethyl polysiloxane (right) phases. Patulin qualitative limit of detection was 1 ng injected for 5% phenylmethyl silicone phase and 8 pg injected for trifluoropropylmethyl polysiloxane phase.

A standard curve of ion intensity for *m/z* 136 extracted from limited mass scans of 60 to 160 Da acquired with the GCQ was linear over a range of 400 femtograms to 400 pg. The patulin qualitative identification limit for GCQ limited mass scan data was 8 pg. Standard deviation (SD) for 4 injections of 400 pg equally spaced over 8 hours of juice analyses was 3.8% and the SD for 4 injections of 40 pg was 12.4% for data recorded during 8 hours of juice analyses. These data are encouraging, but quantitations based on an external standard curve are not recommended for ion trap instrumentation (Roach et al., 2000). The presence or absence of co-eluting materials significantly affects the analyte signal recorded by an ion trap. This affects ion trap quantitation based on an external standard curve. An isotopically labeled internal standard represents a more prudent approach to GC/MS quantitation.

REFERENCES

Brause, A., Trucksess, M., Thomas, F., Page, S., 1996, Determination of patulin in apple juice by liquid chromatography: collaborative study, *J. Assoc. Off. Anal. Chem.* 79:451.

Forbito, P.R., Babsky, N.E., 1985, Rapid liquid chromatographic determination of patulin in apple juice, *J. Assoc. Off. Anal. Chem.* 68:950.

Official Methods of Analysis, 1995, 16th Ed. March 1996 Supplement, AOAC International, Gaithersburg, MD, sec 995.10.

Ralls, J.W., Lane, R.M., 1977, Examination of cider vinegar for patulin using mass spectrometry, *J. Food Sci.* 42:1117.

Roach, J.A.G., White, K.D., Trucksess, M.W., Thomas, F.W., 2000, Capillary gas chromatography/mass spectrometry with chemical ionization and negative ion detection for confirmation of identity of patulin in apple juice, *Int. J. Assoc. Off. Anal. Chem.* 83:104.

Rupp, H.S., Turnipseed, S.B., 2000, Confirmation of patulin and 5-hydroxymethylfurfural in apple juice by gas chromatography/mass spectrometry, *Int. J. Assoc. Off. Anal. Chem.* 83:612.

Rychlik, M., Schierberle, P., 1998, Synthesis of ^{13}C-labeled patulin [4-hydroxy-4H-furo[3,2-c]pyran-2(6H)-one] to be used as internal standard in a stable isotope dilution assay, *J. Agric. Food Chem.* 46:5163.

Rychlik, M., Schierberle, P., 1999, Quantification of the mycotoxin patulin by a stable isotope dilution assay, *J. Agric. Food Chem.* 47:3749.

Stray, H., 1978, High pressure liquid chromatographic determination of patulin in apple juice, *J. Assoc. Off. Anal. Chem.* 61:1359.

Van Egmond, H., 1989, Current situation on regulations for mycotoxins. Overview of tolerances and status of standard methods of sampling and analysis, *Food Addit. Contam.* 6:139.

METHODS FOR THE DETERMINATION OF DEOXYNIVALENOL AND OTHER TRICHOTHECENES IN FOODS

Gary A. Lombaert

Health Products and Food Branch
Health Canada
Winnipeg, MB R2J 3Y1 CANADA

ABSTRACT

Trichothecene mycotoxins are secondary metabolites of *Fusarium* moulds that routinely infect cereal crops. Processing can reduce, but not eliminate, trichothecenes from cereal-based foods, and the potential presence of the trichothecenes in cereal foods poses a significant health risk to consumers. Deoxynivalenol (vomitoxin, DON) is the most common of the trichothecenes detected in cereal crops and is subject to government regulation in many countries. Sensitive (ng/g) methods for the detection of trichothecenes in cereal grains and food products are needed to protect consumers, to provide data for dietary exposure estimates, and to support research into the control of moulds and subsequent toxin production. Laboratories require simple, rugged and reliable methods for routine testing, with unequivocal identification of suspect mycotoxins. A method employing gas chromatography-negative ion chemical ionisation / mass spectrometry (GC-NICI/MS) has been developed and used for the routine determination of eight of the most significant trichothecenes in a variety of commodities. This chapter discusses GC, liquid chromatographic (LC) and supercritical fluid chromatographic methods that are currently used for the analysis of trichothecene mycotoxins.

INTRODUCTION

The trichothecenes are secondary metabolites of various moulds including *Fusarium*, *Stachybotrys*, *Myrothecium* and *Trichothecium*. *Fusarium* infections of wheat, corn and barley crops are common throughout the world. In North American crops the disease is called "scab" or the kernels as "tombstone" kernels. More correctly, the disease is designated as *Fusarium* head blight.

The trichothecenes are sesquiterpenoids; they have a C-9,10 double bond, a 12,13-epoxide ring, and various hydroxyl and acetoxy groups. The most common trichothecenes are

Mycotoxins and Food Safety, edited by Trucksess *et al.*
Kluwer Academic/Plenum Publishers, 2002

classified as type A or B depending on whether they do (type B) or do not (type A) have a carbonyl group at position C-8. Common type A trichothecenes include HT-2 toxin (HT-2) and T-2 toxin (T-2); DON belongs to the type B designation.

Health Significance

Acute and chronic ingestion of trichothecenes by humans and animals can elicit a variety of toxic effects, including fever, diarrhea, vomiting, necrosis, hemorrhage, inhibition of protein synthesis, and depletion of bone marrow. Consequently, the potential presence of trichothecenes in cereal foods poses a significant health risk to consumers. Modern agricultural processing methods can reduce, but not eliminate, trichothecenes from cereal foods. Although DON is among the least toxic of the trichothecenes, it is the most common and its occurrence is considered an indicator of the possible presence of other, more toxic trichothecenes.

Regulatory Levels for Trichothecenes in Foods

Several countries have set maximum permissible levels for DON in foods or raw cereal grains intended for food (FAO, 1997). The U.S. Food and Drug Administration permits a maximum of 1000 ng/g of DON in finished wheat products, and Health Canada has set a guideline of 2000 ng/g of DON in uncleaned soft wheat destined for human consumption. Austria has guidelines of 500 ng/g of DON in wheat and rye and 750 ng/g in durum wheat intended for foods. In Russia, the maximum permitted level of DON in wheat, flour and bran is 1000 ng/g, while only 100 ng T-2 toxin/g is permitted in the same commodities.

Economic Impact

In North America, several epidemics of *Fusarium* head blight, especially since the 1980's, have resulted in tremendous economic losses in the agricultural sector. In Canada, total losses have been estimated at more than $1 billion in just the past seven years (Fernando, 1999). The primary losses are due to down-grading of crops intended for human consumption to use as animal feed only, because of the presence of DON. Secondary losses may be attributed to the diminished performance of livestock fed these infected crops. In Canada and the U.S. (Stack, 1999), millions of dollars are appropriated toward research into crop breeding, chemical and biological control, and biotechnology in attempts to gain an upper hand on this devastating agricultural problem.

Need for Methods

Methods for the determination of trichothecenes in cereal grains and cereal-based foods are required for monitoring relative to current government standards, for survey work before exposure estimates and risk assessments, and in support of research activities aimed at control of the effects of *Fusarium* head blight. Methods must have appropriate sensitivity and specificity for individual trichothecenes. Method costs, speed, ease-of-use, and equipment requirements are other factors important to the laboratory.

The multi-trichothecene method presented (Health Canada, 1999) has developed from that originally published by Scott et al. (1986). The original method involved a two-step clean-up prior to GC with electron capture detection (ECD), and was employed for several years in surveys of Canadian wheat (Scott et al., 1989). It has been modified to employ a single clean-up step, and MS as a more definitive detector. The method has been used in the analysis of raw cereal grains, cereal-based infant foods, breakfast cereals and baked goods for the determination of eight common trichothecenes - DON, nivalenol (NIV), fusarenon-X (F-X), 15-monacetoxyscirpenol (MAS), 15-acetyl-deoxynivalenol (15-ADON), diacetoxyscirpenol (DAS), HT-2 and T-2.

DETERMINATION OF TRICHOTHECENES BY GC-NICI/MS (Health Canada, 1999)

Experimental

Equipment. A Hewlett Packard (HP) 5890 GC (Hewlett-Packard (Canada) Ltd., Mississauga, ON, Canada) equipped with an on-column injector is interfaced to either an HP 5988A or a Fisons VG Trio-1000 MS (Finnigan, Bremen, Germany). Other equipment includes laboratory blenders, a centrifuge, rotary evaporators, magnetic stirrers, vortex mixers, a vial heating block equipped with a nitrogen manifold, microliter hand pipettes, and 24-cm filter paper (Whatman #1 or #2V, Whatman Nucleopore Canada Ltd., Toronto, ON, Canada) or equivalent.

GC Operating Conditions. The analytical column is a DB-1701 (J&W Scientific Inc., Folsom, CA, USA) capillary column (15 m x 0.25 mm i.d., 0.25 μm film thickness), fitted to a 1 m x 0.53 mm i.d. fused silica retention gap. Helium carrier gas at 10 psi column pressure; injector temperature 70°C; oven temperature program: 70°C for 3 min, 30°C/min to 175°C, hold 2 min, 10°C/min to 245°C, hold 10 min; GC-MS transfer line 245°C

NICI/MS Operating Conditions. The MS is operated in the NICI selected ion monitoring (SIM) mode. The MS is tuned with perfluorotributylamine (Sigma-Aldrich, Oakville, ON, Canada) according to manufacturer's instructions, and optimized to favour the response of higher mass (m/z) ions, corresponding to the more specific parent ions of the analytes. Specifically, reagent gas (ultra-high purity methane) flow and ion source temperature have the greatest impact on the relative responses of the parent and progeny ions. On the HP 5988A instrument, the methane flow is optimized at a source pressure of about 1 torr. To minimize fragmentation, the ion source temperature is kept as low as reproducibly possible, eg. 75°C (HP 5988A) or 130°C (VG Trio-1000). At the retention window of each toxin, two specific ions are monitored at equal dwell times.

Selection of Quantifying and Qualifying Ions. Two ions are monitored for each toxin. One ion is used to quantify the toxin while the second is used to qualify its identity. The quantifying and qualifying ions are selected based upon four parameters: abundance, specificity, uniqueness and signal/noise response. To judge abundance, μg amounts of derivatized standard in solution are injected in the full scan (200 to 900 m/z) mode. The extracted spectra are examined over the width of the chromatographic peak to identify the major ions of interest (Figures 1-3). From the same injection, spectra of chromatographic baseline are extracted and subtracted from the standard spectra to ensure the specificity of selected ions. From the proposed ions of interest, it is preferable to select those with the highest m/z, particularly unique parent ions or identifiable fragments. The derivatized standard solution is now re-injected and only the ions of interest monitored, in SIM. A signal/noise calculation is performed for each ion and those with the highest signal/noise ratios advance for final consideration. Finally, the lowest dilution of a derivatized calibration multi-standard (see GC-NICI/MS) is injected to confirm that the proposed ions are sufficiently abundant to elicit a useful, reproducible area response. The ion with the greatest area or signal/noise response is chosen as the quantifying ion and the second most abundant as the qualifying ion. Suggested ions for heptafluorobutyrylimidazole (HFBI) derivatized toxins are: NIV (670, 213), DON (884, 458), F-X (516, 213), MAS (696, 716), 15-ADON (458, 213), DAS (480, 542), HT-2 (816, 213), T-2 (642, 580) (Krishnamurthy et al., 1986).

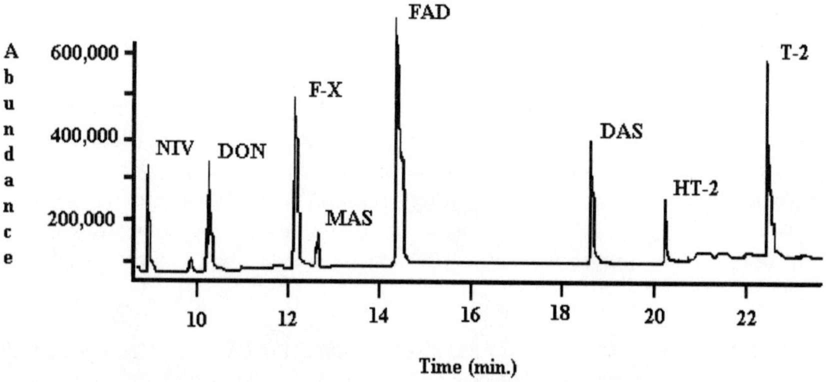

Figure 1. NICI total ion (200-900 m/z) chromatogram of HFBI derivatized trichothecene standards.

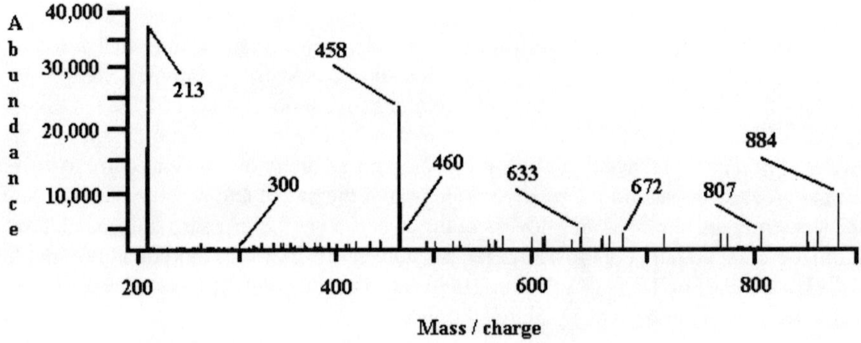

Figure 2. NICI mass spectrum of HFBI derivatized DON.

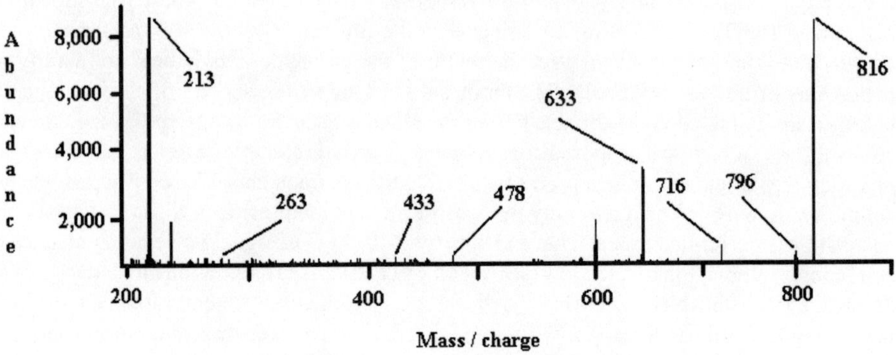

Figure 3. NICI mass spectrum of HFBI derivatized HT-2.

Standards. Standards (Sigma-Aldrich) (5, 10 or 25 mg) are weighed, dissolved in methanol, and then diluted to concentrations of 500, 1000 or 2500 μg/mL. The concentrations of solutions of trichothecenes with significant ultraviolet (UV) chromophores (NIV, DON and F-X) are checked by UV-visible spectrometry, using molar absorptivities described by Cole and Cox (1981) and Sydenham and Thiel (1996).

The relative MS responses of the derivatized trichothecenes will vary depending upon many instrumental parameters: tuning conditions, dwell times, ion selection, etc. To produce chromatograms with approximately equal responses for the trichothecenes, a multi-standard solution is prepared at the following concentrations: NIV (5 μg/mL), DON (2 μg/mL), 15-ADON (12 μg/mL), F-X (12 μg/mL), DAS (16 μg/mL), MAS (2 μg/mL), HT-2 (5 μg/mL) and T-2 (32 μg/mL). 100 μL portions of this multi-standard solution are dispensed into 4 mL glass vials and evaporated to dryness under a gentle stream of nitrogen in a heating block at 60°C. The vials are stored in a freezer until required (*see* **Derivatization**).

Reagents. ChemElut® hydrophilic columns (Varian Inc., Palo Alto, CA, USA). Heptafluorobutyrylimidazole (HFBI) (Pierce Chemical Co., Rockford, IL, USA). Diatomaceous earth (Hyflo Super Cel), ammonium sulfate, sodium hydroxide and potassium dihydrogen sulfate (Fisher Scientific, Ottawa, ON, Canada). Methanol, dichloromethane and iso-octane are of pesticide grade or equivalent.

Phosphate buffer, pH 7.0: Mix 50 mL of 0.1 M (13.6 g/L) potassium dihydrogen phosphate with 29.1 mL of 0.1 M (4 g/L) sodium hydroxide.

Sample Preparation. Grains and dry cereal products are ground in a centrifugal mill to pass a 2-mm screen. Breads and other high moisture baked products, are sliced, cubed and dried at room temperature before grinding.

Extraction. Ground sample (50 g) is weighed into a 500 mL mason jar and extracted by blending for 5 min with 150 mL methanol-water (70:30). For low density commodities, eg. infant cereals, 200 mL of methanol-water is used to promote efficient mixing and extraction, and the calculations are adjusted accordingly. A 50 mL centrifuge tube is filled with the crude sample extract, which is centrifuged for 15 min at 1500 rpm.

Clean-up. 30 mL of the supernatant (equivalent to 10 g sample) is transferred to a 300 mL tall-form beaker. 120 mL of 10% (w/v) aqueous ammonium sulfate and approximately 50 g of diatomaceous earth are added, and the mixture is stirred vigorously on a magnetic stirrer for 2 min. The mixture is then filtered through a fluted filter paper, and 10 mL of filtrate (equivalent to 0.667 g sample) is applied to a ChemElut column. After 2 - 4 min, the trichothecenes are eluted from the column with 8 x 20 mL portions of ethyl acetate, and collected in a 250 mL pear-shaped flask. The ethyl acetate is evaporated to near dryness on a rotary evaporator at a bath temperature of 35°C. The residue is transferred with 3 x 1 mL volumes of dichloromethane-methanol (3:1) to a 4 mL vial. The extract is then completely dried on a heating block at 60°C under a gentle stream of nitrogen.

Derivatization. The dried sample extracts (and dried multi-standard solution) are redissolved in 500 μL of iso-octane. 50 μL of HFBI is added, the vials are sealed with teflon-lined screw-caps, and vortexed for 1 min. The vials are then placed in a 60°C heating block for 60 min. After cooling to room temperature, the contents are again vortexed for 1 min. Subsequently, 1 mL of phosphate buffer is added and the mixture is vortexed for 2 min. The contents are then left undisturbed to allow the layers to separate (about 15 min).

GC-NICI/MS. 40 μL of the organic layer of derivatized sample extract (equivalent to 0.053 g sample) is transferred to a 1.5 mL autosampler vial, and diluted with 1250 μL iso-octane. Final sample concentration is 41.3 μg sample/μL.

For preparation of a four-point calibration curve, 12.5, 25, 50 and 100 μL volumes of the organic layer of a derivatized multi-standard solution are transferred to 1.5 mL autosampler vials, and each is diluted with 1250 μL of iso-octane. If the multi-standard solution is prepared as in **Standards**, the four resulting GC-MS solutions will have concentrations of the individual trichothecenes according to Table 1.

Table 1. Preparation of GC-MS standard curves from multi-standard solution

Trichothecene	Conc. (ng/mL) derivatized soln (100 μL stock soln evaporated and derivatized in 500 μL iso-octane)	Conc. (pg/μL) of varying dilutions (μL derivatized standard + μL iso-octane)			
		12.5 + 1250	25 + 1250	50 + 1250	100 + 1250
NIV	100	9.9	19.6	38.5	74.1
DON	400	4.0	7.8	15.4	29.6
F-X	2400	23.8	47.1	92.3	177.8
MAS	400	4.0	7.8	15.4	29.6
15-ADON	2400	23.8	47.1	92.3	177.8
DAS	3200	31.7	62.7	123.1	237.0
HT-2	1000	9.9	19.6	38.5	74.1
T-2	6400	63.4	125.5	246.2	474.1

Analyses are carried out on the same day as derivatization. 1 μL portions of each dilution of the derivatized multi-standard are injected. For each trichothecene, the ions of interest are monitored in SIM (Figure 4). For each toxin, the areas of both the quantifying and qualifying ions are determined. Calibration curves of area response (quantifying ion) vs. pg injected are constructed, and the ratios of responses for the quantifying and qualifying ions are determined. 1 μL portions of the derivatized sample solution are injected under identical operating conditions (Figure 5). Toxins are presumptively identified by comparison of chromatographic retention times to standards injected under identical conditions. For suspected positive samples, the ratio of responses of the quantifying/qualifying ions is calculated. Ratios must match the mean of the ratios of the calibration standards. For positively confirmed samples, the toxin concentration in the final injection solution, in pg/μL, is determined by comparison to the calibration curve constructed. Derivatized extracts may be diluted with iso-octane and re-injected as required. The concentration of toxin in the final solution is divided by the sample weight in the final solution (41.3 μg/μL) to yield the toxin concentration, in μg/g, in the original sample:

(pg toxin/μL solution) ÷ (μg sample/μL solution) = pg toxin/μg sample = μg toxin/g sample).

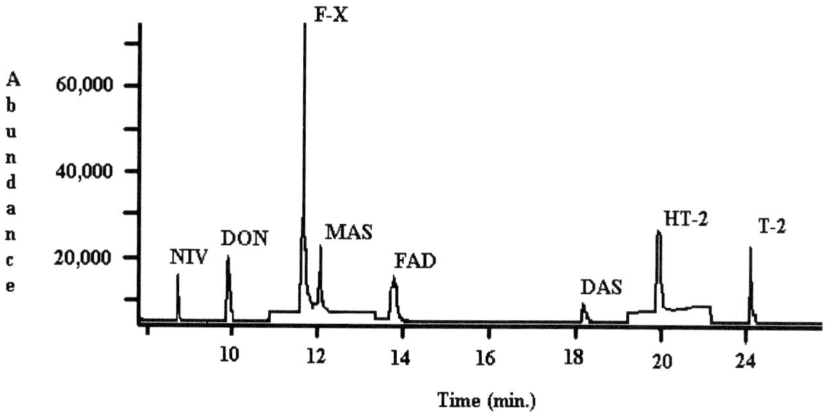

Figure 4. NICI SIM chromatogram of HFBI derivatized trichothecene standards. Toxin: pg injected (ions monitored): NIV: 10 (670, 213), DON: 4 (884, 458), F-X: 24 (516, 213), MAS: 4 (716, 696), 15-ADON: 24, (458, 213), DAS: 32 (542, 480), HT-2: 10 (816, 213), T-2: 64 (642, 580).

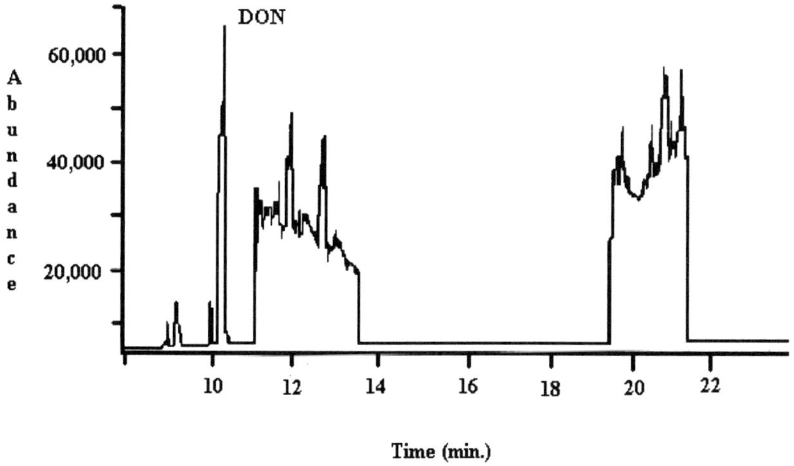

Figure 5. NICI SIM chromatogram of HFBI derivatized sample extract. DON (ions 884 and 458) concentration in the sample is approx. 300 ng/g.

Method Performance

The minimum quantifiable levels (MQL) of DON in cereal products by this method is 20 ng/g. The MQLs of the other toxins range from 20 ng/g (for MAS) to 300 ng/g (for T-2). The method as described has been in use since 1995 for the analysis of raw grains, flour, baked products, breakfast cereals, infant cereals, and various other products. Recoveries of DON spiked into authentic samples at levels from 200 to 900 ng/g, have averaged 90%. Similarly, recoveries of HT-2, at spike levels from 400 to 1000 ng/g, have averaged 91%. Nivalenol recoveries are generally lower and have averaged 77% at spike levels of 500 to 1000 ng/g. Recent reproducibility data (n = 31) on a sample of ground crackers produced a mean of 950 ng/g and a standard deviation of 160 ng/g (18%).

OTHER GC METHODS

GC is the most popular technique for the identification and quantification of multiple trichothecenes in foods. Fused silica capillary columns of medium polarity are routinely used for the separation of these analytes. Derivatization of the extract is favoured prior to the instrumental step to increase the volatility and detectability of the trichothecenes, although a method has recently been published employing MS detection without derivatization (Onji et al., 1998). Preparation of fluorinated propionyl or butyryl esters is the most popular derivatization mechanism, and is applicable to both type A and B trichothecenes. Preparation of the trimethylsilyl (TMS) ethers of the type B trichothecenes is common prior to detection with an ECD. The following section briefly describes some of the available GC methods.

Other GC-MS Methods

Schollenberger et al. (1998) demonstrated the applicability of a combined florisil and cation-exchange solid phase extraction procedure for the analysis of cereal grains and cereal-based foods. Samples were extracted with acetonitrile-water (75 + 25), defatted with hexane and evaporated to dryness. After the two stage clean-up, extracts were derivatized with trifluoroacetic anhydride, evaporated to dryness, redissolved in toluene, washed with water, and dried with sodium sulfate. The GC was interfaced to an ion-trap MS and operated in positive ion chemical ionisation (PICI) SIM with isobutane reagent gas. The authors reported a limit of detection and a minimum quantifiable level for DON of 7 and 23 ng/g, respectively.

A GC-MS method for the determination of DON and several other type B trichothecenes was described by Mirocha et al. (1998). Using a clean-up procedure developed by Tacke and Casper (1996), extracts were derivatized with a mixture of trimethylsilylimidazole and trimethylchlorosilane. The derivatized extracts were injected through a splitless injection system onto a DB-5MS capillary column and into a MS operated in the electron impact (EI), SIM mode. The researchers quantified on the total ion chromatogram, monitoring from five to seven ions per analyte. Recoveries of DON were 88 to 97% at spike levels of 1000 to 10,000 ng/g, and the method sensitivity was reported as 50 ng/g for DON.

Onji et al. (1998) eliminated the usual derivatization step prior to the determination of eight trichothecenes. Barley samples were soaked in water and extracted with acetonitrile. Fatty acids in the crude extract were precipitated with zinc acetate, and the proteins were precipitated with ammonium sulfate. Combined acetonitrile extracts were evaporated to dryness and dissolved in ethyl acetate. After centrifuging, the supernatant was evaporated and redissolved in chloroform-hexane. The extract was further purified on a florisil column, evaporated to dryness and redissolved in acetone, ready for injection onto the GC-MS system. The system employed a cold on-column injector and was operated in the EI mode. Recoveries of 76% were reported for DON, with a quantitative detection limit of 100 to 500 ng/g in SIM.

Yoshizawa and Jin (1995) described the application of GC-MS to the determination of five type B trichothecenes in a survey of wheat and barley. Samples were extracted with acetonitrile-water, defatted with hexane, and cleaned up through a florisil column as described by Luo et al. (1990) before preparation of TMS ethers. The investigators employed capillary GC, operated the MS in the EI, SIM mode, and reported detection limits of about 5 ng/g.

American investigators (Hastings and Stenroos, 1995) have used GC-MS, also in the EI mode, to detect the presence of DON (and other *Fusarium* mycotoxins) in barley, malt and beer. Barley and malt were extracted with acetonitrile-water and cleaned up on a MycoSep 225® (Romer Labs Inc., Union, MO, USA) column, while beer was simply partitioned with ethyl acetate. TMS ethers were separated on a capillary DB-5 column. The MS was operated in SIM, monitoring 512 and 497 m/z for DON. The authors reported typical recoveries of 87 to

99% for barley and malt, and 67% for beer. Detection limits for DON were 190 ng/g for barley and malt, and 11 ng/g for beer.

GC-ECD Methods

GC-ECD has been employed to detect DON (and other trichothecenes) in cereal grains by many authors. Recently, German researchers (Walker and Meier, 1998) analysed wheat flour for the presence and levels of four type B trichothecenes. Samples were extracted with acetonitrile-water and purified on MycoSep® columns. Extracts were derivatized with heptafluorobutyric anhydride and injected in splitless mode onto a capillary column. Detection limits were reported as < 30 ng/g. Recoveries of 81 to 103% were reported, except for NIV, whose recoveries were much lower (45 - 52%).

Kotal and associates (1998) also used GC-ECD for the determination of trifluoroacetylated trichothecenes extracted from wheat. Acetonitrile-methanol extracts were cleaned-up by gel permeation chromatography. The authors reported minimum quantifiable levels of 40 to 200 ng/g, with recoveries ranging from 76 to 100%.

Tacke and Casper (1996) also employed an ECD for the detection of TMS derivatives of DON (and other type B trichothecenes, personal communication) in wheat, barley and malt. They extracted the toxins with acetonitrile-water and cleaned-up the crude extract on a column of C18 and alumina. Excellent recoveries (94 to 105%) were reported from all three commodities, with a limit of quantification of 200 ng/g.

GC-FID Methods

Although the ECD may be the most popular GC detector for trichothecenes, the flame ionization detector (FID) has also been used. Notably, Furlong and Soares (1995) used a FID for the analysis of DON (and other trichothecenes) in wheat. Samples were extracted with methanol-aqueous potassium chloride. The extracts were cleaned-up on a mixed charcoal-alumina-celite column and derivatized by HFBI. Confirmation of identity was accomplished by acetylation and reduction reactions; detection limits were reported as 50 - 100 ng/g.

LC METHODS

The polarity of the trichothecenes lends them readily to reversed phase (RP) LC (Figure 6). However, the weak UV absorbance of the type B trichothecenes, and the lack of a UV chromophore on the type A trichothecenes, limits the use of UV detection. Derivatization of type A trichothecenes can enable their detection with a fluorescence detector (FLD). Advances in LC-MS detectors have turned more researchers toward the inherent advantages of LC for the determination of the trichothecenes. The following is a discussion of only some of the excellent LC methods recently published.

LC-UV

Besides their work with GC-ECD methods, Walker and Meier (1998) also employed LC with a UV diode array detector for the analysis of trichothecenes in wheat flour. Following extraction with acetonitrile-water and purification on MycoSep® columns, the extracts were partitioned into ethyl acetate, evaporated to dryness and redissolved in acetonitrile-water. A gradient mobile phase of acetonitrile and water separated NIV, DON, 15-ADON and 3-ADON on a 250 x 4 mm, 5 μm RP column within 20 minutes. Quantitations were limited to samples containing greater than 1000 ng trichothecene/g. Recoveries by LC-UV were similar to the GC-ECD method, 49 to 55% for NIV and 92 to 105% for the other trichothecenes.

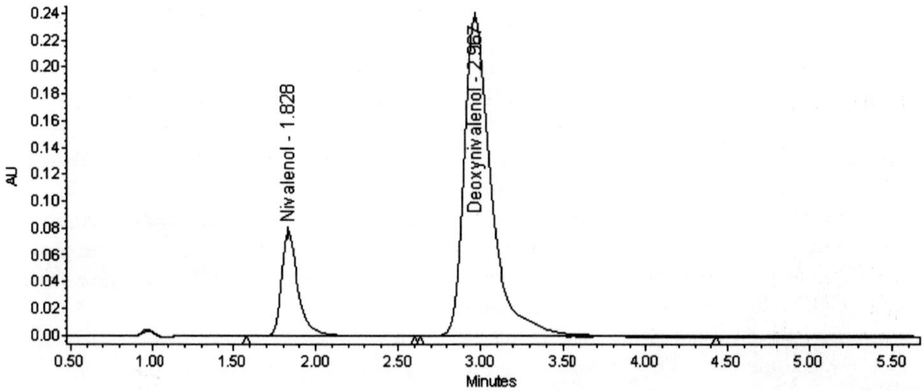

Figure 6. RP LC-UV (200nm) chromatogram of nivalenol (500 ng) and deoxynivalenol (1000 ng): Column - C18, 150 x 3.9 mm, 5μ; Mobile phase - acetonitrile : water : trifluoroacetic acid (1000+9000+1); Flow rate - 0.5 ml/min

LC-FLD

To detect type A trichothecenes, Jiménez and co-workers (2000) attached a fluorescent chromophore to T-2, HT-2, neosolaniol (NEO) and DAS prior to RP-LC. Reagent peaks in the chromatogram were minimised by partitioning with phosphate buffer. An isocratic mobile phase of acidified acetonitrile-water resolved the analytes on a 250 x 4 mm, 5 μm RP column. With the fluorescence detector excitation and emission wavelengths set at 292 and 475 nm, respectively, detection limits of 10 ng/g for T-2 toxin and 15 ng/g for the other trichothecenes were reported. The method was applied to the analysis of rice and corn cultures.

LC-MS

Both type A and B trichothecenes can be detected by MS in either positive or negative ion modes following separation by LC. Berger et al. (1999) used positive ion atmospheric pressure chemical ionization (APCI) for the determination of nine trichothecenes extracted from wheat and rice. Toxins were extract with acetonitrile-water, and isolated with MycoSep® columns. Separation was achieved in under 12 minutes on a RP column using a gradient of methanol/water. Recoveries from wheat ranged from 80% for NIV to 106% for NEO. The minimum quantifiable levels for the method ranged from 10 ng/g (HT-2 toxin) to 100 ng/g (NIV and 3-ADON).

Razzazi-Fazeli and colleagues (1999) applied negative ion APCI to the LC-MS detection of NIV and DON in wheat. The authors extracted with acetonitrile-water and purified the extracts sequentially on two different MycoSep® columns. Chromatographic separation was achieved on a RP column with an isocratic mobile phase of acetonitrile, methanol and water. Using SIM, minimum quantifiable levels were 40 and 50 ng/g for DON and NIV, respectively. The recoveries from wheat were reported as 70% for NIV and 86% for DON.

SUPERCRITICAL FLUIDS

In the past decade, little has been reported on the application of supercritical fluid extraction (SFE) and chromatography (SFC) to the determination of trichothecenes, although each appears to have significant potential. Considering the possible benefits of reduced solvent usage, unattended operation, and the elimination of additional clean-up steps, further research seems warranted.

A carbon dioxide (CO_2) supercritical fluid (modified with methanol) was described (Krska, 1998) for the extraction of DON from wheat. An exciting aspect of this report is that no further clean-up was necessary prior to analysis by either LC-UV or GC-ECD. The method extracted 53% of DON from wheat spiked at levels of 2000 to 16,000 ng/g.

The separation of seven trichothecenes on both packed and capillary columns using supercritical CO_2 has been demonstrated (Young and Games, 1992). Both FID (capillary column) and UV (packed column) detectors were employed. Much earlier, Smith and colleagues (1985) used supercritical CO_2 to separate trichothecenes on bonded phase capillary columns prior to either FID or NICI/MS detection. Sub-picogram detection limits were reported.

IMMUNOAFFINITY COLUMNS AND ENZYME-LINKED IMMUNOSORBENT ASSAYS

Immunoaffinity columns (IAC's) offer unprecedented specificity for clean-up of extracts prior to GC, LC or even UV analysis. Although columns for the isolation of DON from crude extracts are commercially available, their unique specificity reduces their usefulness for analysts interested in multiple trichothecenes. Nevertheless, the presence of DON, generally recognised as the most common of the trichothecenes, can be considered a marker for the potential presence of other trichothecenes. Ideally, an IAC with a broad cross-reactivity to the trichothecene family would be useful for the purification of food extracts prior to analysis by a GC or LC method such as those previously discussed.

Enzyme-linked immunosorbent assays (ELISA's) offer very specific binding and detection of individual trichothecenes without expensive GC or LC instrumentation, and kits are commercially available for at least two of the important trichothecenes (DON and T-2). As with IAC's, however, ELISA's have limited application in laboratories interested in the detection of multiple trichothecenes. An ELISA with wide cross-reactivity to multiple trichothecenes could be useful as a screening tool prior to a GC or LC method. Alternatively, a kit which could test a single sample extract against the antibodies of a number of common trichothecenes would also be of interest and use.

CONCLUSIONS

Developments in the determination of trichothecenes in foods continue to progress, with GC methods leading the way. LC methods, however, are gaining favour and advances in the associated detectors can be expected to contribute to more, and improved, LC methods for trichothecenes. Supercritical fluids offer an exciting alternative to current extraction and chromatographic techniques, though progress is slow to be reported. IAC's and ELISA's have a specific niche in the laboratory for isolating and detecting individual trichothecenes, but require changes in development and format before they will become of great use to investigators interested in the determination of multiple trichothecenes.

ACKNOWLEDGEMENTS

Many thanks to Peter Scott and Val Huzel for their useful comments and input. The opinions expressed in this publication are those of the author and do not necessarily reflect the official views of Health Canada.

REFERENCES

Berger, U., Oehme, M., and Kuhn, F., 1999, Quantitative determination and structural elucidation of type A- and B-trichothecenes by HPLC/ion trap multiple mass spectrometry, *J. Agric. Food Chem.* 47:4240

Cole R.J. and Cox R.H. , 1981, *Handbook of Toxic Fungal Metabolites*, Academic Press, New York, NY

FAO, 1997, *Worldwide Regulations for Mycotoxins, 1995. A Compendium, Food and Nutrition Paper 64*, Food and Agricultural Organisation, Rome

Fernando, D., 1999, Overview of the Fusarium situation in Canada, Canadian Workshop on Fusarium Head blight, Winnipeg

Furlong, E.B. and Valente Soares, L.M., 1995, Gas chromatographic method for quantitation and confirmation of trichothecenes in wheat, *J. AOAC Int.* 78:386

Hastings, D.J. and Stenroos, L.E., 1995, Determination of deoxynivalenol in barley, malt, and beer by gas chromatography-mass spectrometry, *J. Amer. Soc. Brew. Chem.* 53:78

Health Canada, 1999, *WPG-LB-19, Determination of trichothecene mycotoxins in cereal grains and foods by gas chromatography / negative ion chemical ionisation-mass spectrometry*, Health Canada, Health Protection Branch, Winnipeg, Canada

Jiménez, M., Mateo, J.J., and Mateo, R., 2000, Determination of type A trichothecenes by high-performance liquid chromatography with coumarin-3-carbonyl derivatization and fluorescence detection, *J. Chromatogr. A* 870:473

Kotal, F., Holadová, K., Hajšlová, J., Poustka, Z., and Radová, Z., 1998, Determination of trichothecenes in cereals, *J. Chromatogr. A*, 830:219

Krishnamurthy, T., Wasserman, M.B., and Sarver, E.W., 1986, Mass spectral investigations on trichothecene mycotoxins I. Application of negative ion chemical ionization techniques for the simultaneous and accurate analysis of simple trichothecenes in picogram levels, *J. Biomed. Envir. Mass Spectrom.*, 13:503

Krska, R., 1998, Performance of modern sample preparation techniques in the analysis of *Fusarium* mycotoxins in cereals, *J. Chromatogr. A*, 815:49

Luo Y., Yoshizawa, T. and Katayama, T., 1990, Comparative study on the natural occurrence of *Fusarium* mycotoxins (trichothecenes and zearalenone) in corn and wheat from high- and low-risk areas for human esophageal cancer in China, *Appl. Environ. Microbial.* 56:3723

Mirocha, C.J., Kolaczkowski, E., Xie, W., Yu, H. and Jelen, H., 1998, Analysis of deoxynivalenol and its derivatives (batch and single kernel) using gas chromatography/mass spectrometry, *J. Agric. Food. Chem.* 46:1414

Onji, Y., Aoki, Y., Tani, N., Umebayashi, K., Kitada, Y., and Dohi, Y., 1998, Direct analysis of several *Fusarium* mycotoxins in cereals by capillary gas chromatography-mass spectrometry, *J. Chromatogr. A.* 815:59

Razzazi-Fazeli, E., B□hm, J., and Luf, W., 1999, Determination of nivalenol and deoxynivalenol in wheat using liquid chromatography-mass spectrometry with negative ion atmospheric pressure chemical ionisation, *J. Chromatogr. A* 854:45

Scott, P.M., Kanhere, S.R., and Tarter, E.J., 1986, Determination of nivalenol and deoxynivalenol in cereals by electron-capture gas chromatography, *J. Assoc. Off. Anal. Chem.*, 73:503

Scott, P.M., Lombaert, G.A., Pellaers, P., Bacler, S., Kanhere, S.R., Sun, W.F., Lau, P.-Y. and Weber, D., 1989, Application of capillary gas chromatography to a survey of wheat for five trichothecenes, *Food Addit. Contam.*, 6:489

Schollenberger, M., Lauber, U., Jara, H.T., Suchy, S., Drochner, W., and Müller, H.-M., 1998, Determination of eight trichothecenes by gas chromatography-mass spectrometry after sample clean-up by a two-stage solid-phase extraction, *J. Chromatogr. A* 815:123

Smith, R.D., Udseth, H.R. and Wright, B.W., 1985, Rapid and high resolution capillary fluid chromatography (SFC) and SFC-MS of trichothecene mycotoxins, *J. Chromatogr. Sci.* 23:192

Stack, R., 1999, A summary of the USA situation on Fusarium Head blight, Canadian Workshop on Fusarium Head blight, Winnipeg

Sydenham, E.W. and Thiel, P.G., 1996, Physicochemical data for some selected *Fusarium* Toxins, *J. AOAC Int.*, 79:1365

Tacke, B.K. and Casper, H.H., 1996, Determination of deoxynivalenol in wheat, barley, and malt by column cleanup and gas chromatography with electron capture detection, *J. AOAC Int.* 79:472
Walker, F. and Meier, B., 1998, Determination of the *Fusarium* mycotoxins nivalenol, deoxynivalenol, 3-acetyldeoxynivalenol and 15-*O*-acetyldeoxynivalenol in contaminated whole wheat flour by liquid chromatography with diode array detection and gas chromatography with electron capture detection, *J. AOAC Int.* 81:741
Yoshizawa, T. and Zin, Y.-Z., 1995, Natural occurrence of acetylated derivatives of deoxynivalenol in wheat and barley in Japan, *Food Addit. Contam.,* 12:689
Young, J.C. and Games, D.E., 1992, Supercritical fluid chromatography of *Fusarium* mycotoxins, *J. Chromatogr.* 627:247

PROCESSING EFFECTS ON MYCOTOXINS: INTRODUCTION

Lloyd B. Bullerman

Department of Food Science and Technology
University of Nebraska-Lincoln
Lincoln, NE 68583-0919

The mycotoxins of greatest concern in foods are aflatoxins, ochratoxin A, fumonisins, zearalenone and the trichothecenes, particularly deoxynivalenol. The food commodities most commonly contaminated are peanuts and cereal grains, with cottonseed, green coffee and tree nuts also occasionally contaminated. Cereal grains, peanuts and green coffee are utilized for human foods as processed products. If the basic commodity is contaminated with a mycotoxin, subsequent processing of the commodity will have an effect on the stability and concentration of the mycotoxin. In general, mycotoxins tend to be quite stable compounds, but their stability and concentrations can be affected by different processing methods.

Food commodities may be processed by physical, chemical and biological means. Physical processes included physical separation by cleaning and milling processes, physical adsorption and various types of thermal processes. Thermal processes can range form relatively low temperature cooking processes to canning to the high temperatures achieved in roasting and in some extrusion processes. Extrusion processing is used extensively to cook and process cereal grains in the production of breakfast cereals, snack foods and pet foods. Extrusion processing combines high temperatures with high pressures and severe shear forces. These combinations of forces can act to destroy microorganisms, denature proteins and detoxify toxic substances. Extruders are now thought of as high-temperature short-time chemical and bioreactors that have the ability to process relatively dry viscous materials, such as cereal grains, grits and flours at moisture contents around 20-26%. There is evidence that the mycotoxins can be destroyed or transformed under these processing conditions. There is also evidence that certain adsorbent materials such as some clays can selectively adsorb and bind mycotoxins in certain substances and prevent their absorption in the gastrointestinal tracts of animals. Chemical processes that may destroy mycotoxins include reactions with ammonia, calcium hydroxide and certain sulfur containing compounds. Biological processes that may affect mycotoxin stability and concentrations include malting, brewing and fermentations. While some biological processes may reduce the concentrations of some mycotoxins, others such as malting of contaminated barley can cause mycotoxin concentrations to increase.

Mycotoxins and Food Safety, edited by Trucksess *et al.*
Kluwer Academic/Plenum Publishers, 2002

The effects of processing on mycotoxin stability and concentrations vary depending upon the process itself (some processes are more destructive than others), the individual mycotoxin and the effects of the matrix in which the toxin is found. Some mycotoxins are more stable than others and the matrix, especially after being heated, can bind mycotoxins, making recovery and analyses difficult and leading to erroneous results and conclusions in determining losses. All of these factors need to be studied and taken into account when evaluating the ability of processing to reduce mycotoxin concentrations.

In this section authors discuss the effects of various processing methods on mycotoxin stability, removal and reduction. The first paper deals with the use of clay as an enterosorbent to bind aflatoxin in the gastrointestinal tract to reduce absorption of the toxin by the animal body. A second paper discusses the effects of different physical and chemical methods on aflatoxin concentrations and stability in a number of products. In the third paper milling processes and their effects on deoxynivalenol in both wheat and corn are discussed. The fourth paper discusses agricultural practices, storage and transportation practices and processing, particularly roasting on the accumulation and reduction of ochratoxin A in coffee. A fifth paper reviews the effects of milling and thermal processes, including extrusion processing, on the stability of fumonisins in aqueous solutions and corn-based foods. Similarly, the sixth paper discusses the stability of zearalenone in several thermal processes. The final paper in this section discusses the stability of deoxynivalenol during malting, brewing and fermentation.

CHARACTERIZATION OF CLAY-BASED ENTEROSORBENTS FOR THE PREVENTION OF AFLATOXICOSIS

Timothy D. Phillips[1*], Shawna L. Lemke[2], and Patrick G. Grant[3]

[1]Faculty of Toxicology (VAPH), College of Veterinary Medicine, Texas A&M University, College Station, Texas, 77843-4458
[2]Department of Nutrition, University of California, Davis, CA 95616
[3]Lawrence Livermore National Laboratory, Center for Accelerator Mass Spectrometry, Livermore, CA 94550
* to whom correspondence should be addressed

ABSTRACT

Appropriate chemical interventions that can block, retard, or significantly diminish foodborne exposure to aflatoxins are high priorities. A practical and effective approach to the aflatoxin problem has been the dietary inclusion of a processed calcium montmorillonite clay (HSCAS). HSCAS acts as an enterosorbent that rapidly and preferentially binds aflatoxins in the gastrointestinal tract resulting in decreased aflatoxin uptake and bioavailability. In mechanistic studies, we have shown that the intact dicarbonyl system in aflatoxin is essential for optimal sorption by HSCAS. Evidence also suggests that aflatoxins react at multiple sites on HSCAS clay surfaces (especially those within the interlayer region). Due to conceivable risks associated with the dietary inclusion of nonspecific binding agents, all aflatoxin enterosorbents should be tested in sensitive animal models for efficacy, safety, and the potential for nutrient interactions.

INTRODUCTION: AFLATOXIN PROBLEM IN PERSPECTIVE

In historical context, the aflatoxin problem in foods is longstanding, unavoidable and seemingly inextricable. Aflatoxins are harmful by-products of mold growth and (though invisible to the naked eye), are potentially fatal to humans and animals (CAST, 1989; Phillips et al., 1995; 1997; 1998; Phillips, 1999). The aflatoxins are primarily produced by *Aspergillus flavus* and *A. parasiticus* fungi (Diener et al., 1983) and to a lesser extent by *Aspergillus nomius* (Kurtzman et al., 1987). These fungi are widespread and especially a problem during extended periods of drought. Thus, drought stress can be a frequent harbinger of intensified production of aflatoxins in the field. Also, the aflatoxins are heat stable and can survive a variety of food processing procedures and can occur as "unavoidable" contaminants of most foods (particularly those derived from maize and

Mycotoxins and Food Safety, edited by Trucksess *et al.*
Kluwer Academic/Plenum Publishers, 2002

peanuts). Animals can also secrete carcinogenic metabolites of aflatoxin in their milk. Consequently, a variety of dairy products (including cheese and ice cream) can be contaminated with these hazardous chemicals (CAST, 1989; Phillips et al., 1994).

Aflatoxin B_1 (AfB_1), the most toxic of four naturally occurring aflatoxins, is a very potent carcinogen (IARC, 1976, 1987), and has been strongly implicated in the etiology of disease and death in man and animals (Figure 1). Aflatoxin B_1 is a direct acting mutagen and has been shown to disrupt genes involved in carcinogenesis (McMahon et al., 1986; 1987) and tumor suppression (Aguilar et al., 1993). In addition to DNA damage, AfB_1 interacts with RNA and intercellular proteins. The latter occurs through a reaction of the phenolate ion resonance form to produce a Schiff base with protein amino groups (e.g., lysine). These interactions may explain some of the non-genotoxic effects of aflatoxin (Eaton et al., 1994). There is substantial evidence that low-level exposure to aflatoxin may cause suppression of the immune system and increased suceptibility to disease (Richard et al., 1978; Peska and Bondy, 1994). Aflatoxin is also excreted in mother's milk and increases the morbidity of children with kwashiorkor (Adhikari, 1994). Clearly, the young of all species are more sensitive than adults to aflatoxin. What is not clear is how these numerous actions of aflatoxins may contribute to commonly perceived health problems in young humans and animals. Consequently, safe and sustainable dietary interventions that can prevent or reduce exposure of humans and animals to the aflatoxins are highly desirable.

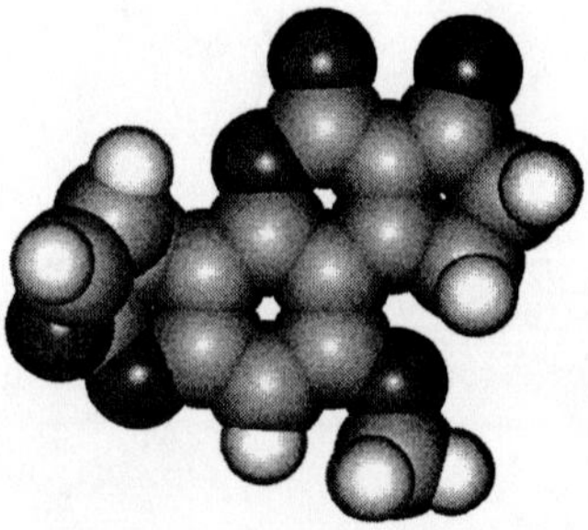

Figure 1. Molecular model of aflatoxin B_1 showing the spacial orientation of functional groups, including the dicarbonyl system (Key: light gray, hydrogen; medium gray, carbon; dark gray, oxygen).

Since earliest recorded history, the unique properties and beneficial uses of clay minerals have been well-documented, and in many instances animals and humans have been reported to eat clay for the prevention of various illnesses (Munn, 1984; Phillips et al., 1995; Ramu et al., 1997; Clark et al., 1998; Phillips, 1999). Many of these clays vary considerably in their structures and ability to sequester toxic chemicals. This chapter will serve to review a simple, clay-based strategy for the prevention of aflatoxicosis in animals and further delineate the surface chemistry associated with aflatoxin sorption to HSCAS clay.

CLAY AND ZEOLITIC MINERALS

Clay Mineral Chemistry

Soil, the loose outer layer of the surface of the earth, is a surprisingly complex mixture. By volume, slightly less than 50% of soil is made up of mineral matter while another 50% is the pore space between particles, filled with water or air. Soil can also contain a variable amount of organic matter (0.5 – 10%) and supports a diversity of living organisms such as bacteria, actinomycetes, fungi, algae and protozoa, which together make up much less than 1% of the soil. The solid particles of soil are classified into three categories based on size: sand (0.05 – 2 mm), silt (0.002-0.05 mm) and clay (less than 2 μm). The relative contribution of each type of particle to a particular soil determines its texture and other physical attributes and is used to name soil classes (Sylvia et al., 1997).

Soil Components

Soil minerals are composed of relatively few elements. Oxygen and silicon are by far the most abundant components; silicates constitute 70-90% of the total mineral mass. Aluminum and iron are the next most abundant elements as they also participate in a variety of complexes with oxygen that form crystalline minerals. Smaller quantities of magnesium, calcium, sodium, potassium, titanium, manganese, nitrogen, phosphorus and sulfur are also present (Sylvia et al., 1997). Mineral classes are divided based on the identity of the dominant anionic group. Major mineral groups include: sulfides, oxides and hydroxides, halides, carbonates, nitrates, borates, phosphates, sulfates, tungstates and silicates. Within these classes, minerals are further subdivided based on structural similarities. The silicates are by far the largest class based on both number of mineral types they form and their overwhelming contribution to total mass of the earth's crust (Schulze, 1989). The basic structural unit for the silicates is a SiO_4 tetrahedron in which Si^{4+} is located at the center and four O^{2-} are positioned at the apices. Tetrahedra can be linked together by sharing O^{2-}, and together can form a variety of more complex structures including rings (cyclosilicates), chains (inosiliates), sheets (phyllosilicates) and three-dimensional arrangements (tectosilicates) (Schulze, 1989).

Phyllosilicate Clay Minerals

The phyllosilicates are a unique class of minerals that have been incorporated into a variety of production processes. Kaolinite is widely used in the ceramic industry and serves as a filler and coating for paper. Vermiculite, which becomes very light and porous when heated, is used as filler for concrete and also serves as a thermal and sound insulator. Montmorillonites have been employed in industry as decolorizing agents and for filtration of water and food products (Millot, 1979). The functionality of this class of minerals is a result of the distinctive structural and chemical properties of silicate layers. Phyllosilicates contain both tetrahedral and octahedral sheets. The tetrahedral sheets are composed of SiO_4 tetrahedra linked together. Each tetrahedron shares three O^{2-} ions with three adjacent tetrahedra. This extends in all directions, forming a plane of basal oxygens. The fourth O^{2-} of each tetrahedron is referred to as the apical oxygen and is free to bond to other elements. The octahedral sheet is comprised of two planes of OH^- groups that form a hexagonal closest packing arrangement. To counter the negative charge of this structure, cations are located in the octahedral spaces between the layers. There are two possible ways to do this: a divalent cation (e.g., Mg^{2+}) must fill every available octahedral space to produce a trioctahedral arrangement, or a trivalent cation (e.g., Al^{3+}) fills two out of every three spaces to produce a dioctahedral arrangement. In a phyllosilicate, apical oxygens from the tetrahedral layer that replace OH- groups from the octahedral layer and coordinate with the metal cation of the octahedral layer link the octahedral and tetrahedral layers. This commonly occurs in one of two ways: the 1:1 layer structure, in which one tetrahedral layer is bonded to one octahedral layer, and the 2:1 layer structure, in which an octahedral layer is bound on either side by a tetrahedral layer (Schulze, 1989). Frequently, cations in either the tetrahedral or octahedral layers are missing or have been replaced through an isomorphous substitution with another cation of lesser charge. The most common substitutions are a replacement of Si^{4+} in the tetrahedral layer with Al^{3+} and a replacement of Al^{3+} or Fe^{3+} in the octahedral layer with Mg^{2+}. These substitutions result in a permanent negative charge on the phyllosilicate mineral (McBride, 1989). The amount of charge is dependent on the amount of isomorphous substitution. Variations in this rate and the identity of the substituted cations create the individual phyllosilicate types. To counteract the negative charge, these minerals attract cations such as Na^+ and Ca^{2+} into the region between the layers (i.e., the interlayer). The amount of charge necessary to balance the negative charge of the clay is referred to as the cation exchange capacity (CEC). The interlayer cations are subject to the formation of layers of hydration, so when wetted, water enters the interlayer regions of the clay and causes it to swell (Bohn et al., 1979).

Phyllosilicates are grouped by the amount of substitution, and therefore, charge per layer. Of the 2:1 phyllosilicates, talc and pyrophyllite have almost no substitution and therefore very little charge; smectites have a layer charge per formula unit of 0.25 to 0.6; vermiculites have 0.6 to 0.9; and micas have complete substitution, resulting in a charge of 1 per formula unit (Bohn et al., 1979) (Table 1). The intermediate charge of smectites gives them optimal swelling properties. These clays attract cations into the interlayer, which hydrate and swell, expanding the space between layers. However, the negative surface charge of the mineral layer is not so high as to limit swelling due to electrostatic attraction to the cations. The montmorillonites are a subclass of smectites that contain isomorphous substitution in the octahedral sheets, resulting in a general formula of $Na_x[(Al_{2-x}Mg_x)Si_4O_{10}(OH)_2]$. These clays are naturally abundant and have surface areas as high as 800 m^2/g, making them ideal sorbent materials (Borchardt, 1989).

Table 1. Classification of 2:1 Phyllosilicate Clays Based on Layer Charge

	Layer Charge per Formula Unit		Predominant Octahedral Cation	
Mineral Group	Tetrahedral Sheet	Octahedral Sheet	Dioctahedral (Al^{3+})	Trioctahedral (Mg^{2+})
Pyrophyllite/Talc	0	0	Pyrophillite	Talc
Smectites	0.25 to 0.6	0	Beidellite	Saponite
	0	0.25 to 0.6	Montmorillonite	Hectorite
Vermiculites	0.6 to 0.9	0	Vermiculite	Vermiculite
Micas	1	0	Muscovite	Biotite

Adapted from Bohn et al. (1979) and Lemke (2000).

Zeolites

Zeolites fall under the class of tectosilicates, containing three-dimensional frameworks of SiO_4 tetrahedra. These minerals also undergo substitution of Si^{4+} with Al^{3+}, resulting in a net negative charge that is balanced by exchangeable cations. Clinoptilolite is a member of the zeolite class and possesses the general structure of $(Na_3K_3)[Al_6Si_{30}O_{72}]24H_2O$, where Na^+ and K^+ are exchangeable cations. Like the phyllosilicates, zeolites are also adsorbent materials due to their large internal surface and void volume (Ming and Mumpton, 1989).

Sorption of Chemicals onto Clay Surfaces

The mechanisms by which compounds bond to or are retained by soil particles, and in particular, clays, have been extensively researched. Sorption can be loosely classified as one of two types: physical/chemical bonding to soil mineral surfaces or hydrophobic partitioning onto soil organic matter (Koskinen and Harper, 1990). A number of physical/chemical bonding mechanisms have been characterized that vary based on the strength of interaction. Van der Waals forces are short-range bonds that result from the interaction of natural or induced dipoles. London forces are a subclass of the van der Waals forces resulting from vibrational dipoles. These two types of interactions are relatively weak (2-4 kJ/mol) and decay rapidly with increasing distance, but can be important for large compounds with a sizeable surface area. A variation on dipole interaction is the hydrogen bond, which results from electrostatic attraction between a hydrogen and the electron pairs of an electronegative atom such as oxygen, nitrogen or sulfur. These forces are slightly stronger than the van der Waals forces (2-60 kJ/mol). The aluminosilicate clay surfaces contain numerous exposed OH groups from octahedral surfaces and edge sites (protonation is pH dependent), making hydrogen bonding an important mechanism for adsorption on these surfaces. A final relatively weak interaction is cation and water bridging. Anionic or polar compounds may interact through electrostatic attraction to exchangeable cations. If the compound cannot displace the waters of hydration surrounding a cation, the interaction is referred to as water bridging (Koskinen and Harper, 1990).

Electrostatic and covalent interactions provide stronger bonds between solutes and surfaces. In anion exchange, an anion becomes electrostatically attracted to an exposed cation at the surface or edge. Ligand exchange involves covalent binding of a compound with a metal ion at the surface. This generally occurs with a carboxylate or hydroxyl functional group that can displace one of the natural surface hydroxyl groups. Cation exchange involves the exchange and retention of a positively charged compound onto the negative clay surface. Protonation involves the same mechanism of exchange, but relies on the mineral surface (assisted by low pH) to transfer a proton to a compound, producing a cation (Koskinen and Harper, 1990).

Alternatively, a variety of compounds become sorbed to soil through partitioning. The organic matter found in soil is a combination of decomposition products from plants, animals and microbes such as carbohydrates, proteins, lignin, fats and waxes. Over time, these materials undergo enzymatic and chemical reactions and become incorporated into humus, a colloidal polymer. This organic matter is important for the structure of the soil as well as buffering pH and increasing the water holding capacity. The identity of humus has not been fully characterized; rather, humus is observationally divided into constituents based on a pH fractionation and relative oxygen and carbon contents (Bohn et al., 1979). These non-humic and colloidal humic substances in the soil are able to interact with other non-polar compounds through hydrophobic sorption (Senesi, 1993; Fetter, 1993).

CLAY-BASED INTERVENTIONS FOR AFLATOXIN CONTROL

The strategy of reducing foodborne exposure to aflatoxins through the inclusion of various binding agents or "detoxifying clays" in the diet has been given considerable attention in the scientific literature. Many of these binding agents have been reported to prevent the deleterious effects of diverse mycotoxins in a variety of animals, apparently by diminishing toxin uptake and distribution to the blood and target organs. As early as 1979, adsorbent clay minerals were reported to bind aflatoxin B_1 in liquids (Masimango et al.,

1979). Also, bleaching clays, that had been used to process canola oil, were found to lessen the effects of T-2 toxin (Carson and Smith, 1983; Smith, 1984).

Enterosorption of Aflatoxins (HSCAS clay)

In the first enterosorbent study with aflatoxins, HSCAS (NovaSil^{TM}), a calcium montmorillonite clay that is used as an anticaking additive for animal feeds, was reported to significantly adsorb aflatoxin B_1 with high affinity and high capacity in aqueous solutions and to protect broiler and Leghorn chicks from the toxic effects of 7,500 ppb aflatoxins in the diet (Phillips et al., 1987, 1988). Since this initial study, HSCAS and other similar montmorillonite clays have been reported to diminish the toxic effects of aflatoxins in a variety of young animals including rodents, chicks, turkey poults, ducklings, lambs, pigs, and mink (Phillips et al., 1990, 1991, 1994, 1995; Colvin et al., 1989; Bonna et al., 1991; Harvey, 1994; Harvey et al., 1989, 1991a, 1991b, 1993; Voss et al., 1993; Kubena et al., 1990a, 1990b, 1991, 1993; Ledoux et al., 1999; Smith et al., 1994; Marquez and Hernandez, 1995; Cerdchai et al., 1990; Lindemann et al., 1993; Abdel-Wahhab et al., 1998; Nahm, 1995; Jayaprakash et al., 1992). Importantly, HSCAS clay has also been shown to decrease the bioavailability of radiolabeled aflatoxins and to reduce aflatoxin residues in poultry (Davidson et al., 1987; Jayaprakash et al., 1992), rats (Sarr et al., 1995) and pigs (Beaver et al., 1990). Levels of aflatoxin M_1 in milk from lactating dairy cattle and goats were also diminished in the presence of HSCAS in the diet (Ellis et al., 1990; Smith et al., 1994; Harvey et al., 1991b).

Specificity of HSCAS for Aflatoxins

Recent studies (*in vivo*) have supported our earlier findings (*in vitro*) that HSCAS has a notable preference (and capacity) for the aflatoxins at levels in the diet at, or below, 0.5% w/w. For example, HSCAS at a level of 0.5% in the diet of poultry, did not impair phytate or inorganic phosphorous utilization (Chung and Baker, 1990). In other poultry nutrition studies, the addition of HSCAS at concentrations of 0.5 and 1.0% did not impair the utilization of riboflavin, vitamin A, or manganese; however, there was a statistically significant reduction in zinc utilization in the presence of 1.0% clay (Chung et al., 1990). It is worthy to note that 0.5% or lower (not 1.0%) of HSCAS is recommended for anticaking activity. Also, in earlier studies, HSCAS (at an inclusion rate of 0.5%) has been shown to protect young chickens from very high levels of aflatoxins (i.e., 7,500 ppb) which represents a "worst-case scenario" for enterosorption.

While clay-based interventions are clearly effective for aflatoxins, an analogous technology is not yet available for other important mycotoxins. For the most part, unmodified HSCAS clays do not "tightly" bind other structurally diverse mycotoxins, e.g., zearalenone, deoxynivalenol, T-2 toxin, ochratoxin A, cyclopiazonic acid, ergota-mine, and fumonisins, nor do they significantly prevent the adverse effects of these mycotoxins when included in the diet of animals. For example, in enterosorbent studies in poultry with mycotoxins (other than the aflatoxins), the inclusion of HSCAS clay in the diet did not significantly prevent the adverse effects of cyclopiazonic acid (Dwyer et al., 1997), T-2 toxin (Kubena et al., 1990a), diacetoxyscirpenol (Kubena et al., 1993), ochratoxin A (Huff et al., 1992), and fumonisins (Lemke, 2000).

The use of HSCAS in mink fed zearalenone helped to alleviate some fetotoxicity but did not reduce the hyperestrogenic effects (Bursian et al., 1992). Also, the addition of HSCAS at 0.5 and 1.0% w/w in the diet, did not influence the average daily gain of pigs exposed to deoxynivalenol. Dilution of the contaminated maize with uncontaminated maize was the only efficacious method for decreasing the toxicity of deoxynivalenol (Patterson and Young, 1993). The possibility of supplementing livestock diets with HSCAS clay to protect from fescue toxicity has also been investigated (Chestnut et al.,

1992). Although *in vitro* experiments predicted good binding of ergotamine to montmorillonite clays in aqueous solution (Chestnut et al., 1992; Huebner et al., 1999), HSCAS (at levels of 2.0% by weight) did not protect rats or sheep from fescue toxicosis and impaired the absorption of magnesium, manganese, and zinc at this high level. Even though lower levels of HSCAS may have less of an effect on these minerals, further work is warranted to determine the dosimetry of this effect and the potential for nutrient interactions in livestock.

MECHANISMS OF AFLATOXIN SORPTION ONTO HSCAS

Earlier studies *in vitro* have assessed the sorption of aflatoxins onto the surface of HSCAS clay (Ellis, 1990, 1991; Sarr, 1992; Ramos and Hernandez, 1996; Grant, 1998; Grant and Phillips, 1998; Phillips et al., 1995; Phillips, 1999). HSCAS, in aqueous solution, was shown to tightly and preferentially bind AfB_1 and similar analogs of AfB_1 that contain an intact β-dicarbonyl system (Phillips et al., 1988; Sarr, 1992).

Prior to isothermal analysis, the physical characteristics of AfB_1, HSCAS, and heat-collapsed HSCAS were confirmed (Grant and Phillips, 1998) (Table 2). The octanol water partition coefficient (K_{ow}) for AfB_1 was estimated by two separate methods, including an energy-minimized molecular model and HPLC analysis. In the scientific literature, the values for the solubility of AfB_1 in water vary from 10-30 μg/ml (Busby and Wogan, 1984). Using estimated log K_{ow} values and the molecular weight and melting point of AfB_1, a solubility range of 11-33 μg/ml was derived from equations described by Meylan and coworkers (1996). The total organic carbon (TOC) in HSCAS was determined to be < 0.05% using a Leco model 523-300 induction furnace. Following the removal of carbonates, organic carbon was converted into CO_2 which was measured with a Horiba IR detector and HP 3369a integrator. The negligible amount of TOC and the specificity of aflatoxin sorption onto the surfaces of HSCAS further confirm that organic partitioning is not an important mechanism in the aflatoxin enterosorption process and protection of animals *in vivo*.

Table 2. Physical characteristics of AfB_1, HSCAS and collapsed HSCAS

Aflatoxin B_1	
Log K_{ow}	1.46 and 1.98
Solubility	11 to 33 μg/mL
ϵ	21,865 [1/(Mcm)]
Vertical cross-sectional area	52.8 $Å^2$
Horizontal cross-sectional area	88.3 $Å^2$
HSCAS	
Total surface area	848 m^2/g
External surface area	70 m^2/g
Total organic carbon	<0.05%
Collapsed HSCAS	
Total surface area	77 m^2/g

HSCAS was found to have a high total surface area of 848 ± 11 m^2/g by the ethylene glycol adsorption method (Dyal and Hendricks, 1950). The surface area was calculated based on the relationship that 3.1 x 10^{-5} g of ethylene glycol covers each square meter of surface. The external surface area of HSCAS was measured by N_2 adsorption and indirectly by measuring heat-collapsed HSCAS with ethylene glycol. Collapsed HSCAS was prepared by heating HSCAS clay to 200°C for 30 minutes and then heating the same sample at 800°C for one hour. The collapsing of the interlamellar region leaves only the exterior surface available for ethylene glycol adsorption. The two methods resulted in values for the external surface area equal to 70 and 77 $\pm$ 2 m^2/g, respectively (Table 2). The near agreement between the two methods confirmed the collapsing of HSCAS and the loss of the interlamellar region in the collapsed clay.

The isotherm shape for the sorption of AfB_1 onto HSCAS can be categorized as an L1 or L2 plot that is reaching (or has reached) a plateau. The maximum amount of AfB_1 that was sorbed by HSCAS was 0.336 mol/kg which was 72.9% of the theoretical maximum capacity (Q_{max}) of 0.461 mol/kg derived from the fitting of the Langmuir model to the data (Grant and Phillips, 1998). The shape of the isotherm plot of AfB_1 binding to collapsed HSCAS was an L2 that had a capacity notably smaller (i.e., 0.0567 mol/kg) than untreated HSCAS (i.e., 0.461 mol/kg). These findings suggested that a significant portion of the binding of AfB_1 to HSCAS is within the interlayer region of the clay. The fitting of the data to the Toth isotherm model gave an exponent of 0.64, indicating more than one type of site for the sorption of AfB_1 and supports the data of Ramos and Hernandez (1996), which suggested that different sites and/or mechanisms of action were involved in aflatoxin binding at the surface of a montmorillonite clay. The Langmuir model was also used to estimate the Q_{max} at different isotherm temperatures and to calculate individual distribution constants (K_ds) to use in enthalpy calculations. The enthalpy of AfB_1 sorption (near or above –40 kJ/mol) showed some variation, suggesting multiple sites on HSCAS with dissimilar thermodynamic properties. Based on this value, it is conceivable that multiple sites on the surface of HSCAS clay act to chemisorb AfB_1.

In previous work, the strength of adsorption to HSCAS had been quantified utilizing an adsorption index (C_α) (Sarr, 1992). The C_α index expressed the ratio of the retained amount of ligand (difference between the amount bound and the amount desorbed) to the initial concentration of ligand. Based on data from these early studies, a proposed mechanism of action was the formation of a stable chelate with metals ions in HSCAS. However, a comparison of C_αs indicated a significant difference in ligand adsorption to HSCAS, even when the ligands possessed a similar structure and the same number of carbonyl functional groups. It was also observed that two pairs of analogs had the same dicarbonyl system, but a slight difference in stereostructure which resulted in a dramatic change in the C_α values. These included: AfB_1/aflatoxin M_1 and aflatoxicol I (AfTI)/aflatoxicol II (AfTII)) (Figure 2). These findings suggested a mechanism involv-ing more than the dicarbonyl system and led to further investigation of AfB_1 and analogs to determine if there was a structural correlation between this set of compounds and the C_α data (Grant, 1998). The partial charges of AfB_1 and the various analogs were estimated by drawing the chemical structures in ISIS Draw 2.0 (MDL Information Systems, San Leandro, CA) and then importing them into HyperChem 4.5 (HyperCube, Waterloo, Ontario, Canada). The structures were energy-minimized using the semi-empirical quantum mechanical AM1 method, which is an improvement of the MNDO method (Dewar, 1985; HyperChem, 1994). The structural information was then imported into the ChemPlus module for the determination of K_{ow}. ChemPlus utilizes previously derived atomic parameters to estimate each individual atom's contribution to the molecule's Log K_{ow} (ChemPlus, 1993; Ghose et al., 1988; Vellarkad et al., 1989). These analogs of aflatoxin included: aflatoxin B_2, G_1, G_2, G_{2a}, M_1, P_1, Q_1 and aflatoxicol I and II, coumarin, dimethoxycylopentenon[2,3-*c*]coumarin, dimethoxycyclopenteno[*c*]cou-marin, esculetin, 4-methylumbelliferone, tetrahydrodeoxy-aflatoxin B_1 and xanthotoxin. ChemPlus was used to measure the cross-sectional areas of AfB_1 and AfM_1. The van der Waals radii of C, O, and H were set to 1.85, 1.40, and 1.20 Å, respectively (Emsley, 1991). The structures were oriented on edge with the dicarbonyl in view and planar with the dihydrofuran in view. The cross-sectional area method was modified by the use of a carbon atom as the reference for the area calculations (Gray et al., 1995). The relative surface coverages were constructed by orienting the test ligand in the proposed orientation with a *periodic box* scribing out the relative surface area. Since the unit cell coordinates for HSCAS clay are not available in the literature, the model for HSCAS was constructed from unit cell coordinates of muscovite, a structurally similar 2:1 phyllosilicate clay (Radoslovich, 1960). The unit cell coordinates were converted to orthogonal coordinates and replicated to form an individual platelet. The platelet was duplicated and arranged to a d_{001} spacing (19 Å) of fully hydrated calcium substituted montmorillonite clay (MacEwan, 1980).

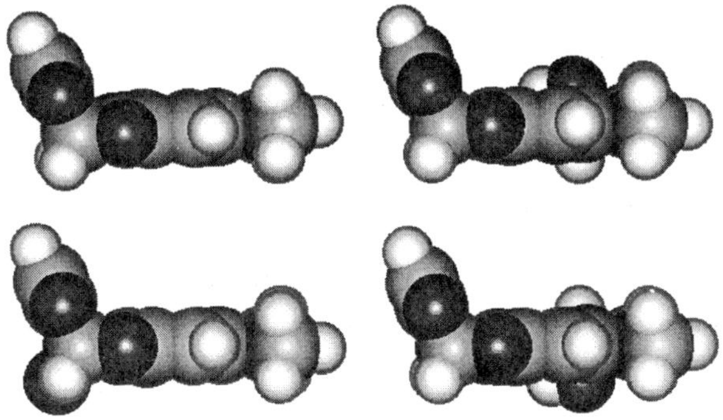

Figure 2. Molecular models of AfB_1 (upper left), AfM_1 (lower left), AfTI (upper right), and AfTII (lower right) illustrating important stereochemical differences.

In structure-activity studies, a log-linear plot of the index of adsorption and the calculated K_{ow} for individual aflatoxins and analogs indicated a lack of correlation. These results confirmed that hydrophobicity of AfB_1 and similar ligands was not critical to the mechanism of sorption onto HSCAS. Molecular models were constructed to estimate the relative surface coverage of the adsorbent based on the possible locations and orientations of the adsorbed ligand. The Q_{max} which was obtained from fitting the Langmuir equation in isotherms, was applied to the total, internal and external surface areas in order to calculate the amount of coverage of AfB_1 on HSCAS clay. These values were used to predict the significance of adsorption in the interlamellar region and to indicate whether a multi-layer formation was possible. The surface area of interest was divided by the number of molecules adsorbed by 1.0 kg of either HSCAS, collapsed-HSCAS or the difference between HSCAS and collapsed-HSCAS to yield the available surface area per adsorbed molecule. This value was then compared to the cross-sectional areas of the probable orientations of adsorbed molecules. The vertical orientation was defined as the dicarbonyl system bound to the clay, since this moiety was found to be important in previous research (Sarr, 1992). The horizontal orientation was defined as the molecule lying planar on the surface with the dihydrofuran away from the surface (resulting in minimal steric hindrance to docking). The total surface area of HSCAS relative to the capacity of the adsorbed AfB_1 molecules was calculated to be 305 Å^2/molecule. This was the area that could bind the the AfB_1 molecule in either orientation without requiring a multi-layer coverage (Grant and Phillips, 1997). Even though the vertical orientation of AfB_1 would be less than a mono-layer, its insertion in the interlamellar space would require a slight tilt from the basal surface. The outer area relative to the amount of AfB_1 bound is 25.2 Å^2/molecule which is smaller than the area required by either orientation (Grant and Phillips, 1998). This confirms that an intact interlamellar region is important in the adsorption of AfB_1 which is also supported by the findings with collapsed-HSCAS. The relative surface coverage for the interlamellar region was calculated by difference between the surface areas and capacities of HSCAS and collapsed-HSCAS. The results of this calculation show that the interlamellar region would have enough surface area to adsorb AfB_1 based on its estimated capacity.

Insight into potential mechanisms for the adsorption of AfB_1 onto the surface of HSCAS came from the observation that stereochemical differences between some of the aflatoxin analogs resulted in a significant effect on the C_α (even though the carbonyl functional groups were identical). AfB_1 and analogs are relatively planar compounds except for the dihydrofuran groups. Interestingly, the analogs that possess a functional group extending out of the major plane of the molecule in the opposite direction of the furan group have a significantly lower C_α and Q_{max} versus similarly structured compounds. For example, AfB_1 ($C_\alpha = 0.93$, $Q_{max} = 0.461$ mol/kg) and AfM_1 ($C_\alpha = 0.71$, $Q_{max} = 0.157$ mol/kg) have the same dicarbonyl structure, and aflatoxicol I ($C_\alpha = 0.43$) and aflatoxicol II ($C_\alpha = 0.18$) differ only in the orientation of the hydroxyl group (Sarr, 1992; Grant and Phillips, 1997; Ellis, 1991; Ellis, 1994). Aflatoxicol (like AfM_1) contains a hydroxyl group pointing out of the plane away from the dihydrofuran group. These results also suggest that the molecular mechanism for the adsorption of aflatoxins onto HSCAS may favor an optimal orientation where the furan is aligned away from the surface (Figure 2).

AfB_1 is strongly bound (chemisorbed) to HSCAS. A potential chemical reaction that may explain these results is an electron donor acceptor (EDA) mechanism. This mechanism involves sharing electrons from the negative surface of the clay with atoms in the adsorbed molecule that are partially positive (Haderlein, 1996). The carbons comprising the dicarbonyl system in aflatoxins are partially positive (electron poor) and have also been shown to be essential to the adsorption process. When the summation of partial charges of the carbons of the carbonyl functional groups for each ligand was plotted versus the C_α there was a significant correlation (Figure 3). When the ligands that were not planar on the side of the molecule opposite the dihydrofuran functional group were removed from the set of test compounds, the correlation was significantly improved (Figure 4).

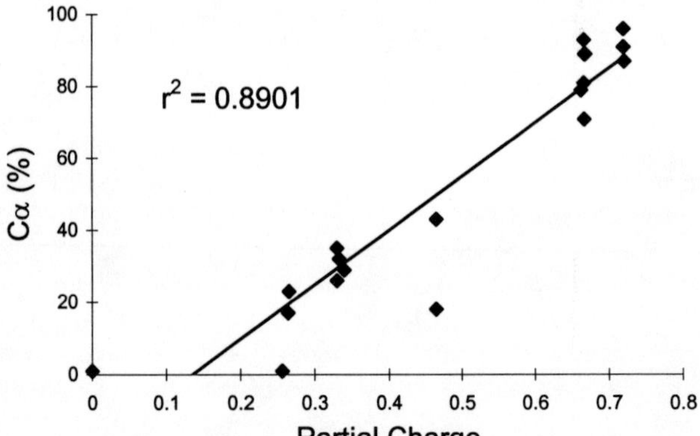

Figure 3. Plot of % chemisorption of ligand onto the surface of HSCAS clay versus the summation of partial charges of the carbons of the carbonyl functional groups of ligands, i.e., aflatoxin B_1 and various analogs. Partial charges: AfB_1 (.67); AfB_2 (.67); AfG_1 (.72); AfG_2 (.72); AfG_{2a} (.72); AfM_1 (.67); AfP_1 (.67); AfQ_1 (.66); AfTI (.47); AfTII (.47); Coumarin (.33); DMCI (.67); DMCII (.26); Esculetin (.33); 4MU (.34); $THDAfB_1$ (.27); Xan (.33). Adapted from Grant, 1998.

Interference from compounds with stereochemical restrictive groups could also play an important role in the adsorption process. For the analogs that contain functional groups that make them thicker than AfB_1, their insertion, docking and adsorption at surfaces in the interlamellar channel might be restricted. In summary, HSCAS clay has been previously shown to protect animals from aflatoxins when included in contaminated feed. With molecular modeling we have demonstrated that this protection may be attributed to the chemisorption of aflatoxins at surfaces within the interlamellar region of HSCAS; the exterior surfaces of the clay were responsible for only minor sorption of aflatoxins. The optimal orientation of the AfB_1 molecule is probably planar on the interlamellar surface. Our results also indicate a good correlation between the magnitude of partial positive charges on carbons C11 and C1 of the ß-dicarbonyl system and the strength of adsorption of planar ligands, suggesting an EDA mechanism with the surface of the clay. Functional groups on the aflatoxin analogs may cause steric hindrance to the adsorption at the surface of HSCAS or may block adsorption by interacting across the interlamellar region. Other mechanisms of AfB_1 sorption to HSCAS surfaces involve the potential chelation of interlayer cations (especially Ca^{2+}) and various edge-site metals (data not shown) (Grant, 1998; Phillips, 1999).

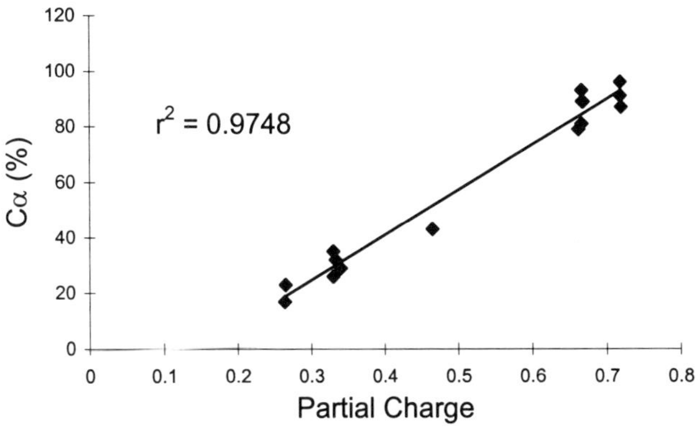

Figure 4. Plot of % chemisorption of ligand onto the surface of HSCAS clay versus the summation of partial charges of the carbons of the carbonyl functional groups of ligands, excluding compounds that are not planar. Adapted from Grant (1998).

CONCLUSIONS

Many concerns about the aflatoxins have originated from the strong implications of their involvement in disease and death in man and animals. Even, at the beginning of a third millennium, scientists and clinicians are still seeking practical ways to control these elusive toxins. The prevention of toxicity and carcinogenicity of aflatoxins through clay-based dietary interventions (i.e., aflatoxin-selective enterosorbents) shows great promise and significant advantages over other methods of detoxification. Evidence suggests that aflatoxins react tightly at multiple sites on HSCAS clay surfaces (especially those within the interlayer region). However, clay and zeolitic minerals comprise a broad family of diverse aluminosilicates and are not created equal; all aflatoxin binding agents should be rigorously tested for efficacy, paying particular attention to their effectiveness and safety in aflatoxin-sensitive animals and their potential for interactions with nutrients and/or synergy with aflatoxins before inclusion in diets.

ACKNOWLEDGEMENTS

This research is based in part upon work supported by USDA Animal Health Grant 9703230, NIH (P42-ES04917), the Texas Agricultural Experiment Station (Grant H-6215), and Center Grant ES09106.

REFERENCES

Abdel-Wahhab, M.A., Nada, S.A., Farag, I.M., Abbas, N.F. and Amra, H.A., 1998, Potential protective effect of HSCAS and bentonite against dietary aflatoxicosis in rat: with special reference to chromosomal aberration. *Nat. Toxins* 6:211.

Adhikari, M., Ramjee, G. and Berjak P., 1994, Aflatoxin, kwashiorkor and morbidity. *Natural Toxins* 2:13.

Aguilar, F., Hussain, S. P., and Cerutti, P., 1993, Aflatoxin B_1 induces the transversion of G $\rightarrow$ T in codon 249 of the p53 tumor suppressor gene in human hepatocytes. *Proc. Natl. Acad. Sci. USA* 90:8586.

Beaver, R.W., Wilson, D.M., James, M.A., and Haydon, K.D., 1990, Distribution of aflatoxins in tissues of growing pigs fed an aflatoxin-contaminated diet amended with a high affinity aluminosilicate sorbent. *Vet. Hum. Toxicol.* 32:16.

Bohn, H.L., McNeal, B.L., and O'Connor, G.A., 1979, *Soil Chemistry*. John Wiley & Sons, New York, NY.

Bonna, R.J., Aulerich, R. J., Bursian, S.J., Poppenga, R.H., Braselton, W.E., and Watson, G. L., 1991, Efficacy of hydrated sodium calcium aluminosilicate and activated charcoal in reducing the toxicity of dietary aflatoxin to mink. *Arch. Environ. Contam.Toxicol.* 20:441.

Borchardt, G., 1989, Smectites, in *Minerals in Soil Environments*, J. B. Dixon and S. B. Weed, Eds., Soil Science Society of America Inc., Madison, WI.

Bursian, S.J., Aulerich, R.J., Cameron, J.K., Ames, N.K. and Steficek, B.A., 1992, Efficacy of hydrated sodium calcium aluminosilicate in reducing the toxicity of dietary zearalenone to mink. *J. Appl. Toxicol.* 12:85-90.

Busby, W.F. Jr. and Wogan, G.N., 1984, Aflatoxins, in *Chemical Carcinogens* Volume 2, Searle, C.E., Ed., pp. 945-1136.

Carson, M.S. and Smith, T.K., 1983, Role of bentonite in prevention of T-2 toxicosis in rats. *J. Anim. Sci.* 57:1498.

Cerdchai, R., Paisansarakit A. and Khajarern, J., 1990, Effect of hydrated sodium calcium aluminosilcate (NovaSil) on reducing aflatoxicosis in ducks. Proc. 7th FAVA Congress, Pattaya, pp. 391.

ChemPlus:Extension for HyperChem; Hypercube: Waterloo, Ontario, Canada, 1993, p.1.

Chestnut, A.B., Anderson, P.D., Cochran, M.A., Fribourg, H.A. and Gwinn, K.D., 1992, Effects of hydrated sodium calcium aluminosilicate on fescue toxicosis and mineral absorption. *J. Anim. Sci.* 70:2838.

Chung, T.K., Erdman, J.W. Jr., and Baker, D.H., 1990, Hydrated sodium calcium aluminosilicate: Effects on zinc, manganese, vitamin A and riboflavin utilization. *Poultry Sci.* 69:1364.

Chung, T.K. and Baker, D.H., 1990, Phoshorus utilization in chicks fed hydrated sodium calcium aluminosilicate. *J. Anim. Sci.* 68:1364.

Clark, K., Sarr, A.B., Grant, P.G., Phillips, T.D. and Woode, G.N., 1998, In vitro studies on the use of clay, clay minerals and charcoal to adsorb bovine rotavirus and bovine coronavirus. *Vet. Microbiol.* 63:137.

Colvin, B.M., Sangster, L.T., Hayden, K.D., Bequer, R.W., and Wilson, D.M., 1989, Effect of high affinity aluminosilicate sorbent on prevention of aflatoxicosis in growing pigs. *Vet. Hum. Toxicol.* 31:46.

Council for Agricultural Science and Technology 1989, *Mycotoxins: Economic and Health Risks.* CAST, Task Force Report No. 116. Ames, Iowa.

Davidson, J.N., Babish, J.G., Delaney, K.A., Taylor, D.R., and Phillips, T.D., 1987, Hydrated sodium calcium aluminosilicate decreases the bioavailability of aflatoxin in the chicken. *Poultry Sci.* 66:89.

Dewar, M.J.S., Zoebisch, E.G., Healy, E.F., Stewart, J.J.P., 1985, AM1: A new general purpose quantum mechanical molecular model. *J. Am. Chem. Soc.* 107:3902.

Diener, U.L. 1981, Unwanted biological substances in foods: Aflatoxins, in *Impact of Toxicology on Food Processing*, J. C. Ayres and J.C. Krishnamachari, Eds., AVI Publishing Company, Inc., Westport, Connecticut.

Dwyer, M.R., Kubena, L.F., Harvey, R.B., Mayura, K., Sarr, A.B., Buckley, S., Bailey, R.H. and Phillips, T.D., 1997, Effects of inorganic adsorbents and cyclopiazonic acid in broiler chickens. *Poultry Sci.* 76:1141.

Dyal, R.S. and Hendricks, S.B., 1950, Total surface of clays in polar liquids as a characteristic index. *Soil Sci.* 69:421.

Eaton, D.L., Ramsdell, H.S and Neal, G.E., 1994, Biotransformation of aflatoxins, in: *The Toxicology of Aflatoxins, Human Health, Veterinary Agricultural Significance*, L.D. Eaton and J. D. Groopman, Eds., Academic Press, NY.

Ellis, J.A., Harvey, R.B., Kubena, L.F., Bailey, R.H., Clement, B.A., and Phillips, T.D., 1990, Reduction of aflatoxin M_1 residues in milk utilizing hydrated sodium calcium aluminosilicate. *Toxicologist* 10:163. [Abstract]

Ellis, J.A., Bailey, R.H., Clement, B.A., and Phillips, T.D., 1991, Chemisorption of aflatoxin M_1 from milk by hydrated sodium calcium aluminosilicate. *Toxicologist* 11; 96. [Abstract].

Emsley J. *The Elements*, 2nd ed.; Clarendon Press: Oxford, 1991; 1p.

Fetter, C., 1993, Transformation, retardation, and attenuation of solutes, in: *Contaminant Hydrogeology*, R. A. McConnin, Ed., Macmillan Publishing Co., New York, NY.

Ghose, A.K., Pritchett, A, Crippen, G.M., 1988, Atomic physicochemical parameters for three dimensional structure directed quantitative structure-activity relationships III: Modeling hydrophobic interactions. *J. Comp. Chem.* 9:80.

Grant, P.G., 1998, Investigation of the mechanism of aflatoxin B_1 adsorption to clays and sorbents through the use of isothermal analysis, Ph.D. Dissertation, Texas A&M University, College Station, TX.

Grant, P.G. and Phillips, T.D., 1998, Isothermal adsorption of aflatoxin B_1 on HSCAS clay. *J. Agric. Food. Chem.* 46:599.

Grant, P.G., Lemke, S.L. Dwyer, M.R. and Phillips, T.D., 1998, Modified Langmuir equation for s-shaped and multisite isotherm plots. *Langmuir*:14:4292.

Gray, M.J., Mebane, R.C., Womack, H.N., Rybolt, T.R., 1995, Molecular mechanics and molecular cross-sectional areas: A comparison with molecules adsorbed on solid surfaces. *J.Colloid Interface Sci.,* 170:98.

Haderlein, S.B., Weissmahr, K.W., Schwarzenbach, R.P., 1996, Specific adsorption of nitroaromatic explosives and pesticides to clay minerals. *Environ. Sci. Technol.* 30:612.

Harvey, R.B., Kubena, L.F., Huff, W.E., Corrier, D.E., Rottinghaus, G.E. and Phillips, T.D., 1990, Effects of treatment of growing swine with aflatoxin and T-2 toxin. *Am. J. Vet. Res.* 51:1688.

Harvey, R.B., Kubena, L.F., Elissalde, M.H., and Phillips, T.D., 1993, Efficacy of zeolitic ore compounds on the toxicity of aflatoxin to growing broiler chickens. *Avian Diseases* 37:67.

Harvey, R.B., Kubena, L.F., Phillips, T.D., Corrier, D.E., Elissalde, M.H., and Huff, W. E., 1991a, Diminution of aflatoxin toxicity to growing lambs by dietary supplementation with hydrated sodium calcium aluminosilicate. *Am. J. Vet. Res.* 52:152.

Harvey, R.B., Phillips, T.D., Ellis, J.A., Kubena, L.F., Huff, W.E., and Petersen, D.V., 1991b, Effects of aflatoxin M_1 residues in milk by addition of hydrated sodium calcium aluminosilicate to aflatoxin-contaminated diets of dairy cows. *Am. J. Vet. Res.* 52:1556.

Harvey, R., Kubena, L.F., Elissalde, M., Corrier, D. and Phillips, T.D., 1994, Comparison of two hydrated sodium calcium aluminosilicate compounds to experimentally protect growing barrows from aflatoxicosis. *J. Vet. Diagn Invest* 6:88.

Huebner, H.J., Lemke, S.L. , Ottinger, S.E., Mayura, K., and Phillips, T.D., 1999, Molecular characterization of high affinity, high capacity clays for the equilibrium sorption of ergotamine. *Food Additives and Contam.* 16:159.

Huff, W.E., Kubena, L.F., Harvey, R.B. and Phillips, T.D., 1992, Efficacy of a hydrated sodium calcium aluminosilicate to reduce the individual and combined toxicity of aflatoxin and ochratoxin A. *Poultry Sci.* 71:64.

HyperChem: Computational Chemistry; Hypercube Inc: Waterloo, Ontario, Canada, 1994, pp.1-285.

International Agency for Research on Cancer, 1976, *IARC Monographs on the evaluation of the carcinogenic risk of chemicals to man: Some naturally occurring substances*, Vol. 10, IARC, Lyon, France.

International Agency for Research on Cancer, 1987, *IARC Monograph on the evaluation of carcinogenic risk to humans, Aflatoxins.* Suppl. 7, IARC, Lyon, France.

Jayaprakash, M., Gowda, R.N.S., Vijayasarathi, S.K. and Seshadri, S.J., 1992, Adsorbent efficacy of hydrated sodium calcium aluminosilicate in induced aflatoxicosis in broilers. *Indian J. Vet. Pathol.* 16:102.

Koskinen, W.C., and Harper, S.S., 1990, The retention process: Mechanisms, in *Pesticides in the Soil Environment: Process, Impacts, and Modeling* SSSA Book Series, no. 2 (H.H. Cheng, Ed.), Soil Science Society of America, Madison, WI.

Kubena, L.F., Harvey, R.B., Phillips, T.D. Corrier, D.E. and Huff, W.E., 1990a, Diminution of aflatoxicosis in growing chickens by dietary addition of a hydrated sodium calcium aluminosilicate. *Poultry Sci.* 69:727.

Kubena, L.F., Harvey, R.B., Huff, W.E., Corrier, D.E., and Phillips, T.D., 1990b, Ameliorating properties of a hydrated sodium calcium aluminosilicate on the toxicity of aflatoxin and T-2 toxin. *Poultry Sci.* 69:1078.

Kubena, L. F., Huff, W.E., Harvey, R.B., Yersin, A.G., Elissalde, M.H., Witzel, D.A., Giroir, L. E., and Phillips, T.D., 1991, Effects of hydrated sodium calcium aluminosilicate on growing turkey poults during aflatoxicosis. *Poultry Sci.* 70:1823.

Kubena, L.F., Harvey, R.B., Huff, W. E., Yersin, A.G., Elissalde, M.H., and Witzel, D. A., 1993, Efficacy of a hydrated sodium calcium aluminosilicate to reduce the toxicity of aflatoxin and diacetoxyscirpenol. *Poultry Sci.* 72:51.

Kurtzman, C.P., Horn, B.W., and Hesseltine, C.W., 1987, *Aspergillus nominus*: A new aflatoxin-producing species related to *Aspergillus flavus* and *Aspergillus tamarii*. Antonie van Leeuwenhoek 53:147.

Ledoux, D.R., Rottinhaus, G.E., Bermudez, A.J. and Alonso-Debolt, M., 1999, Efficacy of a hydrated sodium calcium aluminosilicate to ameliorate the toxic effects of aflatoxin in broiler chicks. Poultry Sci. 78:204.

Lemke, S.L., 2000, Investigation of clay-based strategies for the protection of animals from the toxic effects of selected mycotoxins. PhD Dissertation, Texas A&M University, College Station, Texas.

Lindemann, M.D., Blodgett, D.J., Kornegay, E.T., Schurig, G.G., 1993, Potential ameliorators of aflatoxicosis in weanling/growing swine. *J. Anim. Sci.* 71:171.

MacEwan, D.M.C.; Wilson, M.J., 1980, Interlayer and Intercalation Complexes of Clay Minerals, in *Crystal Structures of Clay Minerals and Their X-Ray Identification,* G.W. Brindleyand G. Brown, Eds., Mineralogical Society, London.

McMahon, G., Hanson, L., Lee, J.J. and Wogan, G.N., 1986, Identification of an activated c-Ki-ras oncogene in rat liver tumors induced by aflatoxin B_1. *Proc. Natl. Acad. Sci.* USA 83:9418.

McMahon, G., Davis, E. and Wogan, G.N., 1987, Characterization of c-Ki-ras oncogene alleles by directing sequencing of enzymatically amplified DNA from carcinogen-induced tumors. *Proc. Natl. Acad. Sci.* USA 84:4974.

Marquez, R.N. and de Hernandez, I.T., 1995, Aflatoxin adsorbent capacity of two Mexican aluminosilicates in experimentally contaminated chick diets. *Fd. Ad. Contam.* 12:431.

Masimango, N., Remacle, J. and J. Ramaut, 1979, Elimination, par des argiles gonflantes, de L'aflatoxine B_1 des milieus contamines. *Ann. Nutr. Alim* 33 (1):137.

Mayura, K., Abdel-Wahhab, M.A., McKenzie, K.S., Sarr, A.B., Edwards, J.F., Naguib, K. and Phillips, T.D., 1998, Prevention of maternal and developmental toxicity in rats via dietary inclusion of common aflatoxin sorbents: Potential for hidden risks. *Toxicol. Sci.* 41:175.

McBride, M. B., 1989, Surface chemistry of soil minerals. An introduction to soil mineralogy, in: *Minerals in Soil Environments* (J. B. Dixon and S. B. Weed, Eds.), Soil Science Society of America Inc., Madison, WI.

Meylan, W.M., Howard, P.H., and Boethline, R.S., 1996, Improved method for estimating water solubility from octanol/water partition coefficient. *Environ. Toxicol. Chem.* 15:100.

Millot, G., 1979, *Clay. Sci. Am.* 240(4):109.

Ming, D.W., and Mumpton, F.A., 1989, Zeolites in soils. An introduction to soil mineralogy, in: *Minerals in Soil Environments,* J. B. Dixon and S. B. Weed, Eds., Soil Science Society of America Inc., Madison, WI.

Munn, C.A., 1994, Macaws: Winged rainbows. *Natl. Geograph.* 185 (Jan):118.

Nahm, K.H., 1995, Prevention of aflatoxicosis by addition of antioxidants and hydrated sodium calcium aluminosilicate to the diet of young chicks. *Japanese J. Poultry Sci.* 32:117.

Patterson, R. and L.G. Young, 1993, Efficacy of hydrated sodium calcium aluminosilicate, screening and dilution in reducing the effects of mold contaminated corn in pigs. *Canadian J. Anim. Sci.* 73:616.

Peska, J.J. and Bondy, G.S. 1994, Immunotoxic effects of mycotoxins, in: *Mycotoxins in Grain: Compounds other than Aflatoxin*, J.D.Miller and H.I. Trenholm (eds), Eagan Press, St. Paul, MN.

Phillips, T.D., Kubena, L.F., Harvey, R.B., Taylor, D.R., and Heidelbaugh, N.D., 1987, Mycotoxin hazards in agriculture: New approach to control. *J. Am. Vet. Med. Assoc.* 190:1617. [Abstract]

Phillips, T.D., Kubena, L.F., Harvey, R.B., Taylor, D.R., and Heidelbaugh, N.D., 1988, Hydrated sodium calcium aluminosilicate: A high affinity sorbent for aflatoxin. *Poultry Sci.* 67:243.

Phillips, T.D., Clement, B.A., Kubena, L.F., and Harvey, R.B., 1990, Detection and detoxification of aflatoxins: Prevention of aflatoxicosis and aflatoxin residues with hydrated sodium calcium aluminosilicates. *Vet. Human. Toxicol.* 32:15.

Phillips, T.D., Sarr, A.B., Clement, B.A., Kubena, L.F., and Harvey, R.B., 1991, Prevention of aflatoxicosis in farm animals via selective chemisorption of aflatoxin, in: *Mycotoxins, Cancer and Health* (G. A. Bray and D. H. Ryan, Eds.), Vol 1, Louisiana State University Press, Baton Rouge.

Phillips, T.D., Clement, B.A.., and Park, D.L., 1994, Approaches to reduction of aflatoxin, in: *The Toxicology of Aflatoxins, Human Health, Veterinary Agricultural Significance*, L.D. Eaton. and J. D. Groopman, Eds., Academic Press, NY.

Phillips, T.D., Sarr, A.B., and Grant, P.G., 1995, Selective chemisorption and detoxification of aflatoxins by phyllosilcate clay. *Natural Toxins* 3:204.

Phillips, T.D.,1999, Dietary clay in the chemoprevention of aflatoxin-induced disease. *Toxicol. Sciences* 52:118.

Radoslovich, E.W., 1960, The Structure of Muscovite $KAl_2(Si_3Al)O_{10}(OH)_2$. *Acta Cryst.*, 13:919.

Ramos, A.J. and Hernandez, E., 1996, In vitro aflatoxin adsorption by means of a montmorillonite silicate. A study of adsorption isotherms. *Anim. Feed Sci. Tech.* 62:263.

Ramu, J., Clark, K., Woode, G.N., Sarr, A.B. and Phillips, T.D., 1997, Adsorption of cholorea and heat-labile Escherichia coli enterotoxins by various adsorbents: An in vitro study. *J. Food Protection* 60:358.

Richard, J.L., Thurston, J.R. and Pier, A., 1978, Effects of mycotoxins on immunity, in: *Toxins: Animal, Plant and Microbial*, P. Rosenberg, Ed., Pergamon Press, New York.

Sarr, A. B., 1992, Evaluation of innovative methods for the detection and detoxification of aflatoxin, Ph.D. Dissertation, Texas A&M University, College Station, TX.

Sarr, A.B., Mayura, K., Kubena, L.F., Harvey, R.B., Phillips, T.D., 1995, Effects of phyllosilicate clay on the metabolic profile of aflatoxin B_1 in Fischer-344 rats. *Toxicol. Lett.* 75:145.

Schulze, D.G., 1989, An introduction to soil mineralogy, in: *Minerals in Soil Environments*, J. B. Dixon and S. B. Weed, Eds., Soil Science Society of America Inc., Madison, WI.

Senesi, N. 1993, Organic pollutant migration in soils as affected by soil organic matter: Molecular and mechanistic aspects, in: *Migration and Fate of Pollutants in Soils and Subsoils* (D. Petruzzelli and F.G. Helffrich, Eds, Springer, Berlin.

Smith, T.K. 1984, Spent canola oil bleaching clays: Potential for treatment of T-2 toxicosis in rats and short-term inclusion in diets of immature swine. *Can. J. Anim. Sci.* 64: 725.

Smith, E.E., Phillips, T.D., Ellis, J.A., Harvey, R.B., Kubena, L.F., Thompson, J. and Newton, G.M. 1994, Dietary hydrated sodium calcium aluminosilicate reduction of aflatoxin M_1 residue in dairy goat milk and effects on milk production and components. *J. Anim. Sci.* 72: 677.

Southern, L.L., Ward, T.L., Bidner, T.D. and Hebert, L.G. 1994, Effect of sodium bentonite and hydrated sodium calcium aluminosilicate on growth performance and tibial mineral concentrations in broiler chicks fed nutrient-deficient diets. *Poultry Sci.* 73: 848.

Sylvia, D., Fuhrmann, J., Hartel, P. and Zuberer, D. 1997, Principles and Applications of Soil Microbiology, Prentice Hall Publishers, Upper Saddle River, N.J.

Vellarkad, N., Viswanadhan, V.N., Ghose, A.K., Revankar, G.N. and Robins, R.K., 1989, Atomic physiochemical parameters for three dimensional structure directed quantitative structure activity relationships. 4. Additional parameters for hydrophobic and dispersive interactions and their applications for an automated superposition of certain naturally occurring nucleoside antibiotics. *J. Chem. Inf. Comput. Sci.* 29:163.

Voss, K.A., Dorner, J.W. and Cole, R.J., 1993, Amelioration of aflatoxicosis in rats by volclay NF-BC, microfine bentonite. *J. Food Prot.* 56:595.

EFFECT OF PROCESSING ON AFLATOXIN

Douglas L. Park

Division of Natural Products
Center for Food Safety and Applied Nutrition
Food and Drug Administration
Washington, DC 20204

ABSTRACT

Naturally occurring toxicant contamination of foods with mycotoxins is unavoidable and unpredictable and poses a unique challenge to food safety. Aflatoxins are toxic mold metabolites produced by toxigenic strains of Aspergillus species. Primary commodities susceptible to aflatoxin contamination include corn, peanuts and cottonseed and animal-derived foods such as milk when the animal is fed aflatoxin-contaminated feed. Risks associated with aflatoxin-contaminated foods can be reduced through the use of specific processing and decontamination procedures. Factors, which influence the effectiveness of a specific process or procedure, include the chemical stability of the mycotoxin(s), nature of the process, type and interaction with the food/feed matrix and interaction with multiple mycotoxins if present. Practical decontamination procedures must: 1) inactivate, destroy, or remove the toxin, 2) not produce or leave toxic residues in the food/feed, 3) retain the nutritive value of the food/feed, 4) not alter the acceptability or the technological properties of the product, and, if possible, 5) destroy fungal spores. For aflatoxins, multiple processing and/or decontamination schemes have been successful in reducing aflatoxin concentrations to acceptable levels. Physical cleaning and separation procedures, where the mold-damaged kernel/seed/nut is removed from the intact commodity, can result in 40-80% reduction in aflatoxins levels. Processes such as dry and wet milling result in the distribution of aflatoxin residues into less utilized fractions of the commodity. The ammoniation of aflatoxin-contaminated commodities has altered the concentrations as well as toxic and carcinogenic effects of aflatoxin by greater than 99%. Nonbiological materials such as selected anticaking agents covalently bind aflatoxins from aqueous suspensions, diminish aflatoxin uptake by animals, prevent acute aflatoxicosis, and decrease aflatoxin residues in milk. Ultimately, the best processing or decontamination process is one that is approved by regulatory agencies, cost-effective, and reduces the mycotoxin concentration to acceptable levels.

Mycotoxins and Food Safety, edited by Trucksess *et al.*
Kluwer Academic/Plenum Publishers, 2002

INTRODUCTION

Aflatoxins are naturally occurring secondary mold metabolites produced primarily by *Aspergillus flavus* and *A. parasiticus*. The primary agricultural commodities associated with aflatoxin contamination include corn, peanuts, tree nuts, cottonseed and dairy products. Toxins of primary public health concern include aflatoxin B_1, B_2, G_1 and G_2 and an animal metabolite, aflatoxin M_1, which occurs in milk when lactating dairy cows are fed rations containing aflatoxin B_1. Aflatoxins are potent liver carcinogens and toxins. The International Agency for Research on Cancer (IARC) has classified aflatoxin as a probable human carcinogen (Stoloff, 1982). In fact, there is epidemiological evidence that humans are not immune to aflatoxicosis, as reported in India and Kenya (Park and Stoloff, 1989); and, where other factors are present such as hepatitis B virus, the carcinogenic event can occur (Henry et al., 1999). All of these factors have highlighted the importance of establishing appropriate food safety management programs for aflatoxins (Park and Stoloff, 1989; Lopez-Garcia and Park, 1998; Park et al., 1999; Park, 1993; Park and Liang, 1993).

FOOD SAFETY MANAGEMENT PROGRAMS

Factors crucial to the effectiveness of food safety management programs for aflatoxin include the establishment of regulatory limits and monitoring programs, the control of alflatoxin formation through good agricultural practices and proper storage and handling, reduction of aflatoxin levels in contaminated commodities through processing and decontamination procedures, and adequate education programs for agricultural producers, food and feed processors and consumers (Lopez-Garcia et al., 1999; Phillips et al., 1994). The primary goal of these efforts is to reduce human exposure to aflatoxin to the lowest practical level while at the same time provide for an adequate food supply. Information necessary for the establishment of regulatory limits and monitoring programs include the toxicological properties of the toxin and metabolites, the major commodities affected, the levels of the toxin in affected commodities, the dietary intake, the availability of analytical methods, and the impact of regulatory limits on the availability of the food/feed supply. It is inappropriate to enact strict programs that would restrict marginally aflatoxin contaminated food/feed products in the absence of a clear public health benefit.

PHYSICAL METHODS OF AFLATOXIN REMOVAL AND DETOXIFICATION

Cleaning and Segregation

The first option in aflatoxin reduction strategies is the physical separation of the mold-damaged kernel/seed/nut from the intact and apparently non-contaminated product. These procedures such as cleaning, sorting and handpicking (Dickens and Whitaker, 1975) are non-evasive and do not alter the product significantly. Flotation and density segregation have been reported to be useful in separating aflatoxin-contaminated corn and peanuts (Cole, 1989). However, complete removal of the contaminated product or aflatoxin cannot be expected with physical methods of separation. Should there be high residual levels of contamination, other procedures must be used to reduce aflatoxin concentrations in the final product to acceptable levels. The peanut industry uses a combination of segregation and other techniques to reduce aflatoxin levels in peanut products such as peanut butter (Table 1). Physical separation is a good alternative for the food industry. An initial investment to purchase adequate equipment is necessary; however, maintenance expenses are minimal.

Table 1. Effectiveness of aflatoxin management strategies at the processing level for peanut products

Technology	*Aflatoxin Concentration (µg/kg)*	*Reduction (%)*	*Cumulative Reduction (%)*
Farmers' stock	217	-	-
Belt separator	140	35	35
Shelling plans	100	29	54
Color sorting	30	70	86
Gravity table	25	16	88
Blanching/color sorting	2.2	91	99.0
Re-color sorting	1.6	27	99.3

From: Park and Liang (1993)

Wet Milling

The wet milling process is widely used for the preparation of corn products. It is, however, necessary to clearly identify the distribution patterns of the toxin in processing by-products when using this process. The contaminated fractions can be diverted to less-risk uses or subjected to decontamination procedures. In experiments studying the fate of aflatoxins during wet milling of corn, aflatoxin B_1 was found to partition primarily into the steep water (39-42%) and fiber (30-38%) fractions. The remainder of the aflatoxin was found in gluten (13-17%), germ (6-10%) and starch (1%) fractions (Wood et al., 1982; Bennett and Anderson, 1978). Relatively little of the aflatoxin was destroyed with this process. It is necessary to consider the significance of the aflatoxin level when determining appropriate uses for the aflatoxin-containing fractions.

Dry Milling

Like wet milling, dry milling will also fractionate the aflatoxin found in contaminated grain. For corn, the highest levels of aflatoxin B_1 following the milling process were found in the germ and hull fractions. Grits, low-fat meal and low-fat flour contained only 6-10% of the original aflatoxin B_1. Studies with rice (Achroder et al., 1968) and wheat (Scott, 1984) showed similar patterns, with the bran containing the highest concentration of the toxin. On an industrial perspective, dry milling can be a cost-effective way of reducing aflatoxin levels in the primary product, flour. Management of the toxic fractions can be effective through alternative lower-risk uses.

Thermal Inactivation

Aflatoxins are relatively heat-stable and are not completely destroyed when boiled in aqueous solution, autoclaved and processed using a variety of different methods used to prepare foods and feeds (Christensen et al., 1977). Several studies have shown that roasting is a good method for reducing aflatoxin levels in certain commodities, i.e., oil and dry-roasted peanuts (Peers, 1975), microwave roasted peanuts (Luter et al, 1982), corn, coffee (Conway et al., 1978; Levi, 1980; Levi et al., 1975), etc. Many of these studies were performed on samples prepared by artificial spiking; therefore, it is important to confirm processing results on naturally incurred toxins since the matrix effects may be different with spiked versus natural contamination (Wood et al., 1982).

Irradiation

Gamma irradiation (2.5 Mrad) did not significantly degrade aflatoxin in contaminated peanut meal (Feuell, 1966). Due to the cost of equipment, limited positive results, and lack of consumer acceptance of the irradiation process, irradiation does not appear to be a good commercial method.

Absorption from Solution and Covalent Binding

The use of nonbiological materials to adsorp or covalently bind aflatoxins is a good method for aflatoxin decontamination. A variety of adsorbent materials, i.e., activated carbon and clays, have been shown to bind aflatoxins from aqueous solutions, and aminosilicates were shown effective with oils and animals feeds (Decker, 1980; Machen et al., 1988). Although specific clays have been reported to be effective in binding aflatoxin from aqueous solutions, decreasing aflatoxin uptake in animals, preventing aflatoxicosis in farm animals and decreasing levels of aflatoxin M_1 in milk, their use for removal of aflatoxin from feed is not permitted by regulatory agencies and the long-term effects and safety of these agents have not been determined (Phillips et al., 1994).

BIOLOGICAL DECONTAMINATION

Biological methods demonstrating effective decontamination properties are usually the result of specific compounds produced by microorganisms. When this occurs, it is usually more economical to add the active agent directly. For the fermentation industry, however, biological methods are a good option. During the beer brewing process, there is approximately 70-80% reduction of aflatoxin B_1 in the starting materials (Chu et al., 1975). After cooking and fermentation of corn, wheat, and corn milo, aflatoxin B_1 was reduced by 47% (Dam et al., 1977). Aflatoxin does not partition into the alcoholic distillate. However, the toxin concentrates in the spent grain. This highlights the need for contamination procedures for spent grain used for animal feed.

CHEMICAL INACTIVATION

Ammonia has been shown to be effective in altering the toxic and carcinogenic effects of aflatoxin in corn, peanuts, cottonseed and meals. The results of these studies have been summarized by Park et al. (1988) and Park and Price (2001). The ammoniation process has been successfully used for over twenty years in the U.S. and France. The procedure has also been used in Senegal, Sudan, Brazil, Mexico and South Africa.

Although several ammonia-based procedures have been developed and studied, the high pressure/high temperature method that utilizes as ammonia (0.5-2.0%) under controlled conditions of moisture (12-16%), pressure (45-55 psi), and temperature (80-100^{0}C) for 20-60 minutes is the most efficient and produces a safer product. During ammoniation, the aflatoxin molecule is chemically modified to compounds possessing reduced or non-detectable toxic or mutagenic potentials. Exhaustive extraction, isolation, and purification studies have shown the toxic/mutagenic potentials and fate of aflatoxin/ammonia reaction products to be non-detectable or many orders of magnitude lower than the parent aflatoxin. The presence of identified aflatoxin/ammonia reaction products, i.e., aflatoxin D_1 and a compound with a molecular weight of 206, in animal feeds have no health significance. In studies determining the distribution and formation of aflatoxin/ammonia reaction products in cottonseed and corn, approximately 12-14% of the original aflatoxin contamination was lost as volatile compounds, 20-24% of the reaction products were extractable with organic solvents and 6-13% of reaction products were extractable with methanol. Following treatment of the corn with acid and base and proteolytic enzymes, an additional 19-22% of the reaction products were detected. The residual cottonseed or corn matrix contained only 37% of the original aflatoxin concentration as ammonia/reaction products.

Metabolic and excretion studies in laboratory and farm livestock have shown that over 98% of the feed-bound ammonia/aflatoxin products were excreted in the urine and feces. Between 0.25-1.6% of the original contaminant was excreted in the milk. Exhaustive animal feeding studies have shown that aflatoxin-ammonia reaction products have minimal if any affect on the health of animals receiving rations containing ammonia-treated corn, peanut, and cottonseed meals.

Another effective ammoniation procedure, the ambient temperature method, usually requires a 3-6 week treatment of the contaminated material. This process also requires close monitoring to confirm effective results. Little research has been done to demonstrate the safety of the ammoniated product.

Nixtamalization, a traditional alkaline treatment of corn used in the manufacture of tortillas, has been shown to significantly reduce aflatoxin levels in corn (Ulloa and Herrera, 1970; Ulloa-Sosa and Shroedes, 1969). Subsequent studies have shown that aflatoxin B_1 is reformed upon acidification (Price and Jorgensen, 1985). The addition of selected food additives alone or in combination with H_2O_2 during the nixtamalization process resulted in significant aflatoxin reductions in contaminated corn (Trujillo, 1997; Lopez-Garcia, 1998; Hagler et al., 1982; Burgos-Hernandez, 1998) and corn co-contaminated with fumonisin B_1 (Park et al., 1996).

CONCLUSIONS

Aflatoxins are naturally occurring toxicants that can contaminate a significant number of staple food commodities, occasionally at very high levels. Obviously, complete avoidance of risks posed by aflatoxins would be total abstinence of affected foods. Commodities such as corn, peanuts and dairy products are important sources of energy, protein, fiber, and oil for humans and livestock. In many developing regions of the world these commodities are food staples. Procedures that prevent the formation of mycotoxins is the best way to avoid aflatoxin exposure; however, should the contamination occur, procedures must be instituted that reduce toxin levels to an acceptable exposure level. Numerous processing and decontamination procedures have been developed to separate the contaminated material, remove or chemically modify the toxin, or reduce the toxic/mutagenic potentials of the toxin.

Processing procedures such as cleaning and separation of the mold damaged kernel/seed/nut from the intact product, wet and dry milling, thermal inactivation and fermentation are useful in producing highly acceptable food products with minimal levels of aflatoxins from highly contaminated starting material. Decontamination procedures including the use of ammonia have been used successfully worldwide to inactive aflatoxin-contaminated animal feed ingredients and rations. Promising procedures but requiring additional study to confirm the efficacy of the process and safety of the treated product include the use of adsorbent clays, ozone, and extraction solvents. With appropriate test results, some of the decontamination procedures could be used for human foods and commodities contaminated with aflatoxins.

REFERENCES

Achroder, H.W., Boller, R. A., and H. Hein, Jr., 1986, Reduction in aflatoxin contamination of rice by milling procedures. *Cereal Chem.* 45:574.

Bennett, G.A., and Anderson, R.A., 1978, Distribution of aflatoxin and/or zearalenone in wet-milled corn products: A review. *J. Agric. Food Chem.* 26:1055.

Burgos-Hernandez, A., 1998, Evaluation of chemical treatments and intrinsic factors that affect the mutagenic potential of aflatoxin B_1-contaminated corn. Louisiana State University, Baton Rouge, Louisiana, United States. (Ph.D. dissertation)

Chu, F.S., Chang, C.C., Ashoor, S.H., and Prentice N., 1975, Stability of aflatoxin B_1 and ochratoxin A in brewing. *Applied Microbiol* 29:313.

Christensen, C.M., Mirocha, C.I., Meronuck, R.A., 1977, Mold, mycotoxin and mycotoxicoses. *In*: Agricultural Experiment Station, Report 142, St. Paul; University of Minnesota.

Cole, R.J., 1989, Technology of aflatoxin decontamination. *In*: S. Natori, K. Hashimoto, Y. Ueno, eds. *Mycotoxins and Phycotoxins '88*. Amsterdam; Elsevier Scientific, p 177.

Conway, H.F., Anderson, R.A., and Bagley, E.B., 1978, Detoxification of aflatoxin-contaminated corn by roasting. *Cereal Chem.* 55:115.

Dam, R.S., Tam, S.W., and Satterlee, L.D., 1977, Destruction of aflatoxins during fermentation and by-product isolation from artificially contaminated grains. *Cereal Chem.* 54:705.

Decker, W.J., 1980, Activated charcoal absorbs aflatoxin B_1. *Vet Human Toxicol* 22:388.

Dickens, W.J., and Whitaker, T.B. , 1975, Efficacy of electronic color sorting and hand picking to remove aflatoxin contaminated kernels from commercial lots of shelled peanuts. *Peanut Sci.* 2:45.

Feuell, A.J., 1966, Aflatoxin in groundnuts. IX. Problems of detoxification. *Trop Sci.* 8:61.

Hagler, W.M., Jr, Hutchings, J.E., and Hamilton, P.B., 1982, Destruction of aflatoxin in corn with sodium bisulfate. *J. Food Prot* 45:1287.

Henry, S.H., Bosch, F.X., Troxell, T.C. and Bolger, P.M., 1999, Reducing liver cancer – Global control of aflatoxin. *Science* 286:2453.

Levi, C.P., 1980, Mycotoxins in coffee. *J. AOAC* 63:1282.

Levi, C.P., Ternk, H.L., and Yeransianm, J.A. , 1975, Investigations of mycotoxins relative to coffee. *Colloq Int. Chim. Cafes* (CR) 7:287.

Lopez-Garcia, R. and Park, D.L., 1998, Effectiveness of post-harvest procedures in management of mycotoxin hazards In . *Mycotoxins in Agriculture and Food Safety*, D. Bhatnagar and S. Sinha, eds, New York, Marcel Dekker, pp. 407-433.

Lopez-Garcia, R., 1998, Aflatoxin B_1 and fumonisin B_1co-contamination: Interactive effects, possible mechanisms of toxicity, and decontamination procedures. Louisiana State University, Baton Rouge, Louisiana, United States. (Ph.D. dissertation).

Lopez-Garcia, R., Park, D.L. and Phillips, T.D., 1999, Integrated mycotoxin management systems, *Food, Nutrition and Agriculture* FAO 23:38

Luter, L., Wyslouzil, W., and Kashyap, S.C., 1982, The destruction of aflatoxins in peanuts by microwave roasting. *Can Inst. Food Sci. Technol. J.* 15:236.

Machen, M.D., Clement, B.A., Shepherd, E.C., Sarr, A.B., Pettit, E.W., and Phillips, T.D., 1988, Sorption of aflatoxins from peanut oil by aluminosilicates. *Toxicologist* 8:265.

Park, D.L., 1993, Controlling aflatoxin in food and feed. *Food Technol.* 47: 92.

Park, D.L., Lee, L.S., Price, R.L., and Pohland, A.E. , 1988, Review of decontamination of aflatoxin by ammoniation: Current status and regulation. *J. AOAC* 71:685.

Park, D.L. and Liang, B., 1993, Perspectives on aflatoxin control for human food and animal feed. *Trends Food Sci. Technol.*, 4:334.

Park, D.L., Lopez-Garcia, R., Trujillo-Preciado, S. and Price, R.L., 1996, Reduction of risks associated with fumonisin contamination in corn. In *Fumonisins in Food*, L.S., Jackson, J.W. DeVries and L.B. Bullerman, eds., Plerum Press, New York, pp.335-344.

Park, D.L., and Price, W.D. , 2001, Reduction of aflatoxin hazards using ammoniation. *Rev. Environ. Contam. Toxicol.* (In Press).

Park, D.L. and Stoloff, L., 1989, Aflatoxin control – How a regulatory agency managed risk from an unavoidable natural toxicant in food and feed. *Regul. Toxicol. Pharmacol.*, 9:109.

Park, D.L., Njapau, H. and Boutrif, E., 1999, Minimizing risk posed by mycotoxins utilizing the HACCP concept. *Food, Nutrition and Agriculture, FAO*, 23:49.

Peers, F.G., and Linsell, C.A.. 1975, Aflatoxin contamination and its heat stability in Indian cooking oils. *Trop Sci.* 17:229.

Phillips, T.D., Clement, B.A., and Park, D.L., 1994, Approaches to reduction of aflatoxins in foods and feeds. In: *The Toxicology of Aflatoxins-Human Health, Veterinary and Agricultural Significance*, D.L. Eaton, J.D. Groopman, eds., San Diego; Academic Press. p 383.

Price, R.L., and Jorgensen, K.V., 1985, Effects of processing on aflatoxin levels and on mutagenic potential of tortillas made from naturally contaminated corn. *J. Food Sci.* 50:347.
Scott, P.M., 1984, Effects of food processing on mycotoxins. *J. Food Prot.* 47: 489.
Stoloff, L., 1982, Mycotoxins, potential environmental carcinogens. In *Carcinogens and Mutagens in the Environment*, H.F. Stich, ed., Boca Raton: CRC Press, p 97.
Trujillo, S., 1997, Reduction and management of risks associated with aflatoxin and fumonisin contamination in corn, Louisiana State University, Baton Rouge, Louisiana (Ph.D. dissertation).
Ulloa Sosa, M. and Herrea, T., 1970, Persistencia de las aflatoxinas durante la fermentacion del pozol. *Rev. Lat-Am Microbiol* 12:19.
Ulloa Sosa, M., and Shroeder, H.W., 1969, Note on aflatoxin decomposition in the process of making tortillas from corn. *Cereal Chem* 46:397.
Wood, G.M., Cooper, S.J. and Chapman, W.B., 1982, Problems associated with laboratory simulation of effects of food processes on mycotoxins. Proceedings of V. Int. IUPAC Symp Mycotoxins, Vienna, Austria, p 142.

EFFECT OF PROCESSING ON DEOXYNIVALENOL AND OTHER TRICHOTHECENES

Dionisia M. Trigo-Stockli

Kansas State University
Dept. of Grain Science & Industry
Manhattan, KS 66506

ABSTRACT

Deoxynivalenol (DON, vomitoxin) and other trichothecene toxins may contaminate crops as weather conditions during the growing season sometimes favor the growth and toxin production by the *Fusarium* fungus. Several processing procedures for the reduction of DON in contaminated wheat and corn have been studied. Although total elimination of toxin is not usually possible, processing methods such as cleaning and milling can reduce toxin to acceptable levels. In addition, more countries have guidelines or advisory levels to follow for the utilization of DON-contaminated grains such as wheat. In recent years, a few studies on grain processing have been conducted, however, there is a continuing effort to find a more effective method or a combination of methods for the reduction of DON in grains and grain products.

INTRODUCTION

The trichothecenes, a group of toxins that includes DON, nivalenol (NIV), T-2 toxin, HT-2, and diacetoxyscirpenol, are produced by several species of *Fusarium*. In animals, the trichothecenes cause feed refusal, dermal necrosis, gastrointestinal effects, and coagulopathy (Osweiler, 1986). Deoxynivalenol (Figure 1), the most commonly studied trichothecene, is produced by *Fusarium graminearum* and *Fusarium culmorum*. Deoxynivalenol causes feed refusal and emetic syndrome in sensitive animals such as pigs. Other naturally occurring trichothecenes, such as T-2 toxin, diacetoxyscirpenol, and nivalenol, have also been experimentally shown to cause feed refusal. Feed refusal is reflected in decreased weight gains and slower growth rates. Emesis may occur in animals that consume small quantities of contaminated cereal. Cases of vomiting, nausea, abdominal pain, and diarrhea in humans following consumption of scabby grain have been reported in Japan (Marasas and Nelson, 1987).

During the growing season, weather conditions may favor the growth of *Fusarium* fungi and toxin formation. Trichothecene toxin occurrence seems to be a continuing

Mycotoxins and Food Safety, edited by Trucksess *et al.*
Kluwer Academic/Plenum Publishers, 2002

problem since certain species of *Fusarium* are favored when weather conditions are favorable for crops while other *Fusarium* species are favored by adverse weather conditions, such as drought. For example, a four-year survey showed heavy infection of Kansas Hard Red Winter Wheat with *F. graminearum,* a head blight fungus, in 1993 and 1995 when the amount of rainfall was high and when temperatures were cool. The level of DON generally correlated with the level of *F. graminearum* infection (Trigo-Stockli et al., 1998).

Use of resistant cultivars would be the ultimate solution to prevent the occurrence of DON and other mycotoxins in general. However, resistant cultivars usually take a long time to develop, and when a resistant cultivar is developed, certain disadvantages may occur. An example is the recently developed barley cultivar, which is resistant to *F. graminearum*, the head blight fungus, but which also has poor malting qualities. Because of this problem, other means of reducing the level of mycotoxins in grain and grain products become more important. Some processing methods have been shown to reduce the level of DON and other trichothecenes in processed products. This paper describes the physical and chemical methods that have been shown to reduce the level of DON and other trichothecenes in naturally-contaminated grains.

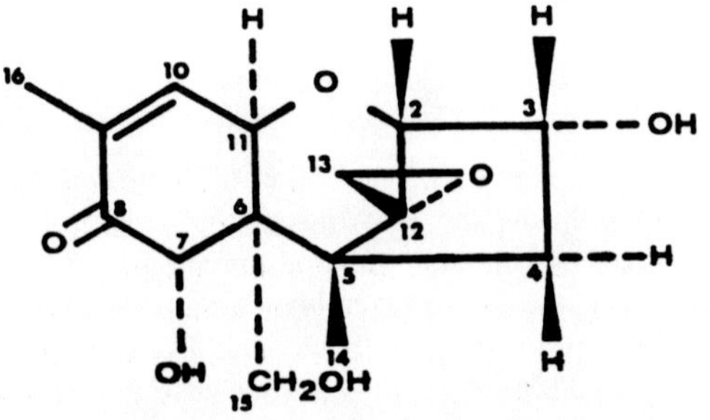

Figure 1. Deoxynivalenol structure.

PHYSICAL PROCESSING METHODS

The common physical methods of processing that have been shown to reduce the levels of DON and NIV are dry milling of wheat and wet milling of corn. Dry and wet milling are processes employed to produce various products from wheat and corn. Other methods that have been used singly or in combination with milling include cleaning, washing, sieving and dehulling, polishing, scarification, baking, feed pelleting, and extrusion processing. The success of these methods depends on the degree of contamination, distribution of the mycotoxin throughout the grain, and the processing methods utilized.

Dry Milling

Products of wheat dry milling generally include bran, shorts, red dog, and flour. Dry milling causes reductions of 15-41% for DON and 20-69% for NIV levels in wheat flour (Table 1). Deoxynivalenol fractionates during the milling process with the highest concentration in the bran and lowest in the flour (Seitz et al., 1985; Lee et al., 1987; Trigo-Stockli et al., 1996). The effectiveness of the milling process to reduce the level of DON in the flour depends on the relative distribution of the toxin within the kernels and the degree of kernel infection (Young et al., 1984; Young, 1986; Seitz et al., 1985). Generally, as DON levels in grain increase, percent reduction of DON during milling decreases. Products of corn dry milling include grits, germ, corn meal, and flour. Dry milling of DON-contaminated corn resulted in the highest DON levels in the germ meal fraction (Scott, 1991).

Table 1. Effect of dry milling on deoxynivalenol (DON) and nivalenol (NIV) in wheat.

Wheat	Toxin level in wheat (mg/kg) DON	NIV	% Toxin reduction DON	NIV	Reference
Hard red spring	8.7	-	15	-	Young et al., 1984
Soft white winter & Hard red spring	0.05-3.0	-	27-31	-	Young et al., 1984; Seitz et al., 1985
Korean wheat	0.31	0.887	20-41	20-69	Lee et al., 1987

Wet Milling

Corn wet milling products include, germ, gluten, fiber, and starch. Wet milling of NIV- and DON-contaminated corn gave the highest DON concentrations in corn steep liquor. Lauren and Ringrose (1997) studied the fate of NIV and DON in New Zealand grown corn in a commercial wet milling process. They reported that the highly water soluble mycotoxins, NIV and DON, were found at high concentrations (up to 8.8 mg/kg) in concentrated steep liquor (CSL) fractions and low levels (< 0.3 mg/kg) were found in the solid (germ, fiber, and gluten) fractions. Fiber, a wet milling by-product, which is composed mainly of pressed fiber, and concentrated steep liquor, contained high levels of the toxins (Table 2). The use of such products for animal feed presents a potential for toxicoses in some animals (Bennett and Richard, 1996).

Table 2. Recovery of nivalenol (NIV) and deoxynivalenol (DON) in corn fractions obtained from wet milling[1]

Process fraction	% Recovery Nivalenol	Deoxynivalenol
Corn	67.5	70.5
Light steep liquor	78.5	74.5
Concentrated steep liquor	86.0	93.0
Germ	72.5	73.5
Fiber	65.0	65.5
Gluten	66.0	70.0
Feed	94.0	95.5
Oil	61.0	73.5

[1]Data are averages of 2 replicate analyses, Lauren and Ringrose (1997)

Lauren and Ringrose (1997) did not collect starch in their study since it could not be reliably related back to a particular batch of input maize. However, Patey and Gilbert (1989) reported that in commercially wet milled corn, approximately 30% of DON partitioned into the starch fraction. Using a bench scale wet milling equipment to produce wet-milled corn products, Collins and Rosen (1981) reported that two-thirds of T-2 toxin initially present in contaminated corn was removed by the steep and process water, 4% was found in starch, and the remainder was distributed between germ, gluten, and fiber.

Other Physical Processing Methods

Cleaning of wheat before the milling process by screening and airflow or by removal of dockage and dusts has been shown to reduce the level of DON up to 35%. In addition, cleaning by aspiration before milling reduced DON levels in spring wheat flour (Bennett and Richard, 1996). Scouring of wheat preceding and following tempering reduced the level of DON in wheat by 22%. Other methods such as sieving, and dehulling also reduce DON levels by 40-83%. Polishing of dehusked and unhusked barley has been shown to reduce DON by 82-100% and 52-100%, respectively (Table 3). Polishing also reduced NIV levels in dehusked and unhusked barley by 48-94% and 37-97%, respectively (Lee et al., 1992).

Washing treatments are effective at reducing DON levels in wheat. However, washing treatments are not usually feasible under commercial situations due to the added costs needed to dry the grain after the treatment. Washing barley and corn three times in distilled water reduced DON concentrations by 65-69% (Trenholm et al., 1992). Moreover, Trenholm et al. (1992) reported that soaking barley, corn, and wheat in a 0.1 M sodium carbonate solution for 24 or 72 h caused a 42-100% reduction in toxin concentration.

Baking has been shown to cause little or no effect on DON levels in flour and dough. A 35% reduction in DON levels was observed in cookies and doughnuts baked from flours containing 0.5 mg/kg DON (Young et al., 1984). These authors observed a decrease only in non-yeast containing doughnuts. In doughnuts containing yeast, Young et al. (1984) speculated that a 118 to 189 increase in DON concentration was due to enzymatic conversion of unidentified precursor into DON. In contrast, Niera et al (1997) observed a 29% difference in DON levels between fermented doughs and baked products. They stated however, that further studies are needed to confirm their results. Baking has also been shown to reduce the DON concentration in dough (0.2-0.9 mg/kg flour) by 20-40% (Seitz et al., 1986).

Little or no reduction in DON concentrations was observed when dough containing flour contaminated with DON levels of 1-7 mg/kg was baked into bread (Scott et al., 1984). Similarly, El Banna et al (1983), reported no reduction in DON levels when wheat flour (2-3.5 mg DON/kg) was baked into Egyptian bread at 350°C, for 2 min. In addition, Tanaka et al. (1986) reported no reduction in DON concentration when flour contaminated with 0.38 mg/kg DON was baked into a sponge cake at 170°C for 30 minutes.

Deoxynivalenol was stable at 98.5°C during a feed pelleting process. Deoxynivalenol levels (0.2, 0.5, and 1.0 ppm) in a naturally contaminated wheat ingredient in the feed mash remained at the same level when they were processed with a CPM model CL-5 laboratory pellet mill into feed pellets (Trigo-Stockli et al., 2000). Roasting of wheat contaminated with 30 mg/kg DON using a commercial gas fired-roaster was shown (Stahr et al., 1987) to reduce DON levels by 50%.

Extrusion processing at 145 and 170°C of DON-contaminated milled flour and whole meal samples that were tempered with water or sodium bisulfite solutions did not cause a significant reduction in DON levels. Soaking DON-contaminated wheat kernels in water or 5% sodium bisulfite solution before extrusion processing lowered DON content by 33 and 89%, respectively (Accerbi et al., 1999). Using DON-spiked corn grits, Wolf-Hall (1995) reported no significant reduction in DON levels after extrusion processing. In addition, she reported that extrusion processing created a matrix effect that reduced recovery of DON during analysis.

Table 3. Effect of physical methods on deoxynivalenol (DON) level in grain.

Method	Grain	DON level (mg/kg)	% Reduction	Reference
Cleaning by removing dockage	Spring wheat	7.1	35	Scott et al., 1983
Cleaning by removing dust screenings	Soft wheat	0.07	7 – 23	Young et al., 1984
Cleaning by screening & airflow	Soft wheat	0.03-2.89	16	Seitz et al., 1985
Scouring	Canadian western red spring	12.5	22	Nowicki et al., 1988
Dehulling	Barley, wheat	5 – 23	40 – 52	Trenholm et al., 1991
Sieving	Barley, wheat, corn	5 – 23	73 – 83	Trenholm et al., 1991
Polishing	Dehusked barley	0.022	82 – 100	Lee et al., 1992
	Unhusked barley	0.025	52 – 100	Lee et al., 1992

CHEMICAL PROCESSING METHODS

Charmley and Prelusky (1994) wrote an excellent review on the effect of chemical methods for the decontamination of trichothecene-contaminated grain or feed. Calcium hydroxide monomethylamine, sodium bisulfite, calcium or sodium hypochlorite, chlorine gas, ascorbic acid, hydrogen peroxide, ammonium hydroxide, hydrochloric acid and many others have been tested to decontaminate grains containing trichothecenes such as T-2, DAS, and DON. A few chemicals have proved successful, while others showed little or no effect (Charmley and Prelusky, 1994). Calcium hydroxide monomethylamine has been shown to effectively decontaminate feeds containing T-2 and diacetoxyscirpenol. Another chemical that has been shown to reduce DON concentration in contaminated corn is sodium bisulfite solution. Moisture content and temperature influence the effectiveness of these chemicals during processing. High moisture content e.g. 25% and high temperatures (100-121°C) during processing cause more toxin reduction.

Sodium bisulfite has also been shown to reduce DON levels in naturally contaminated soft white wheat. Accerbi et al. (1999) reported that soaking contaminated wheat for 1 hr in 5% sodium bisulfite solution reduced DON from 7.3 μg/g to 0.8 μg/g. One concern over the use of sodium bisulfite is the formation of DON-sulfonate adduct (DON-S) that is unstable at high temperatures and high pH (Young, 1986). However, autoclaving and treating corn (7.2 mg DON/kg) with sodium bisulfite appeared to improve feed intake and body weight gain of pigs (Young et al., 1987). Therefore, sodium bisulfite treatment of DON contaminated grain may be useful for decontaminating grains intended for animal feed.

Another chemical that has been shown to reduce DON in wheat is calcium hypochlorite. Chlorine, at gas concentrations greater than 1%, reduced DON levels (Young, 1986). The lack of significant reduction of DON levels, under normal commercial milling and bleaching conditions was speculated to be due to the low levels of chlorine (Young et al., 1984). In our laboratory, the effect of calcium hypochlorite and scarification on DON-contaminated soft white wheat has been studied. The wheat was scarified (0, 10, and 20 sec.) and various levels of calcium hypochlorite (0, 200, 400, and 800 ppm) were added to water during the tempering process. Scarification was accomplished using a 40-grit sandpaper in a 1725 rpm/min scarifier (Forsberg's, Inc). Two-hundred grams of wheat were scarified each time with a total of 500 g for each treatment. Wheat was milled using a Brabender quadromat junior. Deoxynivalenol levels in bran and flour were determined using gas chromatography (Trigo-Stockli et al., 1996).

Our data showed that scarifying DON-contaminated wheat for 20 sec caused a significant reduction in DON level in the bran (Figure 2). In flour, DON levels were not significantly different among scarified and non-scarified wheat. Reduction of DON levels in wheat bran is significant because DON fractionates during the milling process and a greater percentage of DON is usually present in the bran. Wheat middlings which include the bran, and other fractions such as shorts and red dog are utilized for animal feed. Treatment of wheat with different levels of calcium hypochlorite (chlorine), had no effect on DON levels in bran and flour (Figure 3). Although some of the chemicals described above caused some reduction of DON, presently, there are no regulations on the use of chemicals for DON-contaminated grains.

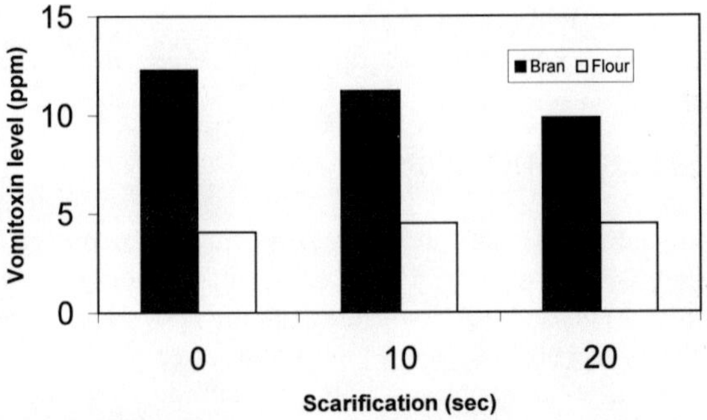

Figure 2. Deoxynivalenol levels in soft wheat bran and flour at various times of scarification.

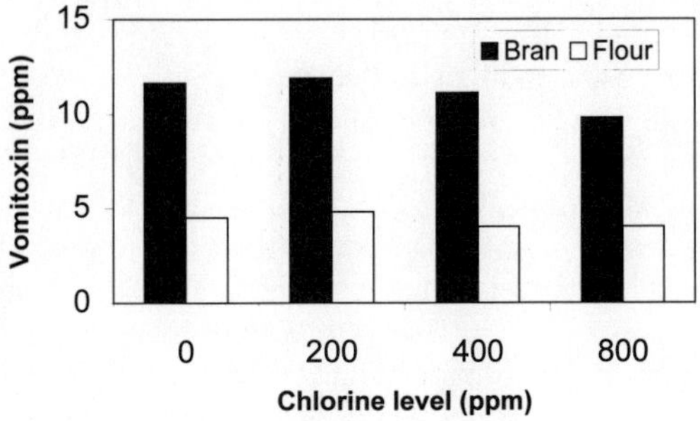

Figure 3. Deoxynivalenol levels in soft wheat bran and flour at various concentrations of calcium hypochlorite.

CONCLUSION

Some processing methods cause some reduction in the levels of DON and other trichothecenes in processed products. The level of reduction, however, depends on the degree of fungal contamination and distribution of DON in the grain, the method of processing utilized, and whether a single or combination of methods are utilized. The reduction of DON through grain processing is important because it allows utilization of processed products containing DON as long as the level meets the Food and Drug Administration's (FDA) advisory levels. The advisory levels for DON in wheat are as follows: 1 ppm in finished wheat products such as flour, bran, and germ intended for human consumption, 10 ppm in grains and grain-by products destined for ruminant beef and feedlot cattle older than 4 months, and chickens such that the ingredients do not exceed 50% of the diet, and 5ppm in grains and grain by-products destined for swine such that the ingredients do not exceed 20% of the diet as well as other animals such that the ingredient do not exceed 40% of their diet. No advisory level has been set for wheat intended for milling. In Canada, guidelines for DON content in grains for domestic use as well as for export have been established at no more than 2 ppm in uncleaned soft white winter wheat for adult consumption, 1.2 ppm for baby food, and 1 ppm for animal feed (Van Egmond, 1989). The guidelines of the European Community (EC) for utilization of DON-contaminated grain are covered in a paper by Van Egmond (1989).

REFERENCES

Accerbi, M., Rinaldi, V.E.A., and Ng, P.K.W., 1999, Utilization of highly deoxynivalenol-contaminated wheat via extrusion processing, *J. Food Prot.* 62:1485-1487.

Bennett, G.A., and Richard, J.L., 1996, Influence of processing on *Fusarium* mycotoxins in contaminated grains, *Food Technol.* 235-238.

Charmley, L.L., and Prelusky, D.B., 1994, Decontamination of *Fusarium* mycotoxins, In *Mycotoxins in Grain, Compounds other than Aflatoxin,* J.D. Miller and H.L. Trenholm, eds., Eagan Press, St. Paul, MN.

Collins, G.J. and Rosen, J.D., 1981, Distribution of T-2 toxin in wet-milled corn products, *J. Food Sci.* 46:877-879.

El Banna, A.A., Lau, P.Y., and Scott, P.M, 1983, Fate of mycotoxins during processing of foodstuffs. II. Deoxynivalenol (vomitoxin) during making of Egyptian bread, *J. Food Prot.* 46:484-488.

Lauren, D.R., and Ringrose, M.A., 1997, Determination of the fate of three *Fusarium* mycotoxins through wet-milling of maize using an improved HPLC analytical technique, *Food Add. Contam.* 14:435-443.

Lee, U.S., Jang, H.S., Tanaka, T., Oh, Y.J., Cho, C.M., and Ueno, Y., 1987, Effect of milling on decontamination of *Fusarium* mycotoxins nivalenol, deoxynivalenol, and zearalenone in Korean wheat, *J. Agric. Food Chem.* 35:126-129.

Lee, U-S., Lee, M-Y., Park, W-Y., and Ueno, Y., 1992. Decontamination of *Fusarium* mycotoxins, nivalenol, deoxynivalenol, and zearalenone in barley by the polishing process, *Mycotoxin Res.* 8:31-36.

Marasas, W.F.O., and Nelson, P.E., 1987, *Mycotoxicology,* The Pennsylvania State University Press, University Park.

Niera, M.S., Pacin, A.M., Martinez, E.J., Molto, G., and Resnik, S.L., 1997, The effects of bakery processing on natural deoxynivalenol contamination, *Int. J. Food Microbiol.* 37: 21-25.

Nowicki, T.W., Gaba, D.G., Dexter, J.E., Matsuo, R.R., and Clear, R.M., 1988, Retention of the *Fusarium* mycotoxin deoxynivalenol in wheat during processing and cooking of spaghetti and noodles, *J. Cereal Sci.* 8:189-202.

Osweiler, G., 1986, Occurrence and clinical manifestations of trichothecene toxicosis and zearalenone toxicoses, In *Diagnosis of Mycotoxicosis,* J.L. Richard and J.R. Thurston, eds., Martinus Nijhoff Publishers, Boston.

Patey, A.L., and Gilbert, J., 1989, Fate of Fusarium mycotoxins in cereals during food processing and methods for their detoxification, in: *Fusarium: Mycotoxins, Taxonomy and Pathogenecity,* J. Chelkowski ed, Elsevier, Amsterdam.

Scott, P.M., Kanhere, S.R., Lau, P.Y., Dexter, J.E., and Greenhalgh, R., 1983, Effects of experimental flour milling and breadbaking on retention of deoxynivalenol (vomitoxin) in hard red spring wheat, *Cereal Chem.* 60:421-424.

Scott, P.M., 1991, Possibilities of reduction or elimination of mycotoxins present in cereal grains, In *Cereal Grain Mycotoxins, Fungi and Quality in Drying and Storage*, J. Chelkowski, ed., Elsevier, Amsterdam.

Scott, P.M., Kanhere, S.R., Dexter, J.E., Brennan, P.W., Trenholm, H.L., 1984, Distribution of the trichothecene mycotoxin deoxynivalenol (vomitoxin) during the milling of naturally contaminated hard red spring wheat and its fate in baked products, *Food Addit. Contam.* 1:313-323.

Seitz, L.M., Eustace, W.D., Mohr, H.E., Shogren, M.D., and Yamazaki, W.T., 1986, Cleaning, milling, and baking tests with hard red winter wheat containing deoxynivalenol, *Cereal Chem.* 63:146-150.

Seitz, L.M., Yamazaki, W.T., Clements, R.L., Mohr, H.E., and Andrews, L., 1985, Distribution of deoxynivalenol in soft wheat mill streams, *Cereal Chem.* 62:467-469.

Stahr, H.M., Osweiler, G.D., Martin, P., Domoto, M., Debey, B, 1987, Thermal detoxification of trichothecene contaminated commodities, in: Biodeterioration Research I, G.C. Llewellyn and C.E. O'Rear, eds, Plenum Press, New York.

Tanaka, T, Hasegawa, A., Yamamoto, Y.M., and Ueno, Y., 1986, Residues of *Fusarium* mycotoxins, nivalenol, deoxynivalenol, and zearalenone, in wheat and processed food after milling and baking, *J. Food Hyg. Soc. Jpn.* 27:653-655.

Trenholm, H.L., Charmley, L.L., Prelusky, D.B., Warner, R.M., 1991, Two physical methods for the decontamination of four cereals contaminated with deoxynivalenol and zearalenone, *J. Agric. Food Chem.* 39:356-360.

Trenholm, H.L., Charmley, L.L., Prelusky, D.B., Warner, R.M., 1992, Washing procedures using water or sodium carbonate solutions for the decontamination of three cereals contaminated with deoxynivalenol and zearalenone, *J. Agric. Food Chem.* 40:2147-2151.

Trigo-Stockli, D.M., Deyoe, C.W., Satumbaga, R.F., and Pedersen, J.R., 1996, Distribution of deoxynivalenol and zearalenone in milled fractions of wheat, *Cereal Chem.* 73:388-391.

Trigo-Stockli, D.M., Sanchez-Marinez, R.I., Cortez-Rocha, M.O., and Pedersen, J.R., 1998, Comparison of the distribution and occurrence of *Fusarium graminearum* and deoxynivalenol in hard red winter wheat for 1993-1996, *Cereal Chem.* 75:841-846.

Trigo-Stockli, D.M., Obaldo, L.G., Dominy, W.G., and Behnke, K.C., 2000, Utilization of deoxynivalenol-contaminated hard red winter wheat for shrimp feeds, *J. World Aquaculture Soc.* 31:247-254.

Van Egmond, H.P., 1989, Current situation on regulations for mycotoxins. Overview of tolerances and status of standard methods of sampling and analysis, *Food Addit. Contam.* 6:139-188.

Wolf-Hall. C.E., 1995, Characterization of deoxynivalenol producing mold, detection of deoxynivalenol and effects of heat on mycotoxins, Ph.D. Dissertation, University of Nebraska, Lincoln.

Young, J.C., 1986, Reduction in levels of deoxynivalenol in contaminated corn by chemical and physical treatment, *J. Agric. Food Chem.* 34:465-467.

Young, J.C., Fulcher, R.G., Hayhoe, J.H., Scott, P.M., and Dexter, J.E., 1984, Effect of milling and baking on deoxynivalenol (vomitoxin) content of eastern Canadian wheats, *J. Agric. Food Chem.* 32:659-664.

Young, J.C., Trenholm, H.L., Friend, D.W., and Prelusky, D.B., 1987, Detoxification of deoxynivalenol with sodium bisulfite, and evaluation of the effects when pure mycotoxin or contaminated corn was treated and given to pigs, *J. Agric. Food Chem.* 35:259-261.

EFFECT OF PROCESSING ON OCHRATOXIN A (OTA) CONTENT OF COFFEE

R. Viani

FAO Consultant
Association Scientifique Internationale du Café (ASIC)
CH 1802 Corseaux, Switzerland

ABSTRACT

Coffee production can be roughly separated into three main steps 1) cherry processing to green coffee beans, 2) storage and transportation of green coffee to the place of consumption, and 3) green coffee processing to roasted and ground coffee and soluble coffee. The mold species which are known to produce ochratoxin A (OTA) in coffee have been identified as *Aspergillus ochraceus*, *A. carbonarius* and occasionally, *A. niger*. The length of time spent at a water activity > 0.80 at any moment until roasting defines the risk of mold growth and OTA production. However, the specific moment and locus of contamination have not yet been clearly identified. Since coffee husks are a significant source of OTA contamination, cleaning and grading of green coffee are effective methods for reducing OTA levels in coffee. During the process of converting green coffee to roasted and soluble coffees, up to 90% reduction in OTA levels can occur.

INTRODUCTION

The coffee travel from production to consumption can be roughly separated into three main sections: 1) cherry ripening, picking and processing to green beans in the producer countries, 2) storage and transportation, and 3) green coffee processing to (decaffeinated), roasted and ground and soluble coffee. This chapter reviews current knowledge on the risk of mold proliferation/ochratoxin A (OTA) contamination and on the reduction of contamination during green coffee processing (Viani, 2000).

The influence of microflora on the organoleptic characteristics (Corte dos Santos et al., 1971) and on the risk of mycotoxin formation (Poisson et al., 1975) in coffee has been recognized since the 1970's, and OTA contaminated commercial green coffee samples were found (Levi et al., 1975), in spite of the poor sensitivity of the analytical methods available at the time, around 20 μg/kg. However, no residue was detected in roast coffee, so that a risk to health was deemed unlikely. When Tsubouchi et al. (1988) reported that traces of OTA were still present in commercial roast coffees sold on the Japanese market, the European coffee industry, through its Institute for Scientific Information on Coffee (ISIC)

Mycotoxins and Food Safety, edited by Trucksess *et al.*
Kluwer Academic/Plenum Publishers, 2002

submitted the problem to the Institute of Toxicology of the Swiss Federal Technical High School in Zurich for an evaluation of possible risks to consumers (Studer-Rohr et al., 1995). This research was followed by the examination of the OTA content of several hundreds commercial roast and soluble coffees selected according to market share in eight European countries (van der Stegen et al., 1997). In addition, a pilot investigation, conducted by the University of Surrey in seven producing countries, examined the mycological situation of coffee production during the years 1996-1998 (Frank, 1998).

A Nestlé team has done research both in a producer country (Thailand), and in the pilot plant, with the following findings:

Storage of fresh coffee cherries (particularly ripe and overripe ones) before processing increases the risk of mold growth (Bucheli et al., 2000). More than 90% of the total OTA load is located in the husks of sun dried dry processed coffee (Bucheli et al., 1998). No increase of OTA content occurred in green Robusta coffee beans stored for up to one year at a relative humidity around 80%, corresponding to a coffee moisture content of 13.5% and and a_w of 0.72 (Bucheli et al., 1998). An 85% reduction in the OTA load of green coffee occurs during processing of green coffee to roasted and soluble coffees (Blanc et al., 1998). Soluble coffee adulterated by the addition of husks may contain relatively large amounts of OTA (Pittet et al., 1996).

Two workshops on prevention of mould growth in coffee were organized by the *Association Scientifique Internationale du Café* (ASIC), in Nairobi in 1997 (Anonymous, 1997) and in Helsinki in 1999 (Anonymous, 1999), where the question of OTA toxicity was also discussed (Walker, 1997; Zepnik et al., 1999). A third workshop was planned for the 19th ASIC Conference, held in Trieste, Italy, in May 2001.

Regulatory authorities in a few European countries have proposed limits for the OTA content of coffee products, and, as reported by the World Trade Organization (WTO) (WHO, 2000), the European Union is considering if and what limits should be adopted.

ON-GOING WORK

FAO Project in Uganda

A program of intervention at coffee origin was initiated in May 1999 by FAO in Uganda with the cooperation of the Uganda Coffee Development Authority (UCDA). Fieldwork by international and national experts has identified the socio-economical and technical parameters controlling green coffee production and export, and is now checking their impact on the OTA load: Coffee is the most important commodity of Uganda employing 15% of the population and accounting for 65% of foreign exchange earnings. Average yearly production during 1996-1999 was 230,000 tons, or 5.7% of total world exports. Production included 10% Arabica and 90% Robusta coffees produced on 272,000 ha, in very small farms averaging in size 0.5 ha. Productivity was around 700 kg of clean coffee per ha. All coffee is sun dried at the farm over 7-14 days. The production chain goes from farmer (through buyer) to primary processor and exporter. Since Uganda is a land-locked country, green coffee must be either trucked (7-10 days) or transported by rail (up to two weeks) to the harbors of Mombasa (Kenya) and Dar es-Salam (Tanzania).

The main points of potential microbiological hazard have been identified as at the farm and at the traders (now often by-passed by farmers) and include the following: storage of fresh cherries before drying; drying on bare soil without protection from rain and dew (75 to 90% of farmers); and marketing of wet coffee (ca. 50% of the coffee has a moisture as high as 20%). The reasons for these situations are high costs of drying equipment, trays or tarpaulins and lack of cash and incentives to the farmers to produce a better quality.

At the primary processor, hazards are linked with the following conditions: 1) poor location of plants which are often erected in swampy areas, 2) inadequate drying facilities for incoming wet cherries, 3) inadequate grading of green coffee, 4) contamination of green coffee by dust, husks and dirty bags, 4) mixed storage of graded export coffee with rejects and husks, and 5) lack of quality control tests for moisture and defects.

Only the introduction of differential pricing, now done in a few districts, would convince farmers and primary processors to apply good agricultural practices and good manufacturing practices.

A workshop, addressed mainly to officials from farmer associations, from the Uganda Coffee Development Authority, and from COREC (the Coffee Research Center) was organized. Teaching material from FAO (Anonymous, 1998) was adapted to train trainers on good agricultural practices/good manufacturing practices/HACCP in coffee production to be used for further training down to those at the village level and for distribution to farmers and processors as a complement to existing manuals (Anonymous, undated). The microbiological problems of coffee production, and the development of a microbiological analytical quality program were reviewed in a second workshop held by an international coffee microbiology expert, and dealt specifically with the appropriate methodology to isolate fungi from coffee, an overview of ochratoxin A in coffee, and tests for ochratoxin A production from fungi isolated from coffee (Taniwaki, 2000).

CFC/ICO/FAO Mold Prevention Program

A four years prevention program is now starting, financed by the Common Fund for Commodities, with funds from the European Coffee Federation and in kind contribution by the participating countries (Brazil, Colombia, Côte d'Ivoire, India, Indonesia, Kenya and Uganda), and by the French agricultural research agency CIRAD, *Centre International de Recherche Agronomique pour le Dévéloppement*. The program's aim is the identification of critical control points for mold infection and OTA contamination at the producer end, to complete implementation of HACCP to green coffee production.

European Coffee Federation (ECF)

In order to complete the evaluation of the coffee production chain from producer to consumer, ECF has conducted green coffee storage and transport trials in containers (Blanc, 2000). The main fluctuations in temperatures, which may lead to condensation, wetting of the top layer, proliferation of mold, and OTA contamination, occur during overland transport to export harbor, and again on unloading at import ports in the consumer countries, particularly during the cold season. The safest way of transporting coffee found to consist of storing coffee in the hold of ships in closed, non-aerated containers, so that the risk of temperature fluctuations, moisture condensation, and subsequent spoilage during the ocean trip are reduced. The top layer in the container is the most exposed to the risk of wetting and spoilage, and therefore, must be protected.

DISCUSSION AND CONCLUSIONS

The consensus from all the work accomplished so far can be summarized as follows. The toxicological status of OTA has not yet been settled. Coffees contaminated by small amounts of OTA can still be found in the trade, but a trend toward lower levels is perceptible since the publication of the European market analysis in 1997 (van der Stegen et al., 1997). The largest part (more than 90%) of the OTA contamination of coffee is usually concentrated in the husks. Cleaning, grading and hygienic storing of green coffee is, therefore, of paramount importance. An indirect confirmation of the importance of husk removal to reduce OTA contamination is the observation that husk addition, a fraudulent practice sometimes encountered in the manufacture of soluble coffee, may lead to the presence of relatively high levels of OTA (Pittet et al., 1996). The main good agricultural, hygienic and manufacturing practices, suitable for reducing the risk of OTA contamination have now been identified, and are already applied whenever there is an incentive in so doing. They are rapid processing of fresh cherries, avoidance of wetting at each stage of the handling chain before roasting, reduction to a level as low as possible of defects such as husks, un-hulled cherries and moldy beans, and segregation of cleaned and graded green coffee from discarded material, such as husk in dry processing or pulp material in wet processing. Precise data are not yet available for establishing a HACCP system: The maximum length of time spent by coffee within an a_w window roughly situated between 0.99-0.80, or 0.95, when protected by cherry microbial flora (Frank, 1999) at the highest activities, tolerated at any moment before roasting is estimated at 4 to 6 days (Frank, 2000).

The mycological situation is beginning to get clarified. The relative humidity maximum for good conservation of Arabica green coffee beans at 25°C, without mold proliferation, has been estimated (Multon et al., 1974) as 76% (*ca.* 18% moisture), while maximum proliferation occurs between 25 and 30°C (Joosten et al, 2001). The main *Aspergilli*, which are known to produce OTA during coffee drying or storage/transportation, have been identified as *A. ochraceus*, *A. carbonarius* and, rarely, *A. niger* (Heenan et al., 1998; Taniwaki et al., 1999). *A. citricus* and *A. lactocoffeatus*, have been isolated on rare occasions from coffee grown in Indonesia and in Venezuela (Frank, personal communication). The exact temperatures and a_w windows, where there is the greatest danger of mold proliferation and OTA formation, must still be defined. a minimum a_w for growth of both *A. carbonarius* and *A. ochraceus* is around 0.80, and a minimum a_w for OTA production is around 0.85, at 25 °C. The growth of *A. carbonarius* stops at 35°C (Joosten, 2001). Refined data shall be useful to optimize drying of coffee. There is a hypothesis that the presence of competing hydrophilic organisms (yeasts) reduces the risk of mold proliferation at the highest a_w's (Frank, personal communication). If the hypothesis is confirmed, the window of OTA production would drastically be particularly where the model would predict maximum formation. The place and moment of mold infection is still unknown. Is contamination due to the presence of ubiquitous spores in the air and in the soil, settling on coffee at some time during cherry processing, which then leads to mold proliferation whenever a_w conditions become favorable? Or, does a vector, such as the coffee pest *Hypothenemus hampei,* carry *A. ochraceus* (Vega et al., 1999)? The first hypothesis is corroborated by OTA accumulation occurring whenever coffee is rewetted during storage and transport, in the producer country, on the sea, or in the consumer country. Work is ongoing to verify the second hypothesis.

The chemical analytical situation for the measure of OTA can be summarized as follows. Elaborate analytical techniques for the determination of OTA down to less than 0.1 µg/kg in all coffee products are available (Pittet et al., 1996). No sensitive field kit for OTA analysis is yet available. Representative sampling at the consumer end is extremely cumbersome and costly.

To be able to implement HACCP standards of quality assurance, particularly during primary processing in the producer countries, production must comply with the Codex General Principles of Food Hygiene and GMPs as described by Anonymous (1998), and this is the objective of the ongoing CFC/FAO/ICO "Coffee quality improvement project through reduction of mold growth".

REFERENCES

Anonymous, undated, *A guide to good harvesting, handling and processing of natural Uganda coffee*, UCDA, Kampala.

Anonymous, 1997, Special workshop on the enhancement of coffee quality by reduction of mold growth, *Proc. 17th Conference Association Scientifique Internationale du Café*, 367.

Anonymous, 1998, Food quality and safety systems. A training manual on food hygiene and the hazard analysis and critical control points (HACCP) system, FAO, Rome.
Anonymous, 1999, 2nd Workshop on Enhancement of Coffee Quality by Reduction of Mould Growth, *Proc. 18th Conference Association Scientifique Internationale du Café*, 219.
Blanc, M., Pittet, A., Muñoz-Box, R., and Viani, R., 1998, Behavior of ochratoxin A during green coffee roasting and soluble coffee manufacturing, *J. Agric. Food Chem.* 46:673.
Blanc, M., 2000, Personal communication.
Bucheli, P., Kanchanomai, C., Meyer, I., and Pittet, A., 2000, Development of ochratoxin A during robusta (Coffea canephora)- coffee cherry drying, *J. Agric. Food Chem.* 40:1358.
Bucheli, P., Meyer, I., Pittet, A., Vuataz, G., and Viani, R., 1998, Industrial storage of green Robusta coffee under tropical conditions and its impact on raw material quality and ochratoxin A content, *J. Agric. Food Chem.* 46:4507.
Corte dos Santos, A., Hahn, D., Cahagnier, B., Drapron, R., Guilbot, A., Lefebvre, J., Multon, J.L., Poisson, J., and Trentesaux, E., 1971, Etude de l'évolution de plusieurs caractéristiques d'un café arabica au cours d'un stockage expérimental effectué à cinq humidités relatives différentes, *Proc. 5th Conference Association Scientifique Internationale du Café ASIC*, 304.
Frank, J.M., 1998, Appendix 4 – Main results of the pilot project and follow-up actions, in *Project Proposal to the Common Fund for Commodities*, International Coffee Organization, London.
Frank, J.M., 1999, Mycotoxin prevention and decontamination, *Third Joint FAO/WHO/UNEP International Conference on Mycotoxins*, Tunis, March 3-6, 1999.
Frank, J.M., 2000, Development of critical control points for preventing ochratoxin A (OTA) accumulation in coffee, *10th International IUPAC Symposium on Mycotoxins and Phycotoxins*, Guaruja (Brazil) May 21-25, 2000.
Frank, J.M., Personal communication.
Heenan, C.N., Shaw, K.J., and Pitt, J.I., 1998, Ochratoxin A production by *A. carbonarius* and *A. niger* isolates and detection using coconut cream agar. *J. Food Mycol.* 1:67.
Joosten, H.M.L.J., Goetz, J., Pittet, A., Schellenberg, M., and Bucheli, P., 2001, Production of ochratoxin A by *Aspergillus carbonarius* in coffee, *Int. J. Food Microbiology*, 65:39.
Levi, C.P., Trenk, H.L. and Yeransian, J.A., 1975. Investigations of mycotoxins relative to coffee, *Proc. 7th Conference Association Scientifique Internationale du Café*, 287.
Multon, J.L., Poisson, J., Cahagnier, B., Hahn, D., Barel, M., and Corte dos Santos, A., 1974, Evolution de plusieurs caractéristiques d'un café arabica au cours d'un stockage expérimental effectué à cinq humidités relatives et quatre températures différentes, *Café Cacao Thé* 18:121.
Pittet, A., Tornare, D., Huggett, A., and Viani, R., 1996, Liquid chromatographic determination of ochratoxin A in pure and adulterated soluble coffee using an immunoaffinity column cleanup procedure, *J. Agric. Food Chem.* 44:3564.
Poisson, J., Cahagnier, B., Multon, J.L., Hahn, D., and Corte dos Santos, A., 1975, La microflore du café. Méthode de dénombrement et influence sur les qualitiés organoleptiques, *Proc. 7th Conference Association Scientifique Internationale du Café*, 311.
Studer-Rohr, I., Dietrich, D.R., Schlatter, J., and Schlatter, C., 1995, The occurrence of ochratoxin A in coffee, *Food Chem. Toxicol.* 33:341.
Tanawaki, M.H., Pitt, J.I., Urgano, G.R., Teixeira, A.A., and Leitão, M.F.F., 1999, Fungi producing ochratoxin A in coffee, *Proc. 18th Conference Association Scientifique Internationale du Café*, 239.
Taniwaki, M.H. 2000. Personal communication.
Tsubouchi, H., Terada, H., Yamamoto, K., Hisada, K., and Sakabe, Y., 1988, Ochratoxin A found in commercial roast coffee, *J. Agric. Food Chem.* 36:540.
van der Stegen, G., Jörissen, U., Pittet, A., Saccon, M., Steiner, W., Vincenzi, M., Winkler, M., Zapp, J., and Schlatter, C., 1997, Screening of European coffees final products for occurrence of ochratoxin A (OTA), *Food Addit. Contam.* 14:211.
Vega, F.E., Mercadier, G., and Dowd, P.F., 1999, Fungi associated with the coffee berry borer *Hypothenemus hamei* (Ferrari) (Coleopter: *Scolytidae), Proc. 18th Conference Association Scientifique Internationale du Café*, 229.
Viani, R., 2000, Coffee, in *Ullmann's Encyclopedia of Industrial Chemistry*, Wiley-VCH, Weinheim (Germany) (available only as CD-ROM).
Walker, R., 1997, Quality and safety of coffee, *Proc. 17th Conference Association Scientifique Internationale du Café*, 51.
WHO Committee on Sanitary and Phytosanitary Measures, 2000, Notification, *G/SPS/EEC/104* of 27 November 2000.
Zepnik, H., Pähler, A., Schauer, U., and Dekant, W., 1999, Biotransformation and lack of mutagenicity of ochratoxin A using combinations of mammalian biotransformation enzymes, *Proc. 18th Conference Association Scientifique Internationale du Café*, 193.

STABILITY OF FUMONISINS IN FOOD PROCESSING

Lloyd B. Bullerman,[1] Dojin Ryu,[1] and Lauren S. Jackson[2]

[1]Department of Food Science and Technology
University of Nebraska-Lincoln
Lincoln, NE 68583-0919
[2]National Center for Food Safety and Technology
Food and Drug Administration
6502 S. Archer Road
Summit-Argo, IL 60501

ABSTRACT

Fumonisins are mycotoxins produced by *Fusarium verticillioides (moniliforme) and Fusarium proliferatum* that are found in corn and processed corn-based food products. Although generally heat stable, fumonisin concentrations appear to decline as processing temperatures increase. At processing temperatures of 125°C or lower, losses of fumonisin are low (25-30%), whereas at temperatures of 175°C and higher, losses are greater (90% or more). Processes such as baking and canning, where product temperatures rarely reach 175°C, result in little or no loss of fumonisin. Processes such as frying and extrusion cooking, where temperatures can exceed 175°C, result in greater losses. Heating fumonisin in the presence of glucose results in an apparent first order loss of the toxin. Adding glucose to corn muffins and extrusion mixes results in high losses of fumonisins during baking and extrusion processing. Little information exists on the effects of chemical and bioprocessing on fumonisins. Alkaline processing of corn, such as in the nixtamalization process, hydrolyzes fumonisins and results in a more toxic product. Additional research is needed to identify and to determine the toxicity of fumonisin decomposition products.

INTRODUCTION

Fumonisins are a group of mycotoxins that include fumonisin B_1 (FB_1), fumonisin B_2 (FB_2) and fumonisin B_3 (FB_3) which are produced mainly by *Fusarium verticillioides* (formerly known as *F. moniliforme* Sheldon), *Fusarium proliferatum* and some related species. These species are some of the most prevalent fungi associated with corn and are found in all corn growing regions of the world (Marasas et al., 1984). *Fusarium verticillioides* is a soil borne plant pathogen that can cause stalk and kernel rot in corn, and can also cause asymptomatic infections of corn plants. It is not uncommon to find lots of shelled corn with 100% kernel infection, but without visible mold damage or deterioration (Marasas et al., 1984). Consequently, food-grade corn can be contaminated without outward signs.

FB_1 has been associated with several fatal diseases in animals, including equine leukoencephalomalacia (Kellerman et al., 1990; Riley et al., 1997) and porcine pulmonary edema (Harrison et al., 1990; Osweiler et al., 1992; Smith et al., 1996). The toxin has also been demonstrated experimentally to be hepatotoxic and hepatocarcinogenic in rats (Gelderblom et al., 1991). Epidemiological evidence indicates a possible correlation between the consumption of fumonisin/*F. moniliforme* (*verticillioides*) contaminated corn and the high incidence of esophageal cancer in countries where corn is a dietary staple (Sydenham et. al., 1990; Franceschi et al., 1990; Sydenham et. al., 1991; Ross et al., 1991; Rheeder et al., 1992; Chu and Li, 1994). It has also been shown that culture material of *F. moniliforme* (*verticillioides*), when added to a carbohydrate diet low in fat, was atherogenic in a non-human primate, raising the possibility that fumonisins may be involved in other human diseases (Fincham et al., 1992). In China the incidences of atherosclerosis and strokes are higher for Northern Chinese, who consume more corn, than for the Southern Chinese, who consume more rice (Chang et al., 1991; Norred and Voss 1994).

Numerous surveys have been conducted on the fumonisin content of corn-based human foods. In a survey of corn-based foods obtained from different regions of the U.S., Castelo et al. (1998a) reported finding FB_1 levels as high 5900 ng/g. Thiel et al. (1992) reported that of 29 commercial corn products surveyed in the U.S., FB_1 concentrations ranged from 0 to 2790 ng/g. Pittet et al. (1992) surveyed 120 corn-based products in Switzerland and found FB_1 (54-792 ng/g) in 44 samples and FB_2 (52-162 ng/g) in 15 samples. Other surveys of human foods obtained in the U.K. (Patel et al., 1997), Uruguay (1997), Spain (Sanchis et al., 1994), Italy (Doko and Visconti, 1994), Brazil (Machinski and Valente, 2000) and other countries (Sydenham et al., 1991) have shown similar fumonisin levels. Gutema et al. (2000) reported the co-occurrence of FB_1 with moniliformin in corn-based foods obtained from the U.S.

Most surveys have indicated that corn-based foods that receive minimum processing treatments, such as corn meal, flour, grits and polenta contain more fumonisin than more highly processed foods (corn flakes, cereals, tortilla chips, etc.). Sydenham et al. (1991) reported that corn meal and corn grits from the U.S. contained from 0 to 2790 ng/g of FB_1 and 0 to 1065 ng/g of FB_2. In a survey of corn products obtained in different regions of the U.S., Castelo et al. (1998a) found FB_1 levels in corn meal as high as 5900 ng/g. Surveys indicate that corn meal and grits from Canada, Germany, Italy, South Africa, Egypt, Peru and other countries contain levels of fumonisins similar to those found in U.S. corn meal (Sydenham et al., 1991; Doko and Visconti, 1994; Patel et al., 1997; Usleber and Martlbauer, 1998; Machinski and Valente, 2000). Lower or non-detectable fumonisin concentrations were found in corn flakes, alkali treated corn tortillas and chips, and breakfast cereals (Pittet et al., 1992; Doko and Visconti, 1994; Patel et al, 1997; Pineiro et al, 1997; Castelo et al., 1998a; Machinski and Valente, 2000).

Data from surveys and toxicological studies using a variety of livestock and experimental animals (Marasas et al., 1988; Harrison et al, 1990; Gelderblom et al., 1991; Wilson et al, 1992; Colvin et al, 1993; Floss et al., 1994; Smith et al., 1996; Riley et al., 1997) suggest that human health risks associated with exposure to fumonisins are possible. Consequently, the U.S. Food and Drug Administration has drafted guidance levels for fumonisin in corn-based products to reduce human exposure to these natural toxins (U.S. Food and Drug Administration, 2000). These values are 2.0 μg/g total fumonisins for degermed dry milled corn products (flaking grits, corn grits, corn meal, corn flour with fat content of < 2.25 %, dry weight basis), 4.0 μg/g for wholly or partially degermed dry milled corn products (flaking grits, corn grits, corn meal, corn flour with fat content of < 2.25 %, dry weight basis), 4.0 μg/g for dry milled corn bran, 4.0 μg/g for cleaned corn intended for masa production, and 3.0 μg/g for cleaned corn intended for popcorn.

EFFECTS OF PROCESSING ON STABILITY OF FUMONISINS IN CORN-BASED FOODS

Processing methods that may affect the stability of fumonisins in corn-based foods include chemical treatment to destroy or change the toxin, biological treatment to remove or destroy the toxin and physical treatments that may cause redistribution, removal or destruction of the toxin in the processed food.

Chemical Processing

Limited information is available on destruction of fumonisins by chemical treatments. Norred et al. (1991) reported that ammoniation of *F. moniliforme (verticillioides)* culture material as well as naturally contaminated corn for 4 days at 50°C and atmospheric pressure resulted in the reduction of FB_1 by 30 and 45%, respectively. However, the toxicity of the culture material to rats was not altered by the ammoniation treatment. Park et al. (1992) reported a 79% reduction in FB_1 when ammoniation was carried out at 20°C and at higher pressure, though the toxicity of the treated corn was not tested. Treating FB_1-contaminated corn with H_2O_2 and $NaHCO_3$ reduced the fumonisin concentrations by up to 100% and reduced the toxicity considerably (Park et al., 1996).

Sydenham et al. (1995a,b) reported that treating corn or culture material with calcium hydroxide and heat (nixtamalization) resulted in the hydrolysis of FB_1. In another report, nixtamalization of *F. proliferatum* cultured corn at 80-100°C for 1 hr with 1.2% calcium hydroxide resulted in the formation of hydrolyzed fumonisins (Hendrich et al., 1993). However, the nixtamalized corn and culture materials were still toxic to rats (Hendrich et al., 1993; Voss et al., 1996). Treating an aqueous solution of FB_1 with ozone gas for 15 sec resulted in the complete loss of FB_1 but the formation of a derivative that was still as toxic to rats as FB_1 (McKenzie et al., 1997). Lemke et al. (2001) reported deamination of FB_1 in aqueous solution at pH 1.0 and 5°C with $NaNO_2$ resulted in the loss of the primary amine from the FB_1 molecule and subsequent reduction of toxicity.

Bioprocessing

Bothast et al. (1992) found little or no loss of FB_1 after three days of a yeast fermentation of corn. When FB_1 and FB_2 were added to wort in the brewing process, both were stable through fermentation and remained in the beer with negligible uptake of the toxins by yeast (Scott et al., 1995). The authors estimated losses of FB_1 and FB_2 to be 3-28% and 9-17%, respectively, during 8-day fermentation period using several strains of *Saccharomyces cerevisiae*. Fumonisins have been detected in commercial beers in the U.S., Canada, and Spain at ranges of 0.3-85.5 ng/ml (Hlywka and Bullerman, 1999; Scott and Lawrence, 1995; Scott et al., 1997; Torres et al., 1998).

Cleaning

Cleaning is one of the first steps in the processing of corn. Removal of fine and broken material from the intact corn kernels by sieving or other means reduces fumonisin levels by 26-29% (Sydenham et al., 1994), since these materials contain higher amounts of fumonisins than whole, intact kernels. Preliminary work by Canela et al. (1996) has shown that washing and steeping corn in water and solutions of sodium bisulfite is effective at moving fumonisin from the matrix. Further processing may result in additional removal of fumonisin from food.

Milling

When *F. verticilliodes* invades corn kernels it most often invades through the tip (Bacon and Williamson, 1992). The hyphae then grow under the seed coat in the tip and germ area and produce fumonisins. In corn kernels that are not highly damaged, the fungus tends to remain near the tip end and does not spread to the endosperm. This allows milling processes to produce fractions that are free of, or have very low concentrations of fumonisins.

Dry milling of corn produces grits of various sizes, bran, germ, corn meal and corn flour. Large (flaking) grits are used for making corn flakes and smaller grits are used to produce extruded cereals and snacks. Katta et al. (1997) studied the fate of fumonisin in dry-milled samples of corn naturally contaminated with three different concentrations (25.4, 3.9 and 0.3 μg/g) of FB_1. Essentially, the same results were obtained with corn of each contamination level. Fumonisins were found in the highest concentrations in the bran and fines with some also being found in the germ and smallest grit fraction. Flaking grits and grit fractions used for extrusion processing contained little or no fumonisin. For this reason, foods made from flaking or extrusion grits are relatively free of fumonisins; the dry milling process removes or separates fumonisins from the fractions used to make these foods. In the wet milling processing, fumonisins partition in the gluten, fiber and germ fractions, but not in the starch (Bennett et al., 1996). The wet milling process removes the fumonisins from starch, the major food component fraction.

Thermal Processing

The fumonisins are fairly heat stable compounds. No loss of FB_1 was observed when *F. moniliforme* (*verticillioides*) culture material was boiled in water for 30 min, then dried at 60°C for 24 hours (Alberts et al., 1990). However, Le Bars et al. (1994) reported that thermal decomposition of FB_1 in dried corn culture followed a first order reaction with half-life times of 10, 38 and 175 min at 150, 125 and 100°C respectively. Dupuy et al. (1993) also reported a 50% reduction in FB_1 levels when dry contaminated corn culture material was heated for 10 min, 30 min, 175 min, and 8 hours at 150, 125, 100, and 75°C, respectively. Castelo et al. (1998b) reported that dry heating (roasting) of artificially contaminated (5000 ng/g) corn meal at 218°C for 15 min resulted in complete loss of FB_1. Heating moist corn meal at 190°C for 60 min and 220°C for 25 min resulted in losses of FB_1 and FB_2 of 40-100% (Scott and Lawrence, 1994). Maragos and Richard (1994) determined that pasteurization (62°C; 30 min) of milk spiked with 50 ng/ml of FB_1 and FB_2 resulted in no loss of either toxin.

In studies on the heat stability of FB_1 and FB_2 using a Parr pressurized heat reactor, the effects of processing time and temperature on FB_1 and FB_2 stability (5000 ng/g) in aqueous buffered solutions at pH 4, 7 and 10 were determined free of any corn matrix effects (Jackson et al., 1996a,b). The rate and extent of FB_1 decomposition increased with processing temperature. Overall, FB_1 was least stable at pH 4 followed by pH 10 and 7, respectively. At temperatures at or above 175°C, over 90% of FB_1 was lost after processing for 60 min, regardless of pH. Basically the same results were found with FB_2. Both compounds were fairly heat stable in aqueous environments at temperatures below 150-175°C. In another study, the effects of cornstarch, zein and glucose on the fate of FB_1 in aqueous solutions heated to 100-150°C were studied (Hlywka, 1997). The results showed that the greatest losses of FB_1 occurred in solutions containing glucose compared with solutions containing cornstarch or zein.

The effects of baking and frying on the stability of FB_1 spiked into corn-based foods were also examined (Jackson et al., 1997). Baking corn muffins spiked with 5000 ng/g FB_1 at 175 and 200°C for 20 min resulted in 16% and 28% loss of FB_1 respectively. At both temperatures, losses of FB_1 were greater at the surface than at the core of the muffins. No significant losses of FB_1 were found when spiked corn masa was fried at 140-170°C for 0-6 min. FB_1 began to degrade at frying temperatures $\geq$ 180°C and times $\geq$ 8 min. Frying chips for 15 min at 190°C resulted in 67% loss of FB_1. Pineiro et al. (1999) also reported that frying of polenta (corn meal) at 160°C for 3 min resulted in approximately 80% reductions of FB_1 and FB_2. In another study, canned (autoclaved) whole-kernel corn, cream style corn, creamed corn for infants and a canned dog food showed minimal reduction in fumonisin levels, as did baked corn bread (Castelo et al., 1998b). Corn-muffin mix artificially contaminated with 5000 ng/g FB_1 and naturally contaminated corn-muffin mix showed insignificant losses of fumonisin upon baking. Pineiro et al. (1999) on the other hand reported that canning (autoclaving) of naturally contaminated whole kernel corn in brine resulted in an approximately 70% loss of FB_1 and FB_2.

In 1996, Murphy et al. (1996) reported that 100 mM fructose or glucose in a model system with 5000 ng/ml (69.3 μM) FB_1 in 50 mM potassium phosphate, pH 7.0 for 48 h at 80°C resulted in the apparent first order rate loss of FB_1. The loss of FB_1 was suggested to be due to the nonenzymatic browning reaction of a reducing sugar with the amino group present in FB_1. Lu et al. (1997) then reported that modifying FB_1 with a reducing sugar such as fructose eliminated FB_1 hepatocarcinogenicity in rats. In a more recent study, N-(carboxymethyl) FB_1 was characterized and identified as the principal reaction product following the heating of FB_1 with reducing sugars (Howard et al., 1998). These studies suggested that thermal processing of corn containing fumonisins with sugars might lead to more significant reductions of FB_1 in processed products.

Castelo (1999) studied the effects of sugars on the stability of FB_1 in spiked and cultured flaking grits during a simulated corn flake process with and without sugars. No significance differences were found when spiked flaking grits were processed without sugars, with sucrose alone and with sucrose in combination with maltose and high fructose corn syrup (HFCS). Corn flake processing of spiked grits without sugars resulted in 53.5% and 48.7% losses of FB_1 after cooking and toasting, respectively. The percentage losses of FB_1 in spiked grits after cooking and toasting with sucrose were 43.8% and 47.6%, respectively. About 44.1% and 48.7% FB_1 were lost when cooked and toasted, respectively, with sucrose in combination with maltose and HFCS. Similar results were found when cultured grits were made into corn flakes. Loss of FB_1 averaged 36% when cultured flaking grits were processed without sugars and 44.7% and 54% when processed with sucrose, alone and in combination with maltose and HFCS, respectively. As expected, more significant losses of FB_1 were found when both spiked and cultured grits were processed with glucose, alone and in combination with maltose and HFCS. Greater reductions of FB_1 in spiked (up to 89.2%) and cultured (up to 86.3%) grits were observed after toasting. Castelo et al. (2000) reported that adding glucose to baked corn muffins resulted in significantly lower FB_1 levels, and Jackson and Bullerman (1999) also found that adding glucose to corn masa dough before frying at 180°C for 8 min significantly reduced fumonisin levels.

Extrusion Processing

Extrusion processing is one of the most versatile technologies available to the food industry (Mulvaney et al., 1992). Extrusion cooking of foods, especially cereal grains, has been used for over 50 years (Harper, 1989; 1992) in the production of breakfast cereals, snack foods and pet foods. One advantage of extrusion processing is the ability to process relatively dry viscous materials, such as cereal grains, grits and flours at moisture contents around 20% (Harper, 1989). During processing through the extruder, a dough-like mixture is forced through a stationary metal tube or barrel by a rotating screw shaft. As this occurs, heat can be added in the form of steam and is also generated by the mechanical energy of the turning screw and the friction of the barrel. As a result, very high temperatures

(>150°C) can be reached (Harper, 1992). In addition, very high pressures and severe shear forces are generated that contribute to chemical reactions and molecular modifications and accomplish the cooking process in a short period of time. Two general types of extruders in use are single screw and twin screw. Different processing conditions can be established in each type of extruder, and each can be used for different purposes.

Extrusion cooking can be used to inactivate microorganisms and toxic substances (Cheftel, 1989). Grehaigne et al. (1983) and Van de Velde et al. (1984) reported 23-66% reduction in aflatoxin B_1 levels in extruded peanut meal. Losses of aflatoxin reached 87% by adding 2.0-2.5% ammonium hydroxide. Similar results were reported when corn meal naturally contaminated with aflatoxin was extruded with or without chemicals (Price, 1999). These studies suggest that extrusion processing by itself or in combination with various chemicals and food additives may offer a means of reducing the concentration of mycotoxins in corn and other cereal products.

Castelo et al. (1998c) studied the effects of extrusion conditions (extruder temperature, type of extruder screw, corn grit moisture level) on fumonisin levels in artificially contaminated (5000 ng/g) corn grits. Extrusion cooking resulted in more apparent loss of FB_1 when the extruder was equipped with mixing screws than nonmixing screws. Losses of recoverable FB_1 were observed at 120°C and 160°C with the mixing screws. A linear decrease in FB_1 levels was observed with the nonmixing screws as the moisture content of the grits was increased. In a second study by Katta et al. (1999) corn grits spiked with FB_1 at a level of 5000 ng/g were extrusion cooked in a co-rotating twin-screw extruder at different temperatures (140, 160, 180, and 200°C) and screw speeds (40, 80, 120, and 160 rpm). Both the barrel temperature and the screw speed significantly affected the extent of fumonisin reduction in extruded grits. As expected, the FB_1 recovered decreased with an increase in temperature and a decrease in screw speed. The amount of FB_1 lost from cooking grits at the different extrusion parameters used in this study ranged from 34 to 95%. About 46-76% of the spiked FB_1 was lost when the grits were cooked at temperatures and screw speeds that resulted in acceptable product expansion and color. Similar results were reported by Pineiro et al. (1999) who found 70-90% losses of FB_1 and FB_2 when corn flour (14% moisture) was extruded using a single screw extruder at temperatures of 150-180°C. When corn flour was extruded (70-105°C for 5 min) then roasted (170-220°C for 50 sec) to produce a commercial corn flake, 71% of the fumonisins in the product were lost (de Girolamo et al., 2001)

Castelo et al. (2001) studied the effects of added sugars on fumonisin levels in extruded corn grits. Factors that were studied included sugar type (glucose, fructose, sucrose) and level (2.5 and 5.0 %) and extruder temperature and speed. Extrusion cooking of the grits resulted in significant reductions of FB_1 in all treatments relative to unextruded controls, but use of glucose resulted in greater reductions (44.8-66.6%) than with fructose (32.4-53.2%) or sucrose (26-42.7%). In the second experiment, Castelo et al. (2001) extruded corn grits at 160°C with glucose concentrations (2.5, 5.0, 7.0, and 10.0%) and screw speeds as the experimental factors. As expected, both the screw speed and glucose concentration significantly affected the extent of FB_1 reduction in extruded grits, with greater reductions of FB_1 (up to 92.7%) observed at lower screw speeds and higher glucose concentrations.

These studies suggest that extrusion cooking and other processing methods that utilize high temperatures (above 150°C) may be very effective in reducing fumonisins in processed corn products, particularly if glucose can be included as an ingredient. Additional studies are needed to prove that the fumonisins are truly destroyed and not simply bound to the corn matrix, and that toxicity is lost and no new toxic breakdown products are formed.

SUMMARY AND CONCLUSIONS

Fumonisins are found in corn from all corn growing regions in concentrations that range from 100 ng/g (ppb) to more than 1 μg/g (ppm). Little information is available on the effects of chemical processing and bioprocessing on the stability of fumonisins in food. Treatment with ammonia, H_2O_2 or $NaHCO_3$ can result in loss of fumonisins in contaminated corn. However, little is known about the toxicity of the reaction products. Treating corn with calcium hydroxide (nixtamalization) produces hydrolyzed fumonisins that may be toxic. Bioprocessing in the form of yeast fermentation seems to have little effect on the fumonisin.

Processing of corn into food products affects the distribution and concentration of fumonisins in the finished product. Cleaning, washing and steeping can remove fumonisins and greatly reduce its concentration in the raw product prior to processing. Dry milling tends to concentrate fumonisins in the germ and bran fractions leaving the grit fractions, which are processed into food, with very low to non-detectable amounts of the toxins. Likewise wet milling tends to concentrate the fumonisins in the gluten, germ and fiber fractions with little or no contamination of the corn starch, the major food component fraction.

While fumonisins are fairly heat stable compounds, losses of fumonisin can occur during thermal processing, particularly if processing temperatures above 150-175°C are used. Processes such as baking, canning, and frying at temperatures below 190°C result in little or no loss of fumonisins. However frying at higher temperatures (above 190°C), roasting or dry heating above 200°C, and extrusion cooking at temperatures above 180°C result in substantial losses of fumonisins. Extrusion processing shows promise for reducing fumonisin levels in corn-based foods. Baking, frying and extrusion cooking of contaminated corn with added glucose results in the greatest reduction of fumonisins in processed products. More work is needed to identify fumonisin decomposition products and to determine actual loss of toxicity of fumonisins in heat processed and extruded foods.

REFERENCES

Alberts, J.F., Gelderblom, W.C.A., Thiel, P.G., Marasas, W.F.O., Van Schalkwyk, D.J., and Behrend, Y., 1990, Effects of temperature and incubation period on production of fumonisin B_1 by *Fusarium moniliforme*, *Appl. Environ. Microbiol.* 56:1729.

Bacon, C.W., and Williamson, J.W., 1992, Interactions of *Fusarium moniliforme*, its metabolites and bacteria with corn, *Mycopathologia* 117:65.

Bagneris, R.W., Carter Jr., L., Guerrero, H.G., and Ware, G.M., 1992, (Abstract), Rapid HPLC detection and survey of fumonisin B_1 in corn and corn screenings using fluorescence detection, Presented at the 106th Annual AOAC International Meeting. August 31-September 2, Cincinnati, OH.

Bennett, G.A., and Richard, J.L., 1996, Influence of processing on *Fusarium* mycotoxins in contaminated grains, *Food Technol.* 50:235.

Bothast, R.J., Bennett, G.A., Vancauwenberge, J.E., and Richard, J.L., 1992, Fate of fumonisin B_1 in naturally contaminated corn during ethanol fermentation, *Appl. Environ. Microbial.* 58:233.

Canela, R., Pujol, R., Sala, N., and Sanchis, V., 1996, Fate of fumonisins B_1 and B_2 in steeped corn kernels, *Food Add. Contam.* 13:511.

Castelo, M.M., Sumner, S.S. and Bullerman, L.B., 1998a, Occurrence of fumonisins in corn-based food products, *J. Food Prot.* 61:704.

Castelo, M.M., Sumner, S.S. and Bullerman, L.B., 1998b, Stability of fumonisins in thermally processed corn products, *J. Food Prot.* 61:1030.

Castelo, M.M., Katta, S.K., Sumner, S.S., Hanna, M.A. and Bullerman, L.B., 1998c, Extrusion cooking reduces recoverability of fumonisin B_1 from extruded corn grits, *J. Food Sci.* 63:696.

Castelo, M.M., 1999, Stability of Mycotoxins in thermally processed corn products, Ph.D. Dissertation, University of Nebraska, Lincoln.

Castelo, M.M., Jackson, L.S., Hanna, M.A., Reynolds, B.H., and Bullerman, L.B., 2001, Loss of fumonisin B_1 in extruded and baked corn-based foods with sugars, *J. Food Sci.* 66:416.

Chang, Y.S., Zhang, C.L., Zhao, P.Z., and Deng, Z.L., 1991, Human aortic proteoglycans of subjects from districts of high and low prevalence of atherosclerosis in China, *Atherosclerosis* 31:9.

Cheftel, J.C., 1989, Extrusion cooking and food safety, in *Extrusion Cooking*, C. Mercier, P. Linko, and J.M. Harper eds., American Association of Cereal Chemists, Inc., St. Paul, MN.

Chu, F.S., and Li, G.Y., 1994, Simultaneous occurrence of fumonisins B_1 and other mycotoxins in moldy corn collected from the People's Republic of China in regions with high incidence of esophageal cancer, *Appl. Environ. Microbiol.* 60:847.

Colvin, B.M., Cooley, A.J., Beaver, R.W. 1993. Fumonisin toxicosis in swine: clinical and pathologic findings. *J. Vet. Diagn. Invest.* 5:232.

Doko, M.B. and Visconti, A. 1994. Occurrence of fumonisins B_1 and B_2 in corn and corn-based human foodstuffs in Italy. *Food Addit. Contam.* 11:433.

Dupuy, J, Le Bars, P., Boudra, H., and Le Bars, J., 1993, Thermostability of fumonisin B_1, a mycotoxin from *Fusarium moniliforme* in corn, *Appl. Environ. Microbiol.* 59: 2864.

Fincham, J.E., Marasas, W.F.O., Taljaard, J.J.F., Kriek, N.P.J., Badenhorst, C.J., Gelderblom, W.C.A., Seier, J.V., Smuts, C.M., Mieke Faber, Weight, M.J., Slarus, W., Woodroof, C.W., Van Wyk, M.J., Marita Kruger, and Thiel, P.G., 1992, Atherogenic effects in a non-human primate of *Fusarium moniliforme* cultures added to a carbohydrate diet, *Atherosclerosis.* 94:13.

Floss, J.L., Casteel, S.W., Johnson, T.C., Rottinghaus, G.E., and Krause, G.F. 1994, Developmental toxicity of fumonisin in Syrian hamsters. *Mycopathologia* 128:33.

Franceschi, S., Bidoli, E., Baron, A.E., and LaVeccia, C., 1990, Maize and risk of cancers of the oral cavity, pharynx and esophagus in Northeastern Italy, *J. Nat'l. Cancer Inst.* 82:1407.

Gelderbom, W.C.A., Kriek, N.P.J., Marasas, W.F.O., and Thiel, P.G. 1991, Toxicity and carcinogenicity of the *F. monilforme* metabolite, FB_1, in rats. *Appl. Environ. Microbiol.* 12:1247.

Gelderblom, W.C.A., Marasas, W.F.O., Thiel, P.G., and Cawood, M.E., 1992, Fumonisins: Isolation, chemical characterization and biological effects, *Mycopathologia* 117: 11.

de Girolamo, A., Solfrizzo, M., and Visconti, A., 2001, Effect of processing on fumonisin concentration in corn flakes, *J. Food Protect.* 64:701.

Grehaigne, B., Chouvel, H., Pina, M., Graille, J., and Cheftel, J.C., 1983, Extrusion cooking of aflatoxin-containing peanut meal with and without addition of ammonium hydroxide, *Lebensm. Wiss. Technol.* 16:317.

Gutema, T., Munimbazi, C., and Bullerman, L.B., 2000, Occurrence of fumonisins and moniliformin in corn and corn-based food products of U. S. origin, *J. Food Prot.* 63:1732.

Harper, J.M., 1989, Food extruders and their applications, in *Extrusion Cooking*, C. Mercier, P. Linko, and J.M. Harper, eds., American Association of Cereal Chemists, Inc., St. Paul, MN.

Harper, J.M., 1992, A comparative analysis of single and twin-screw extruders, in *Food Extrusion Science and Technology*, J.L. Kokini, C.T. Ho, and M.V. Karwe, (eds.), Marcel Dekker, Inc., New York.

Harrison, L.R., Colvin, B.M., Greene, T.J., Newman, L.E., and R.J. Cole. 1990, Pulmonary edema and hydrothorax in swine produced by fumonisin B_1, a toxic metabolite of *Fusarium moniliforme. J. Vet. Diag. Invest.* 2:217.

Hendrich, S., Miller, K.A., Wilson, T.M., and Murphy, P.A., 1993, Toxicity of *Fusarium proliferatum*-fermented nixtamalized corn-based diets fed to rats: effect of nutritional status, *J. Agric. Food Chem.* 41:1649.

Hlywka, J.J., 1997, The thermostability and toxicity of fumonisin B_1 mycotoxin. Ph. D. Dissertation, University of Nebraska, Lincoln.

Hlywka, J.J., and Bullerman, L.B., 1999, The occurrence of fumonisin B_1 and B_2 mycotoxins in commercial beers, *Food Add. Contam.* 16:318.

Howard, P.C., Churchwell, M.I., Couch, L.H., Marques, M.M., and Doerge, D.R., 1998, Formation of N-(carboxymethyl)fumonisin B_1 following the reaction of fumonisin B_1 with reducing sugars, *J. Agric. Food Chem.* 46:3546.

Howard, P.C., Eppley, R.M., Stack, M.E., Warbritton, A., Voss, K.A., Lorentzen, R.J., Kovach, R.M and Bucci, T.J. Carcinogenicity of fumonisin B1 in a two-year bioassay with Fischer 344 rats and B6C3F1 mice. *Environ. Health Perspect.* 109 (suppl. 2):277.

Jackson, L.S., Hlywka, J.J., Senthil, K.R., Bullerman, L.B., and Musser, S.M., 1996a, Effects of time, temperature and pH on the stability of fumonsin B_1 in an aqueous model system, *J. Agric. Food Chem.* 44:906.

Jackson, L.S., Hlywka, J.J., Senthil, K.R., and Bullerman, L.B., 1996b, Effects of thermal processing on the stability of fumonisin B_2 in an aqueous system, *J. Agric. Food Chem.* 44:1984.

Jackson, L.S., Katta, S.K., Fingerhut, D.D., DeVries, J.W., and Bullerman, L.B., 1997, Effects of baking and frying on the fumonisin B_1 content of corn-based foods, *J. Agric. Food Chem.* 45:4800.

Jackson, L.S., and Bullerman, L.B., 1999, Effects of processing on *Fusarium* mycotoxins, in: *Impact of Processing on Food Safety*, L.S. Jackson, M.G. Knize, and J.N. Morgan, eds., Kluwer Academic/Plenum Publishers, New York.

Katta, S.K., Cagampang, A.E., Jackson, L.S., and Bullerman, L.B., 1997, Distribution of *Fusarium* molds and fumonisins in dry-milled corn fractions, *Cereal Chem.* 74:858.

Katta, S.K., Jackson, L.S., Hanna, M.A., and Bullerman, L.B., 1999, Effect of temperature and screw speed on the stability of fumonisin B_1 in extrusion-cooked corn grits, *Cereal Chem.* 76:16.

LeBars, J., LeBars, P., DuPuy. J., Boudra, H, and Cassini, R., 1994, Biotic and abiotic factors in fumonisin production and accumulation, *J. AOAC Int.* 77:517.

Lemke, S.L., Ottinger, S.E., Ake, C.A., Mayura, K., Mc Donald, T., and Phillips, T.D., 2001, Deamination of fumonisin B_1 and biological assessment of reaction product toxicity, *Chem. Res. Toxicol.* 14:11.

Lu, Z., Dantzer, W.R., Hopmans, E.C., Prisk, V., Cunnick, J.E., Murphy, P.A., and Hendrich, S., 1997, Reaction with fructose detoxifies fumonisin B_1 while stimulating liver-associated natural killer cell activity in rats, *J. Agric. Food Chem.* 45: 803.

Machinski, M., Jr. and Valente, L.M. 2000, Fumonisins B_1 and B_2 in Brazilian corn-based food products. *Food Addit. Contam.* 17:875.

Maragos, C.M., and Richard, J.L., 1994, Quantitation and stability of fumonisins B_1 and B_2 in milk, *J. AOAC Int.* 77:1162.

Marasas, W.F.O., Nelson, P.E., and Toussoun, T.A., 1984, *Toxigenic Fusarium Species: Identity and Mycotoxicology*. The Pennsylvania State University Press, University Park, PA.

Marasas, W.F.O., Kellerman, T.S., Gelderblom, W.C.A., Coetzer, J.A.W., Thiel, P.T., van der Lugt, J.J. 1988, Leukoencephalomalacia in a horse induced by fumonisin B_1 isolated *from Fusarium moniliforme*. *Onderstepoort J. Vet Res.* 55:197.

McKenzie, K.S., Sarr, A.B., Mayura, K., Bailey, R.H., Miller, D.R., Rogers, T.D., Norred, W.P., Voss, K.A., Plattner, R.D., Kubena, L.F., and Phillips, T.D., 1997, Oxidative degradation and detoxification of mycotoxins using a novel source of ozone, *Food Chem. Toxicol.* 35:807.

Mulvaney, S.J., Onwulata, C., Lu, Q., Hsieh, F.H., and Brent, J., 1992, Effect of the screw speed on the process dynamics and product properties of corn meal for a twin screw extruder, in *Food Extrusion Science and Technology*, J. L. Kokini, C. T. Ho, and M. V. Karwe, eds., Marcel Dekker, Inc., New York.

Murphy, P.A., Hendrich, S., Hopmans, E.C., Hauck, C.C., Lu, Z., Buseman, G., and Munkvold, G., 1996, Effect of processing on fumonisin content of food, in *Fumonisins in Food,* L.S. Jackson, J.W. DeVries, and L.B. Bullerman, eds., Plenum Press, NY. NY.

Norred, W.P., Voss, K.A., Bacon, C.W., and Riley, R.T., 1991, Effectiveness of ammonia treatment in detoxification of fumonisin-contaminated corn, *Food Chem. Toxicol.* 29:815.

Norred, W. P., and Voss, K. A., 1994, Toxicity and role of fumonisins in animal diseases and human esophageal cancer, *J. Food Prot.* 57: 522.

Osweiler, G.D., Ross, P.F., Wilson, T.M., Nelson, T.M., Witte, S.T., Carson, T.L., Rice, L.G., and Nelson, H.A. 1992, Characterization of an epizootic of pulmonary edema in swine associated with fumonisins in corn screenings. J. Vet. Diagn. Invest. 4:53.

Park, D.L., Rua, S.M., Microcha, C.J., Abd-Alla, E.S.A.M., and Weng, C.Y., 1992, Mutagenic potentials of fumonisin contaminated corn following ammonia decontamination procedure, *Mycopathologia.* 117:105.

Park, D.L., Lopez-Garcia, R., Trujillo-Preciado, S., and Price, R.L., 1996, Reduction of risks associated with fumonisin contamination in corn, in *Fumonisins in Food,* L.S Jackson, J.W. DeVries, and L.B. Bullerman, eds., Plenum Press, NY.

Patel, S., Hazel, C.M., Winterton, A.G.M., and Gleadle, A.E. 1997, Surveillance of fumonisins in UK maize-based foods and other cereals. *Food Addit. Contam.* 14:187.

Pineiro, M., Miler, J., Silva, G., and Musser, S., 1999, Effect of commercial processing on fumonisin concentrations of maize-based foods, *Mycotoxin Res.* 15:2.

Pineiro, M.S., Silva, G.E., Scott, P.M., Lawrence, G.A., and Stack, M.E. 1997, Fumonisin levels in Uruguayan corn products. *J. AOAC Int.* 80:825.

Pittet, A., Parisod, V., Schellengerg, M., 1992, Occurrence of fumonisins B_1 and B_2 in corn-based products from the Swiss market, *J. Agric. Food Chem.* 40:1352.

Price, R. L., 1999, University of Arizona (Personal Communications).

Rheeder, J.P., Marasas, W.F.O., Thiel, P.G., Sydenham, E.W., Shepard, G.S., and Van Schalkwyk, D.J., 1992, *Fusarium moniliforme* and fumonisins in corn in relation to human esophageal cancer in Transkei, *Phytophathol.* 82:353.

Riley, R.T., Showker, J.L., Owens, D.L., and Ross, P.F. 1997, Disruption of sphingolipid metabolism and induction of equine leukoencephalomalacia by *Fusarium proliferatum* culture material containing fumonisin B_2 or B_3. *Environmental Toxicology and Pharmacology* 3:221.

Ross, P.F., Rice, L.G., Plattner, R.D., Osweiler, G.D., Wilson, T.M., Owens, D.L., Nelson, P.A., and Richard, J.L., 1991, Concentrations of fumonisin B_1 in feeds associated with animal health problems, *Mycopathologia* 114:129.

Sanchis, V., Abadias, M., Oncins, L., Sala, N., Vinas, I. and Canela, R., 1994, Occurrence of fumonisins B_1 and B_2 in corn-based products from the Spanish market. *Appl. Environ. Microbiol.* 60:2147.

Scott, P.M., and Lawrence, G.A., 1994, Stability and problems in recovery of fumonisins added to corn-based foods, *J. Assoc. Off. Anal. Chem.* 77:541.

Scott, P.M., and Lawrence, G.A., 1995, Analysis of beer for fumonisins, *J. Food Prot.* 58:1379.

Scott. P.M., Kanhere, S.R., Lawrence, G.A., Daley, E.F., Farber, J.M., 1995, Fermentation of wort containing added ochratoxin A and fumonisins B_1 and B_2, *Food Addit. Contam.* 12:31.

Scott, P.M., Yeung, J.M., Lawrence, G.A., and Prelusky, D.B., 1997, Evaluation of enzyme-linked immunosorbent assay for analysis of beer for fumonisins, *Food Addit. Contam.* 14:445.

Smith, G.W., Constable, P.D., Bacon, C.W., Meredith, F.I., and Haschek, W.M. 1996, Cardiovascular effects of fumonisins in swine. *Fundamental and Applied Toxicology* 31:169.

Sydenham, E.W., Thiel, P.G., Marasas, W.F.O., Shephard, G.S., Van Schalkwyk, D.J., and Koch, K.R., 1990, Natural occurrence of some *Fusarium* mycotoxins in corn from low and high esophageal cancer prevalence areas of the Transkei, Southern Africa, *J. Agric. Food Chem.* 38:1900.

Sydenham, E.W., Shephard, G.S., Thiel, P.G., Marasas, W.F.O., and Stockenström, S., 1991, Fumonisin contamination of commercial corn-based human foodstuffs, *J. Agric. Food Chem.* 39:2014.

Sydenham. E.W., Van der Westhuizen, L., Stockenström, S., Shephard, G.S., and Thiel, P.G., 1994, Fumonisin-contaminated maize: physical treatment for the partial decontamination of bulk shipments, *Food Addit. Contam.* 11:25.

Sydenham. E.W., Stockenström, S., Thiel, P.G., Shephard, G.S., Koch, K.R., and Marasas, W.F.O., 1995a, Potential of alkaline hydrolysis for the removal of fumonisins from contaminated corn, *J. Agric. Food Chem.* 43:1198.

Sydenham. E.W., Thiel, P.G., Shephard, G.S., Koch, K.R., and Hutton, T., 1995b, Preparation and isolation of the partially hydrolyzed moiety of fumonisin B_1, *J. Agric. Food Chem.* 43:2400.

Thiel, P.G., Marasas, W.F.O., Sydenham, E.W., Shephardand, G.S., and Gelderblom, W.C.A., 1992, The implications of naturally occurring levels of fumonisins in corn for human and animal health, *Mycopathologia* 117:3.

Torres, M.R., Sanchis, V., and Romos, A.J., 1998, Occurrence of fumonisins in Spanish beers analyzed by an enzyme-linked immunosorbent assay method, *Int. J. Food Microbiol.* 39:139.

U.S. Food and Drug Administration, 2000, Draft Guidance for Industry. Fumonisin Levels in Human Food and Animal Feeds. Federal Register 65 FR 35945.

Usleber, E. and Martlbauer, E. 1998, A limited survey of cereal foods from the German market for *Fusarium* toxins (deoxynivalenol, zearalenone, fumonisins). *Archiv für Lebensmittelhygiene* 49:25.

Van de Velde, C., Bounie, D., Cugand, J.I., Cheftel, J.C., 1984, Destruction of microorganisms and toxins by extrusion cooking, in *Thermal Processing and Quality of Foods,* P. Zeuthen, J.C. Cheftel, C. Eriksson, M. Jul, H. Leniger, P. Linko, G. Varela and G. Vos, eds., Elsevier Applied Science Publishers, London.

Voss. K.A., Bacon, C.W., Meredith, F.I., and Norred, W.P., 1996, Comparative subchronic toxicity studies of nixtamalized and water-extracted Fusarium moniliforme culture material, *Food Chem. Toxicol.* 34:623.

Wilson, T.M., Ross, R.R., Owens, D., Rice, L.G., Green, S.A., Jenkins, S.J., and Nelson, H.A., 1992, Experimental reproduction of ELEM. *Mycopathologia* 117:115.

EFFECTS OF PROCESSING ON ZEARALENONE

Dojin Ryu,[1] Lauren S. Jackson,[2] and Lloyd B. Bullerman[1]

[1]Department of Food Science and Technology
University of Nebraska-Lincoln
Lincoln, NE 68583
[2]NCFST/FDA
6502 S. Archer Rd
Summit-Argo, IL 60501

ASTRACT

Zearalenone (ZEN), a common contaminant of all major cereal grains worldwide, is produced by some plant pathogenic molds including *Fusarium graminearum* and *F. culmorum*. Thc biological activity of this mycotoxin is mainly attributed to its estrogenic activity that modulates/disrupts endocrine function in animals and possibly humans. Efforts have been made to reduce the level of ZEN by various chemical, physical, and biological processing methods. Some chemical treatments were shown to be effective in reducing zearalenone content in artificially or naturally contaminated foods. During physical processing, the fate of ZEN depended on its distribution in the food matrix and its chemical properties such as heat stability and solubility. For example, wet milling of contaminated corn resulted in starch that was essentially toxin-free. In contrast, animal feed fractions such as bran and germ, by-products of the wet milling process, tended to concentrate ZEN. Extrusion cooking, a complex process where food is subjected to heat, high pressures and shear stress, reduced ZEN levels in food as well as its estrogenic activity. Fermentation of foods with bacteria and yeast resulted in reduction in ZEN levels. However, fermentation can result in the conversion of ZEN to more potent derivatives such as α-zearalenol. Further efforts are needed to identify effective methods for removing/detoxifying ZEN in foods.

INTRODUCTION

Zearalenone (ZEN) is a unique mycotoxin with estrogenic activity that can modulate the endocrine hormone estrogen. This mycotoxin is produced by ubiquitous fungi such as *Fusarium graminearum* and *Fusarium culmorum* (Mirocha et al., 1977). ZEN is frequently found in all major cereal grains worldwide as a result of fungal infection and growth. The spores of *Fusarium* are ubiquitous, and therefore, all cereal grains are susceptible to their

Mycotoxins and Food Safety, edited by Trucksess *et al.*
Kluwer Academic/Plenum Publishers, 2002

invasion and subsequent growth and toxin production (Nelson et al., 1983). Toxic effects observed in animals exposed to ZEN vary with species, but swine are known to be the most sensitive (Mirocha and Christensen, 1974). The toxicity of ZEN is due to its estrogenic activity that can be observed as hyperestrogenism in swine with symptoms of infertility, pseudopregnancy, abnormal lactation, stillbirths, abortions, and rectal or vaginal prolapses (Friend et al., 1990).

Compounds that are capable of modulating endocrine function exist in the environment. They can be either naturally occurring or man-made, but all have the potential for causing adverse effects in humans and/or animals. The significance of endocrine modulation is mainly attributed to altered hormonal balance which may affect the carcinogenic process in endocrine sensitive tissues including adrenals, thyroid, prostate, and breast (Williams and Weisburger, 1991).

Natural Occurrence

Fusarium graminearum Schwabe is the imperfect or anamorphic stage of *Gibberella zeae* (Schw.) Petch. This fungus is well known for its pathogenic potential on a wide variety of plant hosts including corn, wheat, barley, oats and rice (Mirocha et al., 1977). While the natural habitat of *F. graminearum* is the soil, it may also be considered a storage fungus since growth and toxin production may occur under various storage conditions. Corn and wheat are most susceptible to invasion by this fungus. Crown rot (ear rot) on corn and scab or head blight in wheat and barley are caused by *F. graminearum* worldwide, and result in significant economic crop losses. The potential concomitant production of ZEN is a concern in animal and human health. The worldwide incidences of *F. graminearum* and mycotoxins produced by this fungus have been well documented (Hagler et al., 1987; Jelinek et al., 1989; Tuite et al., 1974). Mirocha et al. (1989) tested 114 isolates of *F. graminearum* from the United States, Australia, China, New Zealand, Norway, and Poland and found more than 90% of the isolates were capable of producing ZEN regardless of geographic origin. Large-scale infection by this widespread fungus can occur in corn as observed in the cornbelt in 1966, 1972 and 1975 (Tuite et al., 1974).

The occurrence of ZEN has also been reported in many countries. According to Tanaka et al. (1990), Dutch cereals harvested in 1984/85 were contaminated with ZEN at an average of 61 ng/g in all positive samples. The highest reported level of natural contamination was 3,100 ng ZEN/g in a corncob mix from 1988/89 in the Netherlands (Veldman et al., 1992). A native cereal beer in Nigeria, pito, contained as much as 200 ng/ml of ZEN (Okoye, 1985). ZEN has also been found in processed food products in the U.S., including breakfast cereals, snack foods, popcorn, and corn meal at an average level of 20 ng/g and maximum levels of 120 and 130 ng/g in corn meal and popcorn, respectively (Warner and Pestka, 1987). Lee et al. (1985, 1986) reported that in 1983-84, the most commonly ZEN contaminated crop in Korea was barley. Twenty-five out of 32 unpolished barley and malt samples from the 1983 crop (Lee et al., 1985) and 34 out of 36 samples from the 1984 crop (Lee et al., 1986) were positive for ZEN. Average levels of ZEN contamination were 110 ng/g for barley (1,600 ng/g maximum) and 19 ng/g for malt in 1983. Park et al. (1992) detected ZEN at levels of 183-1416 ng/g and 40-1081 ng/g for husked and polished barley, respectively, from the 1990 crop. At present, no regulations or guidelines are in effect in the U.S. and Canada with regard to ZEN levels in food and feed.

Chemistry and Analysis

ZEN (Figure 1) has been described as 6-(10-hydroxy-6-oxo-*trans*-1-undecenyl)-β-resorcyclic acid lactone (Urry et al., 1966). The biosynthetic pathway of ZEN has been elucidated showing D-(+)-glucose to be the precursor (Zill et al., 1989). A white, crystalline compound with a molecular formula $C_{18}H_{22}O_5$, ZEN is soluble in many organic solvents including benzene, acetonitrile, methanol and acetone, but virtually insoluble in water and slightly soluble in *n*-hexane (Hidy et al., 1977). Despite its large lactone ring, ZEN is very heat stable and exhibits marked heat resistance during food processing (Bennett et al., 1980; Matsuura et al., 1981). However, ZEN can be converted to α-zearalenol and an unknown metabolite by the gut flora of some animals (Kollarczik et al., 1994). This compound, α-zearalenol, is known to have greater estrogenic activity than ZEN (Katzenellenbogen et al., 1979).

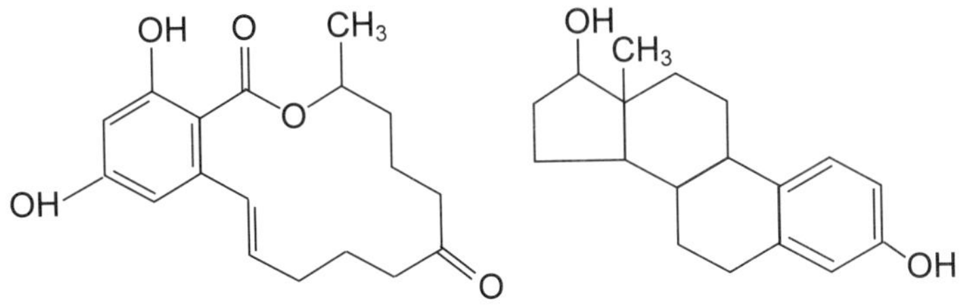

Figure 1. Structure of zearalenone and 17β-estradiol.

ZEN and its metabolites can be quantified by a variety of analytical methods including thin layer chromatography (TLC), gas chromatography (GC), high performance liquid chromatography (HPLC), and enzyme-linked immunosorbant assay (ELISA). An HPLC method (Bennett et al., 1985), using a reverse-phase C18 column and fluorescence detection, is the method of choice for determination of ZEN. This method has also been collaboratively studied (Bennett et al., 1985) and has been adopted as an official method of the Association of Official Analytical Chemists (AOAC), the American Oil Chemists' Society (AOCS), and the American Association of Cereal Chemists (AACC). Detection limits can range between 2 to 50 ng/g, depending on the sample and recovery (Bagnaris et al., 1986; Bennett et al., 1985; Chang and DeVries, 1984; Merino et al., 1993). The detection limit can be lowered if a synthesized internal standard, such as ZEN 6'-oxime is used (Tanaka et al., 1993). Various efforts have been made to improve analysis of ZEN by using post column derivatization (Hetmanski and Scudamore, 1991), and gel permeation chromatography (Dunne et al., 1993). A more sophisticated method of analysis, involving instrumentation such as GC-MS, is also available with high sensitivity (1 ng/g) and accuracy (Schwadorf and Müller, 1992).

In considering the biological activity of ZEN, it is more appropriate to choose a biological method to assess the toxicity of this unique mycotoxin since chemical methods cannot estimate possible interactions in the biological system. The uterotropic assay using rats or mice has proven to be useful to assess estrogenic activity. This assay is based on the measurement of uterine enlargement (Christensen et al., 1965; Thigpen et al., 1987).

There is an increasing tendency to replace animal experiments with *in vitro* methods, such as using sensitive cell lines. In order to assess estrogenic activity, the cell lines must have estrogen receptors (ER) to be able to respond to ZEN. There are several defined cell lines available for this purpose; MCF-7 (Soule et al., 1973), ZR75-1 (Engel et al., 1978), and T47D (Keydar et al., 1979). All of these cell lines are derived from human breast cancer cells. They contain high levels of ER (Horwitz et al., 1978), and their growth is stimulated by estrogens and inhibited by anti-estrogens (Lippman et al., 1976). Among the three, MCF-7 seems to be used most commonly in evaluating estrogenic activity of xenobiotics such as ZEN.

Biological Activity

The toxicity of ZEN is due to its estrogenic activity which can be observed as hyperestrogenism in swine (Friend et al., 1990). Hyperestrogenism in pigs was first described by McNutt et al. (1928) and was attributed to consumption of spoiled corn. Typical signs of hyperestrogenism are prolonged estrus, anestrus, infertility, increased incidence of pseudopregnancy, increased udder or mammary gland development, and abnormal lactation (Mirocha and Christensen, 1974; Etienne and Jemmali, 1982). Stillbirths, abortions, mastitis, vulvovaginitis, and rectal or vaginal prolapses are secondary complications associated with ZEN ingestion (Sundlof and Strickland, 1986). Signs of swollen vulva and mammary glands, and enlargement of the uterus can occur at a level of 250 ng ZEN/g feed in swine. However, infertility, stillbirths and reduced litter size do not occur at levels below 500 ng/g (Friend et al., 1990). ZEN also elicits permanent reproductive tract alterations. In newborn female mice treated with 1 mg ZEN/day for five days, significant ovary-dependent reproductive tract alterations remained eight months after treatment (Williams et al., 1989). The majority of treated mice lacked corpus lutea and uterine glands and exhibited squamous metaplasia of the uterine luminal epithelium. Kumagai and Shimizu (1982) also reported similar observations with persistent anovulatory estrus in the rat exposed to ZEN soon after birth.

Endocrine modulators can be defined as all natural or synthetic compounds that alter activity of the endocrine system in animals or humans. Those compounds generally mimic endocrine hormone(s) to modulate or disrupt the endocrine function(s) and cause various adverse effects. In the early 1980's, unknown exogenous estrogenic compounds in food were suspected as causal agents for an outbreak of precocious pubertal changes in thousands of young children in Puerto Rico in 1978-81(Sáenz de Rodríguez, 1984; Sáenz de Rodríguez et al., 1985). The observed effects included premature pubarche, prepubertal gynecomastia, and precocious pseudopuberty. A similar incidence was reported earlier in Italy (Fara et al., 1979) with a major clinical effect of breast enlargement in young children.

Alteration in hormonal balance may affect the carcinogenic process, since the endocrine system has an important effect on the progression of tumors in endocrine sensitive tissues such as the gonads, adrenals, thyroid, prostate, and breast (Williams and Weisburger, 1991). For example, estrogen is believed to be etiologically important in the development of human breast cancer. The effects of estrogen in target organs are mediated through a specific receptor (ER), and induce proliferation of breast epithelium (Osborne, 1991). Therefore, benign proliferative lesions of the breast indicate a higher risk for subsequent malignancy (Kreiger and Hiatt, 1992; Dupont and Page, 1985). It is possible, therefore, that the presence of ER in benign breast epithelium may be a risk factor for breast cancer (Khan et al., 1994).

ZEN has been shown to competitively bind to ER in a number of *in vitro* systems including rat liver (Tashiro et al., 1983), mouse uterus (Greenman et al., 1979), and human breast cancer cells (Martin et al., 1978). However, the relative estrogenic potency of ZEN and its derivatives was estimated to be 10 to 1000 times less than that of 17β-estradiol, the most active form of estrogenic hormone, depending on the test system used (Fuller et al., 1982; Katzenellenbogen et al., 1979; Kumagai and Shimizu, 1982; Martin et al, 1978; Ueno and Tashiro, 1981). The order of estrogenic potencies of ZEN and its derivatives was α-zearalanol > α-zearalenol > β-zearalanol > ZEN > β-zearalenol (Ueno and Tashiro, 1981). Among these compounds, α-zearalenol was found in both human and swine as a major metabolite (Mirocha et al., 1981).

ZEN was not found to be genotoxic in the Ames assay using a series of different strains of *Salmonella typhimurium* (TA98, TA100, TA1535, TA1537, and TA1538) with or without activation by rat liver microsomes (Wehner et al., 1978; Bartholomew and Ryan,

1980; Ingerowski et al., 1981) and in a eukaryocyte point-mutation assay with *Saccharomyces cerevisiae* (Kuczuk et al., 1978). However, ZEN and its derivatives had a positive DNA-damaging effect in recombination tests with *Bacillus subtilis* (Ueno and Kubota, 1976; Scheutwinkel et al., 1986). Complete inhibition of DNA synthesis was produced in human peripheral blood lymphocyte cultures at a concentration of 30 μg ZEN/ml (Cooray, 1984). Its ability to cause DNA damage in both microbial and mammalian assays indicates that ZEN is possibly weakly genotoxic. This theory was confirmed by Pfohl-Leszkowicz et al. (1995) who found several DNA adducts (12–15) in the kidney and liver of female mice treated with a single dose of ZEN (2 mg/kg i.p. or orally). They also recovered DNA adducts in mouse ovaries after repeated doses of ZEN (1 mg/kg on days 1, 5, 7, 9 and 10). These results also indicate that ZEN may be able to induce hepatocellular adenomas along with the tumors of genital organs in mice.

Marin et al. (1996) reported that ZEN could be immunotoxic as elevated interleukin levels were observed in a thymoma cell line EL-4. No teratogenic effects were seen when mice were given ZEN by gavage once during pregnancy at levels up to 20 mg/kg body weight (b.w.) (Arora et al., 1983). But ZEN given by gavage at 1 to 10 mg/kg b.w. to rats on days 6-15 of pregnancy caused minor skeletal defects considered to be due to delayed ossification (Ruddick et al., 1976). ZEN has low acute toxicity in mice, rats, and guinea pigs. No gross adverse effects were seen in female chicks given a single oral dose of 15 mg/g b.w. (Chi et al., 1980). However, in young pigs, single dose oral administration caused swollen, inflamed vulvae with a no adverse effect level (NOAEL) of less than 3.5 mg/kg b.w. (Farnworth and Trenholm, 1983). In subacute and subchronic toxicity studies, of up to 14 weeks duration, ZEN caused atrophy of seminal vesicles and testes, osteoporosis, myelofibrosis of bone marrow, hyperkeratosis of vagina, and endometrial hyperplasia (Anonymous, 1982). Most of these effects can be attributed to the estrogenic effect of ZEN. In a 2-year oral carcinogenicity study, no effect was observed at levels of 25 or 50 ppm (1 or 2 mg/kg b.w.) in the Fisher (F344/N) rat (Anonymous, 1982). In the same study with B6C3F1 mice, which were given ZEN in the diet at a maximum of 100 ppm (15.8 mg/kg b.w. in males and 18.5 mg/kg b.w. in females), no effect was observed in male mice, whereas estrogen-related effects were seen in several tissues, as well as myelofibrosis in the bone marrow of females.

Chang et al. (1979) established the NOAEL for these effects at less than 25 ng ZEN/g in the diet, corresponding to approximately 0.37 mg ZEN/kg b.w. The pubertal gilt given ZEN at low dietary levels (3 to 9 μg/g) throughout gestation was even more sensitive, and showed an increase in pseudopregnancy, decreased breeding, and decreases in live litters at a NOAEL of 0.06 mg ZEN/kg b.w. (Young and King, 1986). ZEN also dose-dependently inhibits testosterone secretion by isolated testicular cells (Fenske and Fink-Gremmels, 1990).

EFFECT OF FOOD PROCESSING

The fate of ZEN during various food processes is largely dependent on its chemistry especially its solubility and heat stability. Factors such as processing temperature and time, food matrix, and presence of other chemicals including food additives affect the rate of decomposition of ZEN. Chemical, physical and biological treatments have been studied for removing ZEN in contaminated grains.

Chemical Treatments

Several attempts have been made to reduce the level of ZEN in contaminated crops with various chemicals (Table 1). Ammoniation or use of acetic acid, propionic acid, hydrochloric acid, sodium bicarbonate and hydrogen peroxide were not effective in reducing the level of ZEN in contaminated yellow corn (Bennett et al., 1980). However, almost complete (96–100%) destruction of ZEN in both artificially and naturally contaminated corn grits was achieved by using either gaseous or liquid formaldehyde. McKenzie et al. (1997) reported complete destruction of ZEN when an aqueous solution of the toxin was treated with 10% ozone (O_3) for 15 sec. Treatment of corn with 3% ammonium hydroxide and heat (50°C, 16 hrs) resulted in 96% reduction of the toxin (Bennett et al., 1980). According to Matsuura et al. (1981), about 30% of ZEN in wheat flour cake was reduced when heated for 15 min at 100°C in the presence of ammonium persulfate (0.03%) or potassium carbonate (1%). Soaking contaminated kernels of corn, barley, and wheat in 0.1 M sodium carbonate solution for 24–72 hr reduced ZEN by 46-100%, and gentle agitation of kernels in a 1 M sodium carbonate solution for 4 hr reduced the toxin up to 95% (Trenholm et al., 1992).

Table 1. Chemical treatments for reducing zearalenone contents

Reagent	Condition	Initial ZEN Conc.(ppm)	Matrix	% Reduction in ZEN Level
CH_3COOH	3%, Room temp, 3d	3-5	Corn grit	0[1]
CH_3CH_2COOH	3%, Room temp, 3d	3-5	Corn grit	0[1]
HCl	1.85%, Room temp, 3d	3-5	Corn grit, spiked	0[1]
$NaHCO_3$	10%, 50°C, 16 h	3-5	Corn grit, spiked	0[1]
Na_2CO_3	1 M, Soaking or Stirring	0.89-1.58	Corn, barley, wheat	46-100[2]
H_2O_2	3%, Room temp, 3d	3-5	Corn grit, spiked	0[1]
H_2O_2	3%, 50°C, 16 h	3-5	Corn grit, spiked	0[1]
HCHO (*l*)	3.7%, 50°C, 16 h	3-5	Corn grit, spiked	100[1]
HCHO (*g*)	Room temp, 10 d	10	Corn grit, spiked	100[1]
NH_4OH	3%, 50°C, 16 h	3-5	Corn grit, spiked	96[1]
$Ca(OH)_2$	2%, 100°C, 5 m → soaking 12 h	0.2, 4.2	Corn, Natural	100,52[3]
O_3	10%, 15 s	20	Pure solution	100[4]

[1]Bennett et al. (1980); [2]Trenholm et al. (1992); [3]Abbas et al. (1988); [4]McKenzie et al. (1997)

Abbas et al. (1988) studied the effects of alkali treatment on ZEN in naturally contaminated and spiked corn. Tortilla dough (masa) was prepared from two samples of naturally contaminated corn (0.23 and 4.23 ppm ZEN) by boiling the corn in lime water (2% $Ca[OH]_2$). The resultant tortilla dough contained only trace amounts of ZEN (Abbas et al., 1988). Similar lime treatment resulted in 71-74% destruction of ZEN in tortilla dough spiked with 1 and 5 ppm ZEN.

Physical Processing

Efficacies of a number of physical processes have been tested in reducing ZEN levels in foods (Table 2). ZEN is known for its marked heat stability. No change was observed in the amount of ZEN in an aqueous solution after 4 hr at 120°C (Bata and Örsi, 1981). In another study (Bennett et al., 1980), neither pure ZEN nor ZEN present in ground corn was decreased by heating at 150°C for 44 hr. ZEN levels in spiked wheat flour (20 μg ZEN/g), containing 35% moisture, were reduced by 3.2% after heating for 15 min at 100°C and 28.5% after 60 min at 150°C (Matsuura et al., 1981). At 200°C, the toxin levels decreased linearly with heating time; 37% and 69% after 30 min and 60 min, respectively. In addition, ZEN levels decreased 34-40% when dough was baked (190-200°C) into bread, 48-62% during the manufacture of instant noodles, and 16-27% during the preparation of biscuits (Matsuura et al., 1981). Flame roasting of naturally contaminated corn (0.02 and 0.06 μg/g) at temperatures of 110-140°C reduced the concentration of ZEN by 50% (Hamilton and Thompson, 1992).

Sieving through a series of screens following grinding may be employed to reduce ZEN levels in naturally contaminated grain (Trenholm et al., 1991). Trenholm et al. (1991) reported 67-79% removal of ZEN when ground barley and corn was sieved with +9 (barley) and +16 (corn) mesh screens to remove small pieces of the grain. However, it should be noted that this process resulted in losses of 34-69% in grain weight. ZEN in barley (0.6-1.0 μg/g) was reduced by 85% with a loss of 24% of the kernel weight using a commercial dehuller. Using the same dehuller, the toxin was completely removed from barley and wheat containing lower levels (<0.6 μg/g) of ZEN (Trenholm et al., 1991). Similarly, ZEN in naturally contaminated barley (15 ng/g) was removed effectively (100%) by the polishing process while the toxin in bran fractions for animal feed increased 2.7-4.2 fold (Lee et al., 1992). The level of ZEN in naturally contaminated corn and barley could also be reduced to some extent (2–61%) by washing with distilled water (Trenholm et al., 1992).

Several researchers have examined the fate of ZEN in contaminated grain during dry milling operations. Using a laboratory and commercial scale dry milling process, Bennett et al. (1976) reported that all dry milling fractions of corn contained ZEN and that cleaning removed only 3-10% of the toxin from the contaminated grain. The highest levels of ZEN were found in germ and bran, while grits and flour contained 10-22% of the total ZEN (Bennett et al., 1976). Lee et al. (1987) reported a similar distribution of ZEN in fractions obtained from the dry milling of Korean wheat. In study with wheat samples from Kansas, Trigo-Stockli et al. (1996) revealed that 53 and 44% of ZEN were distributed in bran and shorts fractions, respectively, but only 3% in the flour. These results elucidate that the dry milling process is not effective in totally removing ZEN from naturally contaminated field crops. Dry milling results in low levels of ZEN in flour fractions while germ and feed fractions can have high concentrations of the toxin.

Studies on the wet milling of corn naturally contaminated with ZEN showed that ZEN was not destroyed, but rather concentrated in all fractions except for the starch fraction (Bennett et al., 1978). The steeping procedure using SO_2 had no effect on ZEN and the toxin was not bound to gluten, the fraction containing the highest level of ZEN. It was shown that the gluten fraction might contain as much as seven times the amount of ZEN as in the unprocessed grain. The toxin was distributed in the milled fractions in the order of gluten > solubles > fiber > germ. In another study of commercial scale wet milling (Lauren and Ringrose, 1997), ZEN was found in corn germ, fiber, and gluten fractions at a much higher concentration (2.2-4.8 μg/g) than in the concentrated steep liquor (0.6 μg/g). In addition, high concentrations of ZEN (4.6 μg/ml) were found in a sample of corn oil recovered from the study. Compared to the results of Bennett et al. (1978), Lauren and Ringrose (1997) found lower levels of ZEN in the solubles fractions. Both studies indicate that corn starch, the major food fraction, is produced free of ZEN in wet milling, while the animal feed fractions tend to be contaminated.

The effects of extrusion cooking at different moisture contents and temperatures were recently reported by Ryu et al. (1999). Using a laboratory scale twin-screw extruder, the percent reductions of ZEN in spiked corn grits ranged from 66-83% at temperatures 120–160°C. The moisture content of the grits (18–26%) had no significant effect on reduction of ZEN during extrusion. The authors suggested that the high efficacy of extrusion cooking in reducing ZEN might be attributed to the high temperature, high pressure and severe shear. The percent reduction in estrogenic activities of the extruded corn grits ranged from 53% to 72% when tested in a cell culture assay using MCF-7 (Ryu and Bullerman, 1999).

Table 2. Physical processes for reducing zearalenone content

Method	Condition	Initial ZEN Conc.(ppm)	Matrix	% Reduction in ZEN Level
Heating	150°C, 44 h	3-5	Corn grit	0[1]
	150°C, 1 h	20	Wheat flour	29[2]
	200°C, 1 h	20	Wheat flour	69[2]
Baking	190-200°C	20	Wheat flour	34-40[2]
Flame roasting	110-140°C, 6-15 m	0.02, 0.06	Corn	50[3]
Washing	Water	0.89, 1.58	Barley, corn	2, 61[4]
Sieving	Mesh (+9, +16)	0.6-1.0, 1.2	Barley, corn	67, 79[5]
Dehulling	Commercial	0.6-1.0	Barley	85[5]
Polishing	Commercial	0.009, 0.015	Barley	100, 47[6]
Dry milling	Roller mill	0.9-7.8	Corn	75-87 (grit), 2-3X* (germ)[7]
	Bühler test mill	0.001	Wheat	100 (flour), 2X* (shorts, bran)[8]
	Bühler test mill	0.95	Wheat	90 (flour), 1.8X* (shorts) 2.2X* (bran)[9]
Wet milling	Laboratory scale	0.9-9.4	Corn	>99 (starch), 1-2X* (germ) 2-7X* (gluten)[10]
	Commercial	0.5	Corn	100 (steep liquor), 9X* (oil), 2-3X* (germ, fiber, gluten)[11]
Extrusion	120-160°C	4.4	Corn grit	65-83[12]

[1]Bennett et al. (1980); [2]Matsuura et al. (1981); [3]Hamilton and Thompson (1992); [4]Trenholm et al. (1992); [5]Trenholm et al. (1991); [6]Lee et al. (1992); [7]Bennett et al. (1976); [8]Lee et al. (1987); [9]Trigo-Stockli et al. (1996); [10]Bennett et al. (1978); [11]Lauren and Ringrose (1997); [12]Ryu et al. (1999)

* times higher than the initial concentration

Biological Processing

Decomposition or detoxification of ZEN in contaminated grain by different microorganisms has been studied with varying success. Fermentation of yellow corn naturally contaminated with ZEN using *Saccharomyces uvarum* for five days at 32°C resulted in no destruction of the mycotoxin, and concentrations of ZEN in the recovered solids were about twice the levels in the original corn (Bennett et al., 1981). However, the presence of the toxin did not affect ethanol yield and no carry-over residue of ZEN was detected in the distilled ethanol. However, studies on koji extracts spiked with ZEN and incubated with several strains of *Saccharomyces cerevisiae* showed marked conversion to β-zearalenol over an eight day period (Matsuura and Yoshizawa, 1985). Another study on the fermentation of wort containing ZEN by *S. cerevisiae* also resulted in conversion of 69% of the toxin to β-zearalenol, a derivative which has less estrogenic activity than the parent compound (Scott et al., 1992). Boswald et al. (1995) also reported that several yeasts involved in fermentation, contamination, or spoilage of fermented products converted ZEN to α- and β-zearalenol. Yeasts of *Saccharomyces* spp., *Schizosaccharomyces* spp., *Zygosaccharomyces rouxii*, *Bretanomyces* spp., *Candida* spp., *Pichia* spp., *Hansenula anomala*, and *Kloeckera apculata* were included in the test (Boswald et al, 1995). On the other hand, Scott et al. (1993) did not find any ZEN or its metabolites in Canadian and imported beers. Okoye (1986; 1987) reported a 51% carry-over of ZEN into a native Nigerian cereal beer, pito, which is brewed from millet, corn, or mixtures of both. In another fermentation process (ensilaging) using a strain of lactic acid bacteria (*Lactobacillus delbruckii*), the level of ZEN in ground corn was reduced up to 90% after 10 weeks (Wiewiorkówska et al., 1988).

CONCLUSIONS

The significance of the mycotoxin zearalenone (ZEN) is mainly attributed to its worldwide occurrence in all major cereal grains and its ability to disrupt endocrine activity. Due to the health concerns associated with the consumption of ZEN and other mycotoxins, considerable attention has been given to treatments that reduce mycotoxin levels in food and feed. The use of chemical treatments to detoxify ZEN in foods has seen limited success. Formaldehyde and ammonium hydroxide treatments reduced ZEN levels in spiked corn grits but not in whole kernel corn. Other chemical processes using ozone or alkaline solution ($Ca[OH]_2$) were effective in reducing ZEN in corn or in aqueous solution. Physical processing methods have had some success in reducing ZEN levels in food. In general, thermal processing was not effective in reducing ZEN. However, use of heat in combination with pressure during processing (extrusion cooking) resulted in substantial losses of ZEN in corn. Dry milling of wheat and corn resulted in significant reductions in ZEN levels in grit and flour fractions. Germ and bran, by-products of wet milling, tended to concentrate the toxin. These results suggest that ZEN is mainly distributed in the outer portion of the grain while not much is present in the endosperm. Biological processing, treatment of foods with microorganisms such as brewer's yeast (*Saccharomyces cerevisiae*) and lactic acid bacteria (*Lactobacillus delbruckii*), resulted in the metabolism of ZEN to more toxic derivatives such α-zearalenol. More work is needed to develop effective methods for reducing ZEN levels in food.

REFERENCES

Abbas, H.K., Mirocha, C.J., Rosiles, R., and Carvajal, M., 1988, Decomposition of zearalenone and deoxynivalenol in the process of making tortillas from corn, *Cereal Chem.* 65:15.

Anonymous, 1982, *Carcinogenesis Bioassay of Zearalenone in F334/N Rats and B6C3F1 Mice*, National Toxicology Program, Technical Report Series No. 235, Department of Health and Human Services, Research Triangle Park, NC.

Arora, R.G., Frölen, H., and Fellner-Feldegg, H., 1983, Inhibition of ochratoxin and teratogenesis by zearalenone and diethylstilbestrol, *Food Chem. Toxicol.* 21:779.

Bagneris, R.W., Gaul, J.A., and Ware, G.M., 1986, Liquid chromatographic determination of zearalenone and zearalenol in animal feeds and grains, using fluorescence detection, *J. AOAC.* 69:894.

Bartholomew, R.M., and Ryan, D.S., 1980, Lack of mutagenicity of some phytoestrogens in the *Salmonella*/mammalian microsome assay, *Mutat. Res.* 78:317.

Bata, Á., and Örsi, F., 1981, Stability of zearalenone with respect to the possibility of detoxification, *Acta Aliment.* 10:161.

Bennett, G.A., Peplinski, A.J., Brekke, O.L., and Jackson, L.K., 1976, Zearalenone: Distribution in dry-milled fractions of contaminated corn, *Cereal Chem.* 53:299.

Bennett, G.A., Vandegraft, E.E., Shotwell, O.L., Watson, S.A., and Bocan, B.J., 1978, Zearalenone: Distribution in wet-milling fractions from contaminated corn. *Cereal Chem.* 55:455.

Bennett, G.A., Shotwell, O.L., and Hesseltine, C.W., 1980, Destruction of zearalenone in contaminated corn, *J. Amer. Oil Chem. Soc.* 57:245.

Bennett, G.A., Lagoda, A.A., Shotwell, O.L., and Hesseltine, C.W., 1981, Utilization of zearalenone-contaminated corn for ethanol production, *J. Amer. Oil Chem. Soc.* 58: 974.

Bennett, G.A., Shotwell, O.L., and Kwolek, W.F.K., 1985, Liquid chromatographic determination of α-zearalenol and zearalenone in corn: collaborative study, *JAOAC.* 68:958.

Chang, K., Kurtz, J.H., and Mirocha, C.J., 1979, Effects of the mycotoxin zearalenone on swine reproduction, *Amer. J. Vet. Res.* 40:1260.

Chang, H.L., and Devries, J.W., 1984, Short liquid chromatographic method for determination of zearalenone and alpha-zearalenol, *J. AOAC.* 67:741.

Chi, M.S., Mirocha, C.J., Weaver, G.A., and Kurtz, J.II., 1980, Effect of zearalenone on female white leghorn chickens, *Appl. Environ. Microbiol.* 39:1026.

Christensen, C.M., Nelson, G.H., and Mirocha, C.J., 1965, Effect on the white rat uterus of a toxin substance isolated from *Fusarium*, *Appl. Microbiol.* 13:653.

Cooray, R., 1984, Effects of some mycotoxins on mitogen-induced blastogenesis and SCE frequency in human lymphocytes, *Food Chem. Toxicol.* 22:529.

Dunne, C., Meaney, M., Smyth, M., and Tuinstra, L.G.M., 1993, Multimycotoxin detection and clean-up method for aflatoxins, ochratoxin and zearalenone in animal feed ingredients using high-performance liquid chromatography and gel permeation chromatography, *J. Chromatogr.* 629:229.

Dupont, W.D., and Page, D.L., 1985, Risk factors for breast cancer in women with proliferative breast disease, *N. Engl. J. Med.* 312:1461.

Engel, L.W., Young, N.A., Tralka, T.S., Lippman, M.E., O'Brien, S.J., and Joyce, M.J., 1978, Human breast carcinoma cells in continuous culture: establishment and characterization of 3 new cell lines, *Cancer Res.* 38:3352.

Etienne, M., and Jemmali, M., 1982, Effects of zearalenone (F2) on estrous activity and reproduction in gilts, *J. Anim. Sci.* 55:1.

Fara, G.M., Del Corvo, G., Bernuzzi, S., Bigatello, A., Di Pietro, C., Scaglioni, S., and Chiumello, G., 1979, Epidemic of breast enlargement in an Italian school, *Lancet* 11:295.

Farnworth, E.R., and Trenholm, H.L., 1981, The effect of acute administraion of the mycotoxin zearalenone to female pigs, *J. Environ. Sci. Health* 16:239.

Fenske, M., and Fink-Gremmels, J., 1990, Effects of fungal metabolites on testosterone secretion *in vitro*, *Arch. Toxicol.* 64:72.

Friend, D.W., Trenholm, H.L., Thompson, B.K., Hartin, K.E., Fisher, P.S., Asem, E.K., and Tsang, B.K., 1990, The reproductive efficiency of gilts fed very low levels of zearalenone, *Can. J. Anim. Sci.* 70:635.

Fuller, G.B., Bornett, B, Graham, C., and Hobson, W., 1982, A primate model for assessing estrogenicity - The castrate female rhesus monkey, *Int. J. Primatol.* 3:283.

Greenman, D.L., Mehta, R.G., and Wittliff, J.L., 1979, Nuclear interactions of *Fusarium* mycotoxins with estradiol binding sites in the mouse uterus, *J. Toxicol. Environ. Health.* 5:593.

Hagler, Jr., W.M., Bowman, D.T., Babadoost, M., Haney, C.A., and Swanson, S.P., 1987, Aflatoxin, zearalenone, and deoxynivalenol in North Carolina grain sorghum, *Crop Sci.* 27:1273.

Hamilton, R.M.G., and Thompson, B.K., 1992, Chemical and nutrient content of corn (*Zea mays)* before and after being flame roasted, *J. Sci. Food Agric.* 58:425.

Hetmanski, M.T., and Scudamore, K.A., 1991, Detection of zearalenone in cereal extracts using high-performance liquid chromatography with post-column derivatization, *J. Chromatogr.* 588:47.

Hidy, P.H., Baldwin, R.S., Greasham, R.L., Keith, C.L., and McMullen, J.R., 1977, Zearalenone and some derivatives: Production and biological activities, in *Advances in Applied Microbiology*, Vol. 22, Academic Press, New York.

Horwitz, K.B., Zava, D.T., Thilagar, A.K., Jensen, E.M., and McGuire, W.L., 1978, Steroid receptor analyses of nine human breast cancer cell lines, *Cancer Res.* 38:2434.

Ingerowski, G.H., Scheutwinkel-Reich, M., and Stan, H.J., 1981, Mutagenicity studies on veterinary anabolic drugs with the *Salmonella*/microsome test, *Mutat. Res.* 91:93.

Jelinek, C.F., Pohland, A.E., and Wood, G.E., 1989, Review of Mycotoxin Contamination: Worldwide occurrence of mycotoxins in foods and feeds - An update, *JAOAC.* 72:223.

Katzenellenbogen, B.S., Katzenellenbogen, J.A., and Mordecai, D., 1979, Zearalenones: Characterization of the estrogenic potencies and receptor interactions of a series of fungal β-resorcylic acid lactones, *Endocrinology* 105:33.

Keydar, I., Chen, L., Karby, S., Weiss, F.R., Delarea, J., Radu, M., Chaitcik, S., and Breeners, H.J., 1979, Establishment and characterization of a cell line of human breast carcinoma origin *Eur. J. Cancer* 15:659.

Kollarczik, B., Gareis, M., and Hanelt, M., 1994, In vitro transformation of the *Fusarium* mycotoxins deoxynivalenol and zearalenone by the normal gut microflora of pigs, *Nat. Toxins* 2:105.

Kreiger, N., and Hiatt, R.A., 1992, Risk of breast cancer after benign breast diseases: variation by histologic type, degree of atypia, age at byopsy, and length of follow-up, *Am. J. Epidemiol.* 135:619.

Kuczuk, M., Benson, P., Heath, H., and Hayes, A., 1978, Evaluation of the mutagenic potential of mycotoxins using *Salmonella typhimurium* and *Saccharomyces cerevisiae*, *Mutat. Res.* 53:11.

Kumagai, S., and Shimizu, T., 1982, Neonatal exposure to zearalenone causes persistent anovulatory estrus in the rat, *Arch. Toxicol.* 50:279.

Lauren, D.R., and Ringrose, M.A., 1997, Determination of the fate of three *Fusarium* mycotoxins through wet-milling of maize using an improved HPLC analytical technique, *Food Addit. Contam.* 14:435.

Lee, U.S., Jang, H.S., Tanaka, T., Hasegawa, A., Oh, Y.J., and Ueno, Y., 1985, The coexistence of the *Fusarium* mycotoxins nivalenol, deoxynivalenol and zearalenone in Korean cereals harvested in 1983, *Food Addit. Contam.* 2:185.

Lee, U.S., Jang, H.S., Tanaka, T., Hasegawa, A., Oh, Y.J., Cho, C.M., Suguira, Y., and Ueno, Y., 1986, Further survey on the *Fusarium* mycotoxins in Korean cereals, *Food Addit. Contam.* 3:253.
Lee, U.S., Jang, H.S., Tanaka, T., Hasegawa, A., Oh, Y.J., Cho, C.M., and Ueno, Y., 1987, Effect of milling on decontamination of *Fusarium* mycotoxins nivalenol, deoxynivalenol, and zearalenone in Korean wheat, *J. Agric. Food Chem.* 35:126.
Lee, U.S., Lee, M.Y., Park, W.Y., and Ueno, Y., 1992, Decontamination of *Fusarium* mycotoxins, nivalenol, deoxynivalenol, and zearalenone, in barley by the polishing process, *Mycotoxin Res.* 8:31.
Lippman, M., Bolan, G., and Huff, K., 1976, The effects of estrogens and antiestrogens on hormone-responsive human breast cancer in long-term tissue culture, *Cancer Res.* 36:4595.
Marin, M.L., Murtha, J., Dong, W., and Pestka, J.J., 1996, Effects of mycotoxins on cytokine production and proliferation in EL-4 thymoma cells, *J. Toxicol. Environ. Health* 48:379.
Martin, P.M., Horwitz, K.B., Ryan, D.S., and McGuire, W.L., 1978, Phytoestrogen interaction with estrogen receptors in human breast cancer cells, *Endocrinology* 103:1860.
Matsuura, Y., Yoshizawa, T., and Morooka, N., 1981, Effect of food additives and heating on the decomposition of zearalenone in wheat flour, *J. Food Hyg. Soc. Japan* 22:293.
Matsuura, Y., and Yoshizawa, T., 1985, Conversion of zearalenone, an estrogenic mycotoxin, by brewing microorganisms, *J. Food Hyg. Soc. Japan* 26:24.
McKenzie, K.S., Sarr, A.B., Mayura, K., Bailey, R.H., Miller, D.R., Rogers, T.D., Norred, W.P., Voss, K.A, Plattner, R.D., Kubena, L.F., and Phillips, T.D. 1997, Oxidative degradation and detoxification of mycotoxins using a novel source of ozone. *Food Chem. Toxicol.* 35:807.
McNutt, S.H., Purwin, P., and Murray, C., 1928, Vulvovaginitis in swine; preliminary report. *J. Amer. Vet. Med. Assoc.* 73:484.
Merino, M., Ramos, A.J., and Hernández, E., 1993, A rapid HPLC assay for zearalenone in laboratory cultures of *Fusarium graminearum*, *Mycopathologia* 121:27.
Mirocha, C.J., and Christensen, C.M., 1967, Estrogenic metabolite produced by *Fusarium graminearum* in stored corn, *Appl. Microbiol.* 15:497.
Mirocha, C.J., and Christensen, C.M., 1974, Oestrogenic mycotoxins synthesized by *Fusarium*, in: *Mycotoxins*, I.F.H. Purchase, ed., Elsevier Publishing Co., Amsterdam.
Mirocha, C.J., Pathre, S.V., and Christensen, C.M., 1977, Zearalenone, in: *Mycotoxins in Human and Animal Health*, J.V. Rodricks, C.W. Hesseltine, and M.A. Mehlman, eds., Pathotox, Park Forest South, IL.
Mirocha, C.J., Abbas, H.K., Windels, C.E., and Xie, W., 1989, Variation in deoxynivalenol, 15-acetyldeoxynivalenol, 3-acetyldeoxynivalenol, and zearalenone production by *Fusarium graminearum* isolates, *Appl. Environ. Microbiol.* 55:1315.
Nelson, P.E., Toussoun, T.A., and Marasas, W.F.O., 1983, *Fusarium Species: An Illustrated Manual for Identification*, The Pennsylvania State University Press, University Park, PA.
Okoye, Z.S.C., 1986, Zearalenone in native cereal beer brewed in Jos Metropolis of Nigeria, *J. Food Safety* 7:233.
Okoye, Z.S.C., 1987, Stability of zearalenone in naturally contaminated corn during Nigerian traditional brewing, *Food Addit. Contam.* 4:57.
Osborne, M.P., 1991, Breast development and anatomy, in: *Breast Disease*, Harris, J.R., Hellman, S., Henderson, I.C., and Kinne, D.W., eds., J. P. Lippincott Co, New York.
Park, K.J., Park, A.R., and Lee, Y.W., 1992, Natural occurrence of *Fusarium* mycotoxins of the 1990 barley crop in Korea, *Food Addit. Contam.* 9:639.
Ruddick, J.A., Scott, P.M., and Harwig, J., 1976, Teratological evaluation of zearalenone administered orally to the rat, *Bull. Environ. Contam. Toxicol.* 15:678.
Ryu, D., Hanna, M.A., and Bullerman, L.B., 1999, Stability of zearalenone during extrusion of corn grits, *J. Food Prot.* 62:1482.
Ryu, D., and Bullerman, L.B., 1999, Measurement of the estrogenic activity of zearalenone by cell culture technique, *Food Sci. Biotech.* 8:227.
Sáenz de Rodríguez, C.A., 1984, Environmental hormone contamination in Puerto Rico, *N. Eng. J. Med.* 310:1741.
Sáenz de Rodríguez, C.A., Bougovanni, A.M., and Conde de Borrego, L., 1985, An epidemic of precocious development in Puerto Rican children, *J. Pediatr.* 107:393.
Scheutwinkel, M., Hude, W.V.D., and Basler, A., 1986, Studies on the genotoxicity of the anabolic drugs trenbolone and zearanol, *Arch. Toxicol.* 59:4.
Schwadorf, K., and Müller, H.M., 1992, Determination of and zearalenol and zearalenone in cereals by gas chromatography with ion-trap detection, *J. Chromatogr.* 595:259.
Scott, P.M., Kanhere, S.R., Dailey, E.F., and Farber, J.M., 1992, Fermentation of wort containing deoxynivalenol and zearalenone, *Mycotoxin Res.* 8:58.
Scott, P.M., Kanhere, S.R., and Weber, D., 1993, Analysis of Canadian and imported beers for *Fusarium* mycotoxins by gas chromatography-mass spectroscopy, *Food Addit. Contam.* 10:381.
Soule, H.D., Vazquez, J., Long. A., Abert, S., and Brennan, M., 1973, A human cell line from a pleural effusion derived from a breast carcinoma, *J. Natl. Cancer Inst.* 51:1409.

Sundlof, S.F., and Strickland, C., 1986, Zearalenone and zeranol: Potential residue problems in livestock, *Vet. Hum. Toxicol.* 28:242.

Tanaka, T., Yamamoto, S., Hasegawa, A., Aoki, N., Besling, J.R., Sugiura, Y., and Ueno, Y., 1990, A survey of the natural occurrence of *Fusarium* mycotoxins deoxynivalenol, nivalenol and zearalenone in cereals harvested in The Netherlands, *Mycopathologica* 188:19.

Tanaka, T., Teshima, R., Ikebuchi, H., Sawada, J.I., Terao, T., and Ichinoe, M., 1993, Sensitive determination of zearalenone and a-zearalenol in barley and Job's-tears by liquid chromatography with fluorescence detection, *J. AOAC. Int.* 76:1006.

Tashiro, F., Shibata, A., Nishimura, N., and Ueno, Y., 1983, Zearalenone reductase from rat liver, *J. Biochem.* 93:1557.

Thigpen, I.E., Li, L.A., Richter, C.B., Lebetkin, E.H., and Jameson, C.W., 1987, The mouse bioassay for the detection of estrogenic activity in rodent diets. I. A standardized method for conducting the mouse bioassay, *Lab. Anim. Sci.* 37:596.

Trenholm, H.L., Charmley, L.L., Prelusky, D.B., and Warner, R.M., 1991, Two physical methods for the decontamination of four cereals contaminated with deoxynivalenol and zearalenone, *J. Agric. Food Chem.* 39:356.

Trenholm, H.L., Charmley, L.L., Prelusky, D.B., and Warner, R.M., 1992, Washing procedures using water or sodium carbonate solutions for the decontamination of three cereals contaminated with deoxynivalenol and zearalenone, *J. Agric. Food Chem.* 40:2147.

Trigo-Stockli, D.M., Deyoe, E.W., Satumbaga, R.F., and Pedersen, J.R., 1996, Distribution of deoxynivalenol and zearalenone in milled fractions of wheat, *Cereal Chem.* 73:388.

Tuite, J., Shaner, G., Rambo, G., Foster, J., and Caldwell, R.W., 1974, The Gibberella ear rot epidemics of corn in Indiana in 1965 and 1972, *Cereal Sci. Today* 19:238.

Ueno, Y., and Kubota, K., 1976, DNA-attacking ability of carcinogenic mycotoxins in recombination-deficient mutant cells of *Bacillus subtilis*, *Cancer Res.* 36:445.

Ueno, Y., and Tashiro, F., 1981, α-Zearalenol, a major hepatic metabolite in rats of zearalenone, an estrogenic mycotoxin of *Fusarium* species, *J. Biochem.* 89:563.

Urry, W.H., Wehrmeister. J.L., Hodge, E.B., and Hidy, P.H., 1966, The structure of zearalenone, *Tetrahedron Lett.* 27:3109.

Veldman, A., Gorggreve, G.J., Mulders, E.J., and Van de Lagemaat, D., 1992, Occurrence of the mycotoxins ochratoxin A, Zearalenone and deoxynivalenol in feed components, *Food Addit. Contam.* 9:647.

Warner, R.L., and Pestka, J.J., 1987, ELISA survey of retail grain-based food products for zearalenone and aflatoxin B_1, *J. Food Prot.* 50:502.

Wehner, F., Marasas, W.F.O., and Thiel, P., 1978, Lack of mutagenicity of *Salmonella typhymurium* of some *Fusarium* mycotoxins, *Appl. Environ. Microbiol.* 35:659.

Wiewiórkowksa, M., Goliński, P., and Szebiotko, K., 1988, Utilization by low energy processing of plant origin materials contaminated with mycotoxins, *Acta Aliment. Pol.* 14:157.

Williams, B.A., Mills, K.T., Burroughs, C.D., and Bern, H.A., 1989, Reproductive alterations in female C57BL/Crgl mice exposed neonatally to zearalenone, an estrogenic mycotoxin, *Cancer Lett.* 46:225.

Williams, G.M., and Weisburger, J.H., 1991, Chemical carcinogenesis. In: *Casarett and Doull's Toxicology: The basic science of poisons*, M.O. Amdur, J. Doull, and C.D. Klaassen, eds., McGraw-Hill, Inc.

Young, L.G., and King, G.J., 1986, Low concentrations of zearalenone in diets of mature gilts, *J. Anim. Sci.* 63:1191.

Zill, G., Engelhardt, G., Wohner, B., and Wallnöfer, P.R., 1989, Formation of [^{13}C]- and [^{14}C] zearalenone by *Fusarium graminearum* DSM 4529, *Appl. Microbiol. Biotechnol.* 32:340.

MYCOTOXINS AND FERMENTATION - BEER PRODUCTION

Charlene E. Wolf-Hall and Paul B. Schwarz

Department of Cereal Science
North Dakota State University
Fargo, North Dakota 58014

ABSTRACT

Along with food safety issues due to mycotoxins, the effects of *Fusarium* infections on malt and beer quality can be disastrous. While some of the *Fusarium* head blight mycotoxins, such as DON, present in infected barley may be lost during steeping, the *Fusarium* mold is still capable of growth and mycotoxin production during steeping, germination and kilning. Therefore, detoxification of grain before malting may not be practical unless further growth of the mold is also prevented. Methods for reducing the amount of mold growth during malting are needed. Physical, chemical and biological methods exist for inhibiting mold growth in grain. Irradiation is a promising means for preventing *Fusarium* growth during malting, but its effects on malt quality and mycotoxin production in surviving mold need to be evaluated in more detail. Chemical treatments such as ozonation, which do not leave chemical residues in beer, also appear to be promising. Although biological control methods may be desirable, the effects of these inhibitors on malt and beer quality require further investigation. In addition, storage studies are needed to determine the effect of biological control on *Fusarium* viability and malt quality. It may also be possible to incorporate detoxifying genes into fermentation yeasts, which would result in detoxification of mycotoxins present in wort. Development of these types of technological interventions should help improve the safety of products, such as beer, made from *Fusarium* infected grain.

INTRODUCTION

Beer is consumed on a regular basis by a large number of people in the world. Per capita consumption of beer in the United States in 1997 was 21 gallons, or 224, 12 ounce cans (Beer Institute [Online, 1997] http://www.beerinst.org). From the standpoint of food safety, the consumption of this amount of beer could possibly result in significant exposure to mycotoxins. Sources of these mycotoxins could include contaminated malt or adjunct grains used in the beer making process. A previous review by Scott (1996) discusses the fate of several mycotoxins during the malting and brewing processes and reports results of surveys for mycotoxins in beer samples from various countries. Schwarz (2000) recently reviewed the impact of *Fusarium* head blight and associated mycotoxins on the malting and brewing quality of barley. While much of this previous material has been incorporated into the current review, various strategies that have been studied to remove, inactivate or prevent the formation of these toxins prior to, or during the brewing process will also be discussed.

MYCOTOXINS ASSOCIATED WITH *FUSARIUM* HEAD BLIGHT IN BARLEY

Malted barley is the most common grain used in the beer making process. In recent years, small grains, such as barley, have been greatly affected by the plant disease, *Fusarium* head blight (FHB) also known as scab (McMullen, 1997). Trichothecene mycotoxins including, deoxynivalenol (DON or vomitoxin), nivalenol (NIV), T-2 toxin, HT-2 toxin and diacetoxyscirpenol (DAS), as well as the estrogenic mycotoxin, zearalenone (ZEN) have been detected in FHB infected barley from the upper Midwestern states in the United States (Schwarz et al., 1995a). Of these, DON is the most frequently detected and the mycotoxin produced in greatest quantity in the FHB infected grain. Clear et al. (1996) also found DON to be the predominant mycotoxin in FHB infected barley in Manitoba during recent epidemics. In southwest Germany, the predominant mycotoxin found in barley crops from 1982-1992 was also DON (Müller et al., 1997). In a review of surveys looking for mycotoxins in commercially available beers (Scott, 1996), it was shown that DON, NIV, T-2, HT-2, DAS, ZEN, aflatoxins, ochratoxin A, and fumonisins have been detected in beers at trace (ppb) levels. Mycotoxins are also a concern in other by-products of the malting and brewing processes such as various food ingredients and animal feeds (Flannigan et al., 1984).

EFFECTS OF FHB INFECTION IN BARLEY ON MALTING AND BREWING

Malting

Malt is produced from barley by germinating the seed under regulated conditions of moisture and temperature. The germinated seed is the source of amylase enzymes which will hydrolyze the starch in the barley and adjunct grain into simple sugars for the yeast to metabolize during beer fermentation. To germinate the seed, the barley moisture content is raised to roughly 45% by steeping for 36-52 hours at 12-20°C. The steeping process involves several stages of immersion in aerated water followed by air-rests. The process actually rinses the barley during the drain and fill cycles. After steeping, the seed is germinated at 15-20°C in humidified air. The germinated barley or green malt is then dried to roughly 4-5 % moisture in a kilning step. Kilning temperatures and humidity are controlled to prevent inactivation of desired enzymes. Kilning temperatures rarely exceed 90°C. Kilning causes other chemical reactions to occur in the barley such as browning reactions. These chemical products will affect the final beer by contributing to the organoleptic qualities, such as flavor, color, odor and texture.

Mycotoxin production may occur with the growth of *Fusarium* mold during steeping, germination and possibly kilning in the malting process. Schwarz et al. (1995b) micro-malted several barley samples that were naturally infected with *Fusarium graminearum* and contaminated with DON (4.8-22.5 μg/g), zearalenone (1.5 μg/g) and 15-ADON (2.1 μg/g). Steeping lowered DON levels to near or below the detection limit. DON may have been lost due to solubilization or loss of dust or mold with the overflow and fill-drain cycles of the steep. DON is a water soluble toxin, so it may be somewhat washed away with rinse water. Trenholm et al., (1992) demonstrated that washing heavily contaminated barley with distilled water (3 times for 30 min) could reduce DON concentrations by 69%. However, mold growth during germination increased DON levels by 18-114% on green malts compared to the original infected barley following five days of germination (Schwarz et al., 1995b). A significant increase in ergosterol (a molecular marker for fungal biomass) was observed between two and three days of germination, followed by DON production later in germination. This pattern of toxin production after growth of mycelia is typical for the production of secondary metabolites such as mycotoxins (Bu'Lock, 1980). Similarly, increases in 15-ADON and ZEN levels were seen in germinated barley. No change in DON levels was observed during kilning. DON is known to be stable up to 170°C at neutral to acidic pHs (Wolf and Bullerman, 1998). Munar and Sebree (1997) found DON produced during the early stages of kilning. Wolf-Hall and Bullerman (1998) showed greater production of DON and 15-acetyl DON by certain isolates of *Fusarium graminearum* at 35°C versus 25°C. The rise in temperature in the early stages of kilning may actually stimulate increased production of mycotoxins by certain strains of infecting mold. Many factors could be affecting the amount of toxin produced at various stages of malting, including strain of mold present, the viability of the infecting mold (injury, dormancy, or death), the amount of infecting mold present, the location of mycelia or spores within the seed structure, and competing and/or antagonistic organisms.

Other qualities affected by *Fusarium* mold growth during malting include reduced barley germination, gushing, changes in color and formation of off-flavors in beer (Gyllang et al., 1981). Barley used for malt is required to have germination rates above 95%. Below this level problems develop such as low enzyme production and malt extract. The reduced levels of germination in FHB infected barley could be due to the mycelium invading the embryo and/or the presence of mycotoxins which may also inhibit germination (Haikara, 1983; Nummi et al., 1975; Schapira et al., 1989). Changes in color may be due to the production of pigments by the infecting mold (Sloey and Prentence, 1962), or due to increases in the levels of soluble amino nitrogen compounds in the wort due to proteolytic activity by the mold on barley proteins (Haikara, 1983; Prentice and Sloey, 1960; Zhou, 1998).

Brewing

After barley is malted, it is coarsely ground then extracted with hot water. This step is called mashing, where the ground malt is mixed with water and heated to about 70°C. During this process, proteolytic enzymes in the barley are most active and the result is roughly equal proportions of amino acids, peptides, and protein in the final malt extract. The soluble nitrogenous compounds contribute to the beer color, flavor and the yeasts' metabolism during fermentation. Some of the proteins contribute to the foaming characteristics of the final beer. The amylases produced during malting hydrolyze the starch to simple sugars; adjunct grain may be added as an additional source of starch. After mashing, the liquid portion or wort is separated from the solids or spent grains, which are usually used as animal feed. Water soluble mycotoxins tend to partition into the wort. The wort is boiled to kill spoilage organisms, cooled, aerated and then, the yeast inoculation or pitch is added. The boiling process kills the contaminating fungi, but most mycotoxins will survive this process. The fermentation usually takes 5 to 7 days at 8 to 15°C. The beer is then aged, further carbonated and packaged.

Some of the problems associated with using FHB infected barley malt are alterations in wort composition (soluble nitrogen compounds and carbohydrates), the presence of mycotoxins and other mold metabolites, and the propensity for beer gushing. The mold may produce proteases that result in further digestion of proteins in the malt and wort beyond what the barley proteases accomplish (Sloey and Prentence, 1962; Haikara, 1983). This affects color, flavor, texture, and foaming characteristics in the final beer.

Haikara (1983) noted that the use of FHB infected barley results in an increase in beer pH and in the degree of fermentation. This may be due to increased free amino nitrogen for yeast metabolism as well as to higher content of simple sugars.

Yeast fermentation may also be affected by fungal metabolites in the wort. Lafont et al. (1981) noted a 58-80% decrease in the rate of fermentation by *Saccharomyces cervisiae* in fermentation broths containing 10-50 μg/g T-2 toxin. This decrease was a temporary lag, and the yeast appeared to recover. It has been suggested that T-2 toxin inhibits mitochondrial function, causing slower oxygen utilization by the yeast and slowed growth rate during the exponential phase (Koshinsky et al., 1992). This deceased growth rate has also been shown when ZEN was present at 5-50 μg/g (Flannigan et al., 1985), DAS at 5-10 μg/g (Whitehead and Flannigan, 1989), and DON at 50 μg/g (Ryman, 1994). However, Whitehead and Flannigan, (1989) reported that there was no effect with DON at 20 μg/g. Boeira et al. (1999) found no effect of NIV at concentrations below 50 μg/g on strains of ale and lager yeasts. Boeira et al. (2000) studied the interactions of combinations of DON, ZEN and fumonisin B_1 on yeast. They found a synergistic interaction between DON and ZEN, but only at very high concentrations. Even if FHB infected grain was utilized for brewing, it is unlikely that any of these mycotoxins would be found at concentrations this high.

Several studies have shown that DON levels increase during yeast fermentation. Böhm-Schraml et al. (1997) showed an increase in DON within the first 20 hours of fermentation, which then progressively decreased afterward up to 100 hours. The decrease in DON was theorized to be due to absorption by yeast cells or extracellular metabolism. In a baking study using naturally DON contaminated wheat, Young et al. (1984) saw an increase in DON levels in a yeast fermented doughnut and attributed this to possible DON precursors in the flour which were further converted to DON by the yeast.

By increasing inoculum density or pitching rate, Whitehead and Flannigan (1989) were able to decrease the effects of DAS on fermentation. They suggested that certain yeasts may be able to detoxify mycotoxins such as T-2 toxin. The development of these mycotoxin resistant yeasts would be of great benefit to the fuel ethanol industry, as heavily mycotoxin contaminated grain could be better utilized since it would not be acceptable for food/feed uses. It has been reported that there are microorganisms that can detoxify certain trichothecenes (Shima et al., 1997; Westby et al., 1997) which may be sources of genetic material for incorporation into modified ethanol producing microorganisms.

DON appears to be very stable during the brewing process. Niessen (1993), found DON to be carried over into the final beer. This study also showed a four-fold increase in DON concentrations during mashing possibly resulting from release of the toxin from protein conjugates. Wolf-Hall and Bullerman (1999), found that DON recoveries from water extracts of spiked, extruded corn grits could be improved by 26% using an amylase enzyme in the extraction process. This study suggests that the action of proteases and amylases release additional DON from infected grains.

Schwarz et al. (1995b) pilot-brewed beers from naturally infected barley malt containing 1.8-17.3 μg/g DON and 1.6-4.8 μg/g ZEN. The results showed 80-93% of the DON remained in the final beers and only trace amounts were present in the spent grains. No ZEN was detected in the final beers and about 60% of the ZEN was recovered from the spent grains. It has been suggested that ZEN is converted by yeast metabolism to zearalenol (Scott, 1996), however no zearalenol was detected using gas chromatography/mass spectrometry (GC/MS) (P. Schwarz, North Dakota State University, unpublished results).

Besides the effects of and presence of *Fusarium* mycotoxins in beer, there are other qualities of beer affected by using FHB infected grains. Gushing, or sudden over production of foam upon opening a container, is a defect that is closely associated with the use of FHB infected grain in beer making. Gushing can be a serious quality defect resulting in the permanent loss of consumers. Gushing appears to be caused by the formation and stabilization of large amounts of microbubbles in beer (Casey, 1996; Drager, 1996; Fischer et al., 1997). The nucleation centers for these microbubbles seem to be a product of mold growth in the grain, and not just *Fusarium* mold may cause it - others like *Aspergillus* and *Penicillium* may also cause gushing. However, *Fusarium* species seem to be the most problematic (Haikara, 1983). The topic of beer gushing has been reviewed by Casey (1996). Schwarz et al. (1997) induced gushing in beer by adding cell-free extracts of *Fusarium graminearum* grown in liquid culture broth and proposed that the *Fusarium* gushing factor is a water soluble component produced by the mold. The gushing factor(s) has been correlated with the presence of DON in malt (Schwarz et al., 1996), and both the gushing factor(s) and DON are thought to be produced in parallel by the mold during growth in the germinating barley. It may be that the presence and amount of DON can be used to predict gushing propensity, but it is apparent that much more research is needed to develop analytical tests to determine gushing factor content in malt.

METHODS TO CONTROL FHB MYCOTOXINS IN BEER

The simplest means to control the levels of FHB mycotoxins in beer is to avoid the use of contaminated grain, a practice adopted by malting companies/brewers in the upper Midwestern United States. Malting companies have adopted a standard of 0.5 ppm DON or less in purchased grain (McMullen et al., 1997). However, avoidance of FHB infected grain may not always be feasible. This is especially true in developing countries where food and feed supplies may be very limited, or during years of extreme epidemics of FHB in barley crops. The economic costs associated with avoidance can be devastating to barley producing regions. Growers receive severely discounted prices for FHB infected grain, and processors can have the additional costs of testing and importation of grain from other regions. With the current inability to control FHB in the field, it is important that technology be developed in order to be able to permit the utilization of at least a portion of the infected crop. Various techniques have been explored for the utilization of infected grain, and they generally fall into one of three categories: removal or separation of infected kernels, treatments designed to decontaminate or eliminate mycotoxins present in grain, and treatments intended to prevent or inhibit mold growth.

Removal or Separation of Infected Kernels

In wheat it has been determined that levels of DON can be significantly reduced by removing the heavily FHB infected kernels known as tombstone kernels. These kernels are typically shriveled, smaller and less dense compared to non-infected kernels. These tombstone kernels can be removed by use of a gravity table (Clear et al., 1997). However, in barley, if infection occurs later in kernel development, the shriveling of kernels may be minimal (Steffenson, 1996). Attempting to clean barley may not be a very effective means to lower DON concentrations. Unlike wheat, the hull is usually left on barley and seems to be where the highest concentrations of both mold and DON occur (Clear et al., 1997). Dehulling damages the embryo, decreasing germination rates. Consequently, dehulling is not a practical intervention technique. Hulless genotypes of barley may lead to lower mycotoxin concentrations in the grain, but these varieties are not yet practical for use in malting due to problems with germination and clumping (Kunze, 1996).

Decontamination or Elimination of Mycotoxins Present in Grain

There are some options available for treating scabby grain to reduce levels of DON. Some chemical treatments such as sodium bisulfite, chlorine gas, moist ozone, and ammonia have been shown to reduce DON concentrations. However, these harsh treatments can either be too severe on the grain quality, the reaction product too unstable, or undesired residues may be left on the grain. Young et al. (1987;1986b) used sodium bisulfite to decontaminate DON in infected corn and wheat. The reaction product was a DON sulfonate which could be hydrolyzed upon heating, as was evident in baking studies where DON concentrations increased 50-75% in sulfite-treated wheat that was used to make bread.

Chemical destruction of trichothecene toxicity can be achieved rapidly in alkaline solutions of sodium hypochlorite (Faifer et al., 1994; Young et al., 1986a). Trenholm et al. (1992) saw reductions of 74 and 81% in DON and ZEN respectively in FHB infected barley soaked in 1 M sodium carbonate for 30 minutes followed by two, 30 minute distilled water rinses. DON is unstable at alkaline pHs (Wolf and Bullerman, 1998). In fact, a traditional practice used by some maltsters is to add lime to the first steep. This, however, may have no effect on the later growth and toxin production by *Fusarium* during the germination step.

It may also be possible to develop genetically modified yeasts that can reduce the toxicity of mycotoxins that are carried over into the wort. As stated previously, there are microorganisms that can detoxify certain trichothecenes (Shima et al., 1997; Westby et al., 1997). These microorganisms may be sources of genetic material to be incorporated into modified ethanol producing microorganisms.

Prevention or Inhibition of Mold Growth

Methods to prevent or inhibit mold growth include physical, chemical and biological based technologies. There has been very little research reported regarding physical treatments, such as heat, on molds in barley. There may be ways to heat pasteurize grain in a way to kill mold but still allow germination (Berjak et al., 1992).

There are references showing significant effects of chemical treatments including formaldehyde, hypochlorite, and mercuric chloride on *Fusarium* in barley. These, however, would not be acceptable to the brewers who do not want the chemicals or potentially harmful reaction products to be retained in the beer. Fungicides applied in the field may prevent the initial *Fusarium* infection (McMullen, 1997), but treatments can be expensive and the fungicide needs to be applied within narrow periods of time to be effective. Sanitizing agents such as sodium hypochlorite have been shown to eliminate *Fusarium culmorum* from inoculated barley seeds (Ramakrishna, et al., 1991), and may also detoxify the grain. However, high levels of sanitizer are needed due to the rapid depletion of active components by reactions with organic matter in the grain. Sanitizers may also affect desirable competitive microorganisms on the barley. High levels of residual sanitizing agents or by-products may be detrimental to yeast performance during fermentation and may affect the final beer quality. A promising newer method for sanitizing barley is the use of high concentrations of ozone, which rapidly decomposes to oxygen. This method may be effective for sanitizing barley as well as detoxifying mycotoxins. McKenzie et al. (1997) reported that ozone treatment was effective in deoxifying aflatoxins, cyclopiazonic acid, ochratoxin A, secalonic acid D, and ZEN.

The development of FHB resistant barley cultivars is being explored, but may take up to eight years of breeding to achieve results (Beattie et al., 1998). Even if FHB resistance is achievable, these new cultivars will also need to have characteristics that will command malting quality.

There are references to microbiological control methods using competing fungi, such as *Geotrichum candidum* (Boivin and Malanda, 1997), added as starter cultures to the steeping process. Boivin and Malanda (1997) showed a reduction in *Fusarium* infection rates from 86% to 0% in steeped barley. *Geotrichum candidum* is not typically found on barley, but is routinely found in malt (Flannigan et al., 1984). However, the starter cultures of *Geotrichum candidum* would need to be monitored for toxin production as well, since certain strains are capable of producing clavine alkaloids (Flannigan et al., 1984). Some lactic acid bacteria have also shown an ability to produce antifungal compounds. Haikara et al. (1993) found that addition of *Pediococcus pentosaceus* and *Lactobacillus plantarum* to barley during steeping could reduce the levels of *Fusarium* and other spoilage bacteria. Lactic acid bacteria could also be added as starter cultures. The use of starter cultures to out compete or inhibit *Fusarium* molds has also shown beneficial effects on malt and beer quality, such as improved malt modification and homogeneity, lowering of wort glucans and viscosity, and an improvement in mash filterability (Boivin and Malanda, 1997; Haikara and Laitila, 1995). The qualities of the beers produced were rated organoleptically as good. An additional benefit may be the preservative effect of the lactic acid produced on the brewers spent grains (Suomalained et al., 1995), which would increase the quality of the spent grains for animal feed.

Although microbiological methods show promise, the brewing and malting industries are looking for an immediate solution to the problem. Ideally, they would like a physical method that could be placed "on line" in the process. The use of irradiation to pasteurize malting barley may be a way to treat the grain without adverse effects on germination and malt quality. We have seen *Fusarium* survival being decreased by approximately 78% at 10 kilogray (KGy) with only a 20 % reduction in germination in one barley variety using electron beam irradiation (unpublished preliminary research). At 8 KGy there was still significant reduction in *Fusarium* growth, but an actual increase in the germination ability of the treated infected barley. Researchers in the U.K. (Ramakrishna et al., 1991) were able to stop the growth of *Fusarium* molds in infected barley by using gamma irradiation. According to that report, 4 KGy stopped growth of the mold, but did not hamper the germination ability of the barley. The grain they used had a fairly low level of *Fusarium* infection, less than 20%. Higher infection rates, which are more typical in the FHB outbreaks in the Midwest, may take a stronger dosage to stop the growth of the mold due to more mycelium and spores being present in the kernels. Ramakrishna et al. (1991) found barley germination was little affected, or even slightly increased up to 10 KGy, but then steadily decreased with increasing dosage. They did not analyze for mycotoxins.

Time alone may prove to be an effective means to control *Fusarium* growth during malting. It appears that the viability of *Fusarium* mold decreases over time in storage (Lutey and Christensen, 1963). Beattie et al. (1997) were able to show a decrease in *Fusarium* viability in FHB infected barley stored for seven months under different conditions of aeration and temperature. The viability of the *Fusarium* decreased under all conditions tested, but to the greatest extent in aerated samples held at roughly 24°C. The rate of decline in viability was greatest during the first three months of storage. More research should be done to determine if storage is a practical means to improve the utilization of FHB infected barley.

For any method used to prevent mold growth, tests should be done on the affected flora to determine if by injuring the molds, such as *Fusarium,* that mycotoxin synthesis is not induced. Sub-lethal or inhibitory concentrations of chemicals may prevent fungal growth, but actually stimulate mycotoxin production. This would include chemicals such as organic acids and fungicides. It has been shown that application of fungicides can increase the occurrence of mycotoxins (Miller, 1994). Sorbic acid has been found to stimulate the production of aflatoxin B_1 and T-2 toxin (Bauer et al., 1984). Studies evaluating the effects of irradiation on *Aspergillus flavus* and *Aspergillus parasiticus* have shown that surviving isolates with enhanced ability to produce aflatoxins can be isolated from the treated grain (Moss and Frank, 1987). Altering the proportions of the normal microflora on barley could also potentially cause other problems. Fungi and bacteria have chemical interactions that are not well understood, and alteration in their proportions may result in increased mycotoxin production by *Fusarium* or increased mycotoxin production by other genera of fungi (Moss and Frank, 1987). The practicality of combining intervention methods to impose a hurdle effect to mycotoxin production should be investigated.

CONCLUSION

While some degree of FHB mycotoxins, such as DON, present in infected barley may be lost during steeping, the *Fusarium* mold is still capable of growth and mycotoxin production during steeping, germination and kilning. Therefore, detoxification of grain before malting may not be practical unless further growth of the mold is also prevented. Irradiation looks very promising as a means to prevent *Fusarium* growth during malting, but the effect on the surviving mold to produce mycotoxins and the effect on malt quality needs to be further studied. Chemical treatments such as ozonation, which do not leave residual chemical in the beer also appear to be promising. Biological control methods may be desirable due to the use of "natural" inhibition, but the effects on malt and beer quality will need to be further investigated. Storage studies should be done to determine the effect on *Fusarium* viability and malt quality. It may also be possible to incorporate detoxifying genes into fermentation yeasts through biotechnology, which would allow detoxification of the wort when mold growth is no longer a problem. Development of these types of technological interventions should help improve the safety of products, such as beer, made from FHB infected grain.

REFERENCES

Bauer, G.J., A. Montegelas, and Gedek, B., 1984, Stimulation of aflatoxin B_1 and T-2 toxin production by sorbic acid, *Appl. Environ. Microbiol.* 47:416-418.

Beattie, S., Schwarz, P.B., Horsley, R., Barr, J., and Casper, H., 1997, The effect of grain storage conditions on the viability of *Fusarium* and deoxynivalenol production in infested malting barley, *J. Food Protection.* 61:103-106.

Berjak, P., Whittaker, A., and Mycock, D.J., 1992, Wet-heat treatment: a promising method for the elimination of mycoflora from maize grains, *South African J. Sci.* 88:346-349.

Boeira, L.S., Bryce, J.H., Stewart, G.G., and Flannigan, B., 1999, Inhibitory effect of *Fusarium* mycotoxins on growth of brewing yeasts. 2. Deoxynivalenol and nivalenol, *J. Inst. Brew.* 105:376-381.

Boeira, L.S., Bryce, J.H., Stewart, G.G., and Flannigan, B., 2000, The effect of combinations of *Fusarium* mycotoxins (deoxynivalenol, zearalenone and fumonisin B_1) on growth of brewing yeasts, *J. Appl. Microbiol.* 88:388-403.

Böhm-Schraml, M., Stettner, G., and Geiger, E., 1997, Studies into the influence of yeast on *Fusarium*-toxins in wort. *Cereal Res. Comm.* 25:729-730.

Boivin, P. and Malanda, M., 1997, Improvement of malt quality and safety by adding starter culture during the malting process, *MBAA Tech. Quart.* 34:96-101.

Bu'Lock, J.D., 1980, Mycotoxins as secondary metabolites, in: *The Biosynthesis of Mycotoxins: A Study in Secondary Metabolism,* P. S. Steyn, ed., Academic Press, Inc., New York.

Casey, G., 1996, Primary versus secondary gushing and assay procedures used to assess malt/beer gushing potential, *MBAA Tech. Quart.* 33:229-235.

Clear, R.M., Patrick, S.K., Nowicki, T., Gaba, D., Edney, M., and Babb, J.C., 1997, The effect of hull removal and pearling on *Fusarium* species and trichothecenes in hulless barley, *Can. J. Plant Sci.* 77:161-166.

Dickson, J.G., 1962, Diseases of barley and their control, in: *Barley, Biology, Biochemistry, Technology*, A.H. Cook, ed., Academic Press, New York.

Drager, M., 1996, Physical observations on the subject of gushing, *Brauwelt Intl.* 14:363-367.

Duijnhouwer, I.D.C., Grasshoff, C., and Angelino, S.A.G.F., 1993, Kernel filling and malting barley quality, *Eur. Brew. Conv.* 24:121-128.

Faifer, G.C., Velazco, V., and Godoy, H.M., 1994, Adjustment of the conditions required for complete decontamination of T-2 toxin residues with alkaline sodium hypochlorite, *Environ. Contam. Toxicol.* 52:102-108.

Fischer, S. Schwill-Miedaner, A., Illberg, V., and Sommer, K., 1997, Untersuchung von Einflussfaktoren des gushing phänomens, *Brauwelt.* 137:210-214.

Flannigan, B., Day, S.W., Douglas, P.E., and McFarlane, G.B., 1984, Growth of mycotoxin-producing fungi associated with malting of barley, in: *Toxigenic Fungi*, H. Kurata and Y. Ueno, eds., Elsevier Science Publishing Company, Inc., New York.

Flannigan, B., Morton, J.G. Naylor, 1985, *Fusarium* mycotoxins and the malting of barley, in: *Trichothecenes and Other Mycotoxins*, J. Lacey, ed., Wiley, New York.

Gyllang, H. Kjellén, K., Haikara, A., and Sigsgaard, P., 1981, Evaluation of fungal contamination on barley and malt, *J. Inst. Brew.* 87:248-251.
Haikara, A., 1983, Malt and beer from barley artificially contaminated with *Fusarium* in the field. *Proc. Eur. Brew. Conv.* 19:401-408.
Haikara, A. and Laitila, A., 1995, Influence of lactic acid starter cultures on the quality of malt and beer, *Eur. Brew. Conv.* 25:249-256.
Haikara, A., Uljas, H., and Suurnakki, A, 1993, Lactic starter cultures in malting - a novel solution to gushing problems, *EBC Congress*. 24:163-172.
Koshinsky, H., Crosby, R., and Khachatourians, G., 1992, Effects of T-2 toxin on ethanol production by *Saccharomyces cerevisiae. Biotech. Appl. Biochem.* 16:275-286.
Kunze, W., 1996, *Technology Brewing and Malting*, Versuchs-und Lehranstalt für Brauerei, Berlin.
Lafont, J., Romand, A., and Lafont, P., 1981, Effects de mycotoxines sur l activité fermentaire de *Saccharomyces cerevisiae, Mycopathologia.* 74:119-123
Lutey, R.W. and Christensen, C.W., 1963, Influence of moisture content, temperature, and length of storage upon survival of fungi in barley kernels, *Phytopathology.* 53:713-717.
McKenzie, K.S., Sarr, A.B., Mayra, K., Bailey, R.H., Miller, D.R., Rogers, T.D., Norred, W.P., Voss, K.A., Plattner, R.D., Kubena, L.F., and Phillips, T.D., 1997, Oxidative degradation and detoxification of mycotoxins using a novel source of ozone, *Food Chem. Toxicol.* 35:807-820.
McMullen, M., 1997, Scab of wheat and barley: a re-emerging disease of devastating impact, *Plant Disease.* 81:1340-1348.
Miller, J.D., 1994, Epidemiology of *Fusarium* ear diseases, in: *Mycotoxins in Grain*, J.D. Miller and H. L. Trenholm, eds., Eagan Press, St. Paul, Minnesota.
Moss, M.O. and Frank, M., 1987, Prevention: effects of biocides and other agents on mycotoxin production, in: *Natural Toxicants in Food: Progress and Prospects*, D.H. Watson, ed., Ellis Horwood Ltd., Chichester, England.
Müller, H.-M., Reimann, J., Schmacher, U., and Schwadorf, K., 1997, Natural occurrence of *Fusarium* toxins in an area of southwest Germany, *Mycopathologia.* 137:185-197.
Munar, M. and Subree, B., 1997, Gushing - A maltster's view. *J. Am. Soc. Brew. Chem.* 55(3):119-122.
Niessen, L.M., 1993, Entwicklung and anwendung immunchemischer verfahren zum nachweis wichtiger *Fusarium*-toxine bei der bierbereitung sowie mycologische untersuchungen im zuasammenhang mit dem widwerded (gushing) von bieren, *Doctoral Thesis*, Technical University Munick, Program for Technical Microbiology and Brewing Technology 2.
Nummi, M., Niku-Paavola, M.-L., and Enari, T.-M., 1975. Der Einfluss eines *Fusarium*-toxins auf die Gersten-Vermälzung, *Brauwissenschaft.* 28:130-133.
Prentice, N. and Sloey, W., 1960, Studies on barley microflora of possible importance to malting and brewing quality. I. The treatment of barley during malting with selected microorganisms. *Proc. Am. Soc. Brew. Chem.* 28-34.
Ramakrishna, N., Lacey, J., and Smith, J.E., 1991, Effect of surface sterilization, fumigation and gamma irradiation on the microflora and germination of barley seeds, *Int. J. Food Microbiol.* 13:47-54.
Ryman, R. J., 1994, The effect of natural grain contamination by *Fusarium* and associated mycotoxins in malting and brewing, *Master of Science Thesis*, Heriot-Watt University, Edinburgh, Scotland.
Schapira, S.F.D., Whitehead, M.P., and Flannigan, B., 1989, Effects of the mycotoxin diacetoxyscirpenol and deoxynivalenol on malting characteristics of barley, *J. Inst. Brew.* 95:415-417.
Schwarz, P.B., 2000, Impact of head blight on the malting and brewing quality of barley. Chapter 14 in: *Fusarium Head Blight of Wheat and Barley*. K.J. Leonard and W.R. Bushnell, ed. APS Press, St Paul, MN, *in-press.*
Schwarz, P.B., Beattie, S., and Casper, H.H., 1996, Relationship between *Fusarium* infestation of barley and the gushing potential of malt, *J. Inst. Brew.* 102:93-96.
Schwarz, P.B., Casper, H.H., and Barr, J.M., 1995a, Survey of the natural occurrence of deoxynivalenol (vomitoxin) in barley grown in Minnesota, North Dakota and South Dakotta during 1993, *MBAA Tech. Quart.* 3:190-194.
Schwarz, P.B., Casper, H.H., Barr, J., and Musial, M., 1997, Impact of *Fusarium* head blight on the malting and brewing quality of barley, *Cereal Res. Comm.* 25:813-814.
Schwarz, P.B., Casper, H.H., and Beattie, S., 1995b, Fate and development of naturally occurring *Fusarium* mycotoxins during malting and brewing. *J. Am. Soc. Brew. Chem.* 53:121-127.
Scott, P.M., 1996, Mycotoxins transmitted into beer from contaminated grains during brewing, *J. Assoc. Offic. Anal. Chem. Internat.* 79:875-882.
Shima, J., Takase, S., Takahashi, Y., Iwai, Y., Fujimoto, H., Yamazaki, M., and Ochi, K., 1997, Novel detoxification of the trichothecene mycotoxin deoxynivalenol by a soil bacterium isolated by enrichment culture, *Appl. Environ. Microbiol.* 63:3825-3830.
Sloey, W. and Prentice, N., 1972, Effects of *Fusarium* isolates applied during malting on the properties of malt, *Proc. Am. Soc. Brew. Chem.* 25-29.
Steffenson, B.J., 1996, *Fusarium* head blight of barley: Research update, in: *Proceedings, Red River Barley Day*, American Malting Barley Association, Milwaukee, WI.

Suomalainen, T, Storgårds, E., Mäyrä-Mäkinen, A., and Haikara, A., 1995, Lactic acid bacterial starter cultures in the preservation of spent grains used as animal feed, *Eur. Brew. Conv.* 25:733-739.
Trenholm, H.L., Charmley, L.L., Preluskey, D.B., and Warner, R.M., 1992, Washing procedures using water or sodium carbonate solutions for the decontamination of three cereals contaminated with deoxynivalenol and zearalenone, *J. Agric. Food Chem.* 40:2147-2151.
Trenholm, H.L., Friend, D.W., Thompson, B.K., and Preluskey, D.B., 1988, Effects of zearalenone and deoxynivalenol combinations for pigs, in: *Proceeding of the Japanese Association of Mycotoxicology*, Science University, Tokyo.
Young, J.C., Blackwell, B.A., and ApSimon, J.W., 1986a, Alkaline degradation of the mycotoxin 4-deoxynivalenol, Tetrahedron Letters. 27:1019-1022.
Westby, A., Reilly, A., and Bainbridge, Z., 1997, Review of the effect of fermentation on naturally occurring toxins, *Food Control.* 8:329-339.
Whitehead, M.P., and Flannigan, B., 1989, The *Fusarium* mycotoxin deoxynivalenol and yeast growth and fermentation, *J. Inst. Brew.* 95:411-413.
Wolf-Hall, C.E., Hanna, M.A., and Bullerman, L.B., 1999, Stability of deoxynivalenol in heat-treated foods, *J. Food Protect.* 62:962-964.
Wolf, C.E. and Bullerman, L.B., 1998, Heat and pH alter the concentration of deoxynivalenol in an aqueous environment, *J. Food Protect.* 61:365-367.
Wolf-Hall, C.E. and Bullerman, L.B., 1998, Characterization of *Fusarium graminearum* strains from corn and wheat by deoxynivalenol production and RAPD, *J Food Mycol.* 1:171-180.
Young, J.C., Blackwell, B.A., and ApSimon, J.W., 1986a, Alkaline degradation of the mycotoxin 4-deoxynivalenol, *Tetrahedron Letters.* 27:1019-1022.
Young, J.C., Fulcher, R.G., Hayhoe, J.H., Scott, P.M., and Dexter, J.E., 1984, Effect of milling and baking on deoxynivalenol (vomitoxin) content of eastern Canadian wheats. *J. Agric. Food Chem.* 32:659-664.
Young, J.C., Trenholm, H.L., Friend, D.W., and Preluskey, D.B., 1987, Detoxification of deoxynivalenol with sodium bisulfite and evaluation of the effects when pure mycotoxin or contaminated corn was treated and given to pigs, *J. Agric. Food Chem.* 35:259-261.
Zhou, M., 1998, Effects of *Fusarium* infection of barley on malt quality, *Masters of Science Thesis*, North Dakota State University, Fargo, North Dakota.

TOXICITY, RISK ASSESSMENT AND REGULATORY ASPECTS OF MYCOTOXINS: INTRODUCTION

Jonathan W. DeVries
General Mills, Inc.
James Ford Bell Technical Center
9000 Plymouth Avenue North
Minneapolis, MN 55427

The ultimate result of scientific research regarding a food contaminant is a determination of the need to regulate the level of that contaminant in the food supply, and if the need exists, to determine a scientifically appropriate level at which to regulate. Suitable regulatory actions are necessarily based on sound science. Sound science in this case consists of properly designed safety assessments using appropriate models based on reliable and adequate toxicity studies and inputting sufficiently accurate data with regard to rates of contaminant occurrence and level of contamination when it occurs. Adequate toxicity studies in turn depend upon appropriate study design and relevant biological endpoints. Accurate data depends upon appropriate sampling plans and suitable analytical technology to determine the presence or absence of a contaminant and the level present in the food supply when the contaminant is found. Processing will often reduce the exposure level to mycotoxins in the food supply and this effect must be taken into account when considering regulatory action.

Previous chapters of this book covered the activities that must necessarily be completed prior to conducting an adequate and accurate safety assessment. The chapters of this section cover approaches to safety assessment, approaches to contaminant management, and frameworks for setting suitable regulations to protect consumer health and well being within the context of providing an adequate food supply. Views of recognized experts in risk analysis and regulatory activity are presented.

Although the research on aflatoxins, the first mycotoxin identified, has been carried out for nearly four decades, there are continued concerns regarding human and animal exposure on a worldwide basis. These concerns are covered in the first paper of this section.

Trichothecene exposure and toxicity is of concern, particularly amongst young children, in many parts of the world. The second paper provides a European perspective on approaches and evaluations used for assessing trichothecene safety data.

Exchange of goods and fair trade are often inhibited by a lack of uniformity in regulations and standards for contaminants of concern in food commodities. Paper three reviews the efforts and scientific evaluation of an international body, Codex, in arriving at consensus regulations for Ochratoxin A that can be applied worldwide.

Mycotoxins and Food Safety, edited by Trucksess *et al.*
Kluwer Academic/Plenum Publishers, 2002

Regulations for controlling human and animal exposure to mycotoxins vary in different parts of the world. In paper four, a review of the current regulations in place in various countries worldwide is presented.

It is generally understood that regulations that restrict the flow of commodities, or reduce their utility and/or availability as food ingredients come with an associated cost. Paper five provides insight into some of the costs associated with control of mycotoxin exposure and economic changes that result.

The final paper in this series presents the United States perspective on mycotoxins risk management. This is done through control of mycotoxin exposure, with control levels being based on extensive safety assessments and occurrence surveys.

AFLATOXIN, HEPATITIS AND WORLDWIDE LIVER CANCER RISKS

Sara H. Henry[1], F. Xavier Bosch[2], and J. C. Bowers

[1]U.S. Food and Drug Administration
Washington, D.C. 20204
[2]Institut d'Oncologia
Hospitalet del Llobregat
Barcelona, Spain

ABSTRACT

Aflatoxins are among the most potent mutagenic and carcinogenic substances known. Differential potency of aflatoxin among species can be partially attributed to differences in metabolism; however, current information on competing aspects of metabolic activation and detoxification of aflatoxin in various species does not identify an adequate animal model for humans. Risk of liver cancer is influenced by a number of factors, most notably carriage of hepatitis B virus as determined by the presence in serum of the hepatitis B surface antigen (HBsAg+ or HBsAg-). About 50 to 100% of liver cancer cases are estimated to be associated with persistent infection of hepatitis B (or C) virus. The potency of aflatoxin in HBsAg+ individuals is substantially higher (about a factor of 30) than the potency in HBsAg- individuals. Thus, reduction of the intake of aflatoxins in populations with a high prevalence of HBsAg+ individuals will have greater impact on reducing liver cancer rates than reductions in populations with a low prevalence of HbsAg+ individuals. The present analysis suggests that vaccination against hepatitis B (or protection against hepatits C), which reduces prevalence of carriers, would reduce the potency of the aflatoxins in vaccinated populations and reduce liver cancer risk.

INTRODUCTION

This chapter has been written as an update to a risk assessment of aflatoxins performed by The Joint Food and Agriculture Organization/World Health Organization Expert Committee on Food Additives (JECFA) in 1997 (JECFA, 1998). Since the appearance of the JECFA monograph, additional data have appeared on the role of aflatoxin, hepatitis B and C viruses and other factors in the development of human hepatocellular cancer (HCC).

To review briefly, HCC is the most common cancer in the world with 473,000 new cases appearing per year, with 80% of these cases appearing in developing countries. The

Mycotoxins and Food Safety, edited by Trucksess *et al.*
Kluwer Academic/Plenum Publishers, 2002

male female ratio is approximately 3:1. HCC is rapidly fatal with a survival rate after diagnosis of approximately one year. Half of new cases occur in China; 25% of all new cancers in males in West Africa are HCC (Parkin, 1998). Thus, the combined impact of public health costs as well as years of productivity lost is quite significant on a worldwide scale.

RISK FACTORS FOR LIVER CANCER – HBV AND HCV

The major risk factors for HCC vary between Africa-Asia and Europe-U.S. as shown in Table 1 (Bosch et al. 1999).

Table 1. Liver cancer etiology: Attributable fractions in Europe-U.S. and in Africa-Asia (Bosch et al., 1999).

Risk factor	Europe and U.S.	Africa and Asia
Hepatitis B	<15% (4-50%)	60% (40-90%)
Hepatitis C	60% (12-64%)	<10%
Aflatoxin	Limited or none	Not quantified
Tobacco	<15%	Not estimated
Alcohol	<12%	29% (one study)
Oral contraceptives	(10-50%)	Not estimated
Others including hemochromatosis	<5%	<5%

Many studies have confirmed the association between HCC and hepatitis B virus (HBV). An exception, which deserves further study, is Greenland where HCC is rare and chronic HBV is common. HCC was one of the first known virus-associated cancers and is the first cancer to be prevented by vaccination (IARC, 1994). HBV "carriage" is defined as the presence of the hepatitis B surface antigen (HBsAg) in the bloodstream for 6 months or more. Such carriage is associated with an approximately 50-fold increased risk of HCC (Wild and Hall, 2000; Bosch et al., 1999; IARC, 1994).

Vaccination for HBV has differing goals in different geographical regions. In low-incidence countries, the target is acute hepatitis B in adolescents and young adults; in high incidence HCC countries, the goal is to prevent carriage of the virus because age at infection (i.e., under 10 years of age and especially under 5 years of age) is the major determining factor of carriage (Wild and Hall, 2000; Bosch et al., 1999; IARC, 1994)

Vaccination for HBV can be tremendously cost effective. In the U.K., the cost of HBV vaccine is about U.S.$15 as compared to about U.S. $0.75 in the poorest countries. The resources required to prevent one HBV carrier in the U.K. would enable about 20 infants to be vaccinated in a developing country (Gay and Edmonds, 1998).

Hepatitis C, which has been recognized since 1989, is a RNA virus that shows a high mutation rate within individuals as well as a wide geographical and temporal variation in genotype. The virus, unlike HBV, does not integrate into its host genome. Development of a vaccine is not imminent. Education and clean needle use are important in reducing HCV infection (Hall and Wild, 2000). HCV carriage results in about 80% of those infected. Direct transmission by blood contamination, usually through a needle, is the major mode of transmission. Approximately 80% of drug users in some countries have HCV infection. Most countries have a prevalence of about 1%. (Hoofnagle, 1997; Alter, 1996).

In the United States the Veterans Administration has observed an age-adjusted three-fold increase of HCC between 1993-95. Hepatitis C virus accounts for most of the increase in the number of HCC cases. The increase may be related to drug use among

Vietnam era veterans and is especially a problem among African American veterans (El-Serag et al., 2000). A similar increase in HCC, which may be associated with HCV, has been noted in the U.K. and in France (Deuffic et al., 1998).

RISK ASSESSMENT OF AFLATOXIN WITH AND WITHOUT HBV/HCV

The JECFA risk assessment for aflatoxin (JECFA, 1998) estimated the carcinogenic potency of aflatoxin in a number of animal species and from epidemiological studies in humans with and without HBV. Data at that time was lacking for estimating the potency of aflatoxin in conjunction with HCV. In addition, there was no well-studied animal model which would approximate the relationship between HBV and HCV, aflatoxin and HCC. When ranked on a scale of 1 to 0, with 1 representing the most sensitive animal model – the Fisher rat, there was an approximate 30-fold difference between the carcinogenic potency of aflatoxin in humans who were negative for HBV (HBsAg-) as compared to humans who were positive for HBV (HBsAg+). Limitations of the ranking were that some studies were negative and therefore omitted from the modeling. There were also significant problems involved in estimating the dose of aflatoxin received by humans. First, the HBV status had not always been determined with the most accurate methodology. Second, the presence of confounders was not always sufficiently considered.

The JECFA monograph (JECFA, 1998) noted the development of biomarkers beginning in the early 1980's to more accurately assess aflatoxin exposure at the individual level. The binding of aflatoxin to peripheral blood albumin and urinary aflatoxin metabolites has been investigated in laboratory rodents and in human populations. Although measurement of aflatoxin biomarkers gives a more accurate picture of aflatoxin exposure over the long term than dietary questionnaires or direct analysis of foods, biomarker data are still limited because they reflect individual aflatoxin exposure only for the lifetime of the particular marker, *e.g.*, about 22 days in the case of a serum biomarker. A measure of aflatoxin exposure over a third to half a lifetime is still needed.

Biomarkers for aflatoxins have also enabled the identification of a specific mutation in the p53 tumor suppressor gene in HCC cases from regions of the world with high aflatoxin exposure. The same mutation in plasma DNA in HCC cases collected in The Gambia has been noted; this alteration is infrequent in control subjects (Hall and Wild, 2000).

The JECFA monograph concluded that vaccination for HBV must take high priority in preventing HCC. Since there are so many HBV carriers (approximately 360 million worldwide), and HCV carriers, and an incomplete access to vaccine, especially in developing countries, efforts to reduce aflatoxin exposure are important in developing countries (JECFA, 1998; Hall and Wild, 2000).

A very telling example of the importance of vaccination is the reduction of adult liver cancer by HBV vaccination in Korea. In this country, prevalence of HBV infection and HCC are high, and aflatoxin contamination is relatively high. In 370,000 males followed for more than three years, HBV vaccination drastically reduced HCC (incidence of 215 cases/100,000 vs. 8 cases/100,000) (Lee et al., 1998).

Since the JECFA monograph was written, additional data have accumulated which help to elucidate the role of aflatoxin, HBV, HCV, and other risk factors in HCC. Sun et al. (1999) followed 145 men with chronic HBV hepatitis for 10 years. Aflatoxin exposure was measured by aflatoxin M_1 (AFM1) in pooled urine samples, collected at eight monthly intervals before follow-up began. The anti-HCV and HBsAg status of each patient was measured, and the existence of a family history of HCC was recorded. Serum alanine transaminase (ALT) was also measured.

AFM1 was detected in 78 (54%) of the subjects. The risk of HCC was increased 3.3 fold (95% confidence interval of 1.2-8.7) in those with detectable AFM1 (above 3.6 ng/L).

This relative risk was adjusted for age and for HCV status. The relative risk of fatal cirrhosis for those with elevated AFM1 was 2.8 (0.6, 14.3), and the odds of having a persistently elevated ALT were 2.5-fold greater in those with detectable AFM1. Concomitant infection with HCV increased the risk of HCC 5.8 fold (2.0-17), adjusted for age and AFM1 status. A family history of HCC increased the risk of HCC 5.6-fold, adjusted for age and AFM1 status. Four men with detectable AFM1 and HCC al had missense mutation in codon 249 of the p53 gene in cancer tissues. Sun et al. (1999) concluded that this study shows that exposure to AFM1 can account for a substantial part of the risk of HCC in men with chronic HBV hepatitis and adds importantly to the evidence that HCV and family history of HCC increase the risk of HCC in men with chronic HBV hepatitis. Limitations of the study include that it is unable to estimate the dose of aflatoxin received, *e.g.*, high, medium or low, and therefore estimating a dose-response curve is impossible. HCV appeared to increase the risk of HCC in men with HBV; however, there were too few cases to demonstrate this effect conclusively. The familial association may be partly the result of early familiar transmission of HBV, rather than of genetic factors. If the hazard of HCC increases with the duration of HBV infection, then the association of family history with HCC risk may partly represent uncontrolled confounding from longer duration of HBV infection. The study did not include information on the age at which HBV infection occurred.

The authors estimated that aflatoxin intake in this study was not high. Most of the 145 pooled urine samples had aflatoxin concentrations less than 100 ng/L which is equivalent to the intake of less than 7000 ng aflatoxin daily. Taking 65 kg as the average body weight of a Qidong farmer, the intake was below 110 ng/kg/d. To clarify the relationship between aflatoxin consumption and HCC risk and chronic HBV infection, or to detect a two-fold reduction in HCC incidence rates with a power of 0.9 using a 2-sided log rank test would require following about 700 subjects for 5 years. Longer follow-up periods or larger cohorts would be needed to account for the fact that reducing aflatoxin exposure might not produce an immediate lowering in HCC risk. The authors postulate that lowering the aflatoxin exposure might benefit HBsAg carriers with chronic hepatitis.

INTERVENTION WITH OLTIPRAZ

Chemopreventive strategies in humans and animals have been sought to modulate the balance between aflatoxin activation and detoxification. The drug oltipraz, or 4-methyl-5-pyrazinyl-3H-1,2-dithiole-3-thiole, originally prescribed to treat schistosomiasis has been studied (Hall and Wild, 2000). In several studies, Kensler et al. (Wang et al., 1999; Kensler et al, 1998; Wang et al., 1996) have demonstrated that when oltipraz is administered to Chinese people exposed environmentally to aflatoxin there is an increase in the level of GST conjugation of aflatoxin 8,9 epoxide, but also an inhibition of cytochrome P450 1A2 activity which activates aflatoxin to this reactive epoxide. Although such an intervention study should be viewed as controversial in human subjects, a Phase 2B clinical trial (intervention of over 1 year administering 250 or 500 mg oltipraz weekly) is underway in Qidong. A safe and effective dose of oltipraz will be selected on the basis of the Phase 2b trial for the Phase III trial. Minor side effects, including a syndrome involving numbness, tingling and sometimes pain, in the extremities will also be evaluated. Phase III will require a long follow-up period unless aflatoxin exerts its carcinogenic effect late in the progress of the liver cancer.

Oltipraz has been shown to inhibit HBV replication (Chi et al., 1998) and appears to inactivate reverse transcriptase of human immunodeficiency virus (HIV). When tested in 2.2.15 cells (clonal cells derived from HepG2 cells that were transfected with a plasmid containing HBV DNA) *in vitro*, oltipraz had a dose-dependent inhibitory effect on HBV replication, and specifically blocked HBV transcription in 2.2.15 cells. The ramifications of this finding *in vivo* remain to be elucidated.

REFERENCES

Alter, M.J, 1996, Epidemiology of hepatitis C. *Eur. J. Gastroenterol. Hepatol.* 8:319.

Bosch, F.X., Ribes, J., and Borras, J., 1999, Epidemiology of primary liver cancer in: *Seminars in Liver Disease* 19(3):271.

Chi, W.J., Doong, S.-L., Lin-Shiau, S.-Y., Boone, C.W., Kellloff, G.J., and Lin, J., 1998, Oltipraz, a novel inhibitor of hepatitis B virus transcription through elevation of p53 protein. *Carcinogenesis* 19(12): 2133.

Deurric, S., Poynard, T., Buffet, L., and Valleron, A.-J., 1998, Trends in primary liver cancer. *The Lancet* 351:215.

El-Serag, H.B. and Mason, A.C., 2000, Risk factors for the rising rates of hepatocellular carcinoma in the United States. *Gastroenterol.* 118(4): Suppl 2. 3792.

Gay, N.J., and Edmonds, W.J., 1998, Developed countries could pay for hepatitis B vaccination in developing countries. *Brit. Med. J.* 316:1457.

Hoofnagle, J.H., 1997, Hepatitis C: the clinical spectrum of disease. *Hepatol.* 26:15S.

IARC, 1994, Hepatitis viruses, Monographs on the evaluation of carcinogenic risks to humans. IARC Sci. Publ. 59. Lyon, France.

JECFA, 1998, Aflatoxins. Safety evaluation of certain food additives and contaminants. (The forty-ninth meeting of the Joint FAO/WHO Expert Committee on Food Additives.). WHO Food Additives Series, no. 40. World Health Organization, Geneva 1998. pp 359-468.

Kensler,T.W., He, X., Otieno, M., Egner, P.A., Jacobson, L.P., Chen, B., Wang, J.S., Zhu, Y.R., Zhang, B.C., 1998, Oltipraz chemoprevention trial in Qidong, People's Republic of China: modulation of serum aflatoxin albumin adduct biomarkers. *Cancer Epidemiol. Biomarkers Prev.* 7:127.

Lee, M.-S., Kim, D.-H., Kim, H., Lee, H.-S., Kim, C.-Y., Park, T.-S., Yoo, K.-Y., Park, B.-J., and Ahn, Y-O., 1998, Hepatitis B vaccination and reduced risk of primary liver cancer among male adults: a cohort study in Korea. *Int. J. of Epidem.* 27:316.

Parkin, D.M., 1998, The global burden of cancer, *Semin. Cancer Biol.* 8:219.

Sun, Z., Lu, P., Gail, M.H., Pee, D., Zhang, Q., Ming, L., Wang, J., Wu, Y., Liu, G., Wu, Y., and Zhu, Y., 1999, Increased risk of hepatocellular carcinoma in male hepatitis B surface antigen carriers with chronic hepatitis who have detectable urinary aflatoxin metabolite M1. *Hepatol.* 30:379.

Wang, J.S., Qian, G.S., Zarba, A., He, X., Zhu, Y.R., Zhang, B.C., Jacobson, L., Gange, S.J., Munoz, A., Kensler, T.W., 1996, Temporal patterns of aflatoxin-albumin adducts in hepatitis B surface antigen-positive and antigen-negative residents of Daxin, Qidong County, People's Republic of China. *Cancer Epidemiol. Biomarkers Prev.* 5:253-261.

Wang, J.S., Shen, X., He, X., Zhu, B.C., Zhang, B.C., Wang, J.B., Qian, G.S., Kuang, S.Y., Zarba, A., Egner, P.A., 1999, Protective alterations in phase 1 and 2 metabolism of aflatoxin B1 by oltipraz in residents of Qidong, People's Republic of China. *JNCI* 91:347-354.

Wild, C.P., and Hall, A.J., 2000, Primary prevention of hepatocellular carcinoma in developing countries. *Mut. Res.* 462:381.

RISK ASSESSMENT OF DEOXYNIVALENOL IN FOOD: CONCENTRATION LIMITS, EXPOSURE AND EFFECTS

Moniek N. Pieters[1*], Jan Freijer[1], Bert-Jan Baars[1], Daniëlle C.M. Fiolet[1], Jacob van Klaveren[2] and Wout Slob[1]

[1]Center for Substances and Risk Assessment, National Institute of Public Health and the Environment, RIVM, P.O. Box 1, 3720 BA Bilthoven, The Netherlands
[2] State Institute for Quality Control of Agricultural Products (RIKILT), P.O. Box 230, 6700 AE Wageningen, The Netherlands
* to whom correspondence should be addressed

ABSTRACT

The mycotoxin, deoxynivalenol (DON), is produced world-wide by the Fusarium genus in different cereal crops. We derived a provisional TDI of 1.1 μg /kg body weight (bw) and proposed a concentration limit of 129 μg DON/kg wheat based on this TDI and a high wheat consumption of children. In the period September 1998-January 2000, the average DON concentration in wheat was 446 μg/kg (n = 219) in the Netherlands. During this period, the dietary intake of DON exceeded the provisional TDI, especially in children. Eighty percent of the one-year-olds showed a DON intake above the provisional TDI and 20% of these children exceeded twice the provisional TDI. Our probabilistic effect assessment shows that at these exposure levels, health effects may occur. Suppressive effects on body weights and relative liver weight were estimated at 2.2 and 2.7%. However, the large confidence intervals around these estimates indicated that the magnitudes of these effects are uncertain.

INTRODUCTION

The mycotoxin deoxynivalenol (DON) produced by fungi of the Fusarium genus may occur in various cereal crops (wheat, maize, barley, oats, and rye). Chemically it belongs to the trichothecenes: tetracyclic sesquiterpenes with a 12,13-epoxy group (Eriksen and Alexander, 1998). DON is a very stable compound, during both storage/milling and the processing/cooking of food, and does not degrade at high temperatures (Rotter et al., 1996, Ehling et al., 1997).

In humans food poisoning with DON led to abdominal pain or a feeling of fullness in the abdomen, dizziness, headache, throat irritation, nausea, vomiting, diarrhea, and blood in the stool (Rotter et al., 1996; Eriksen and Alexander, 1998). In 1998 and 1999, high contamination levels of DON were detected in wheat and in wheat containing food products in the Netherlands. This urged the Ministry of Public Health, Welfare and Sports (VWS) to request for the derivation of concentration limits for DON in wheat and wheat containing food products. To that end an evaluation of toxicity data and an estimation of the wheat intake in the Dutch population was needed. The first part of this paper deals with the derivation of a provisional TDI and concentration limit of DON in wheat using the standard approach. Since the derived concentration limit appeared to be low compared to the DON levels in food at the time, the second part of the paper discusses a more detailed exposure and effect assessment. Possible human health risks of exposure to DON at levels encountered in the period 1998-2000 are estimated based on a probabilistic risk assessment approach.

METHODOLOGY

Derivation of Concentration Limits

To derive concentration limits for DON in food products, we evaluated the toxicological literature on DON. Taking into account the quality of the studies evaluated and the relevance of the toxicological endpoints, we selected a NOAEL and applied uncertainty factors for interspecies and intraspecies variation to derive a provisional TDI (EHC, 1994).

Subsequently, the wheat intake of the Dutch population was estimated. Food consumption data were obtained from the Dutch National Food Consumption Survey (VCP). The survey includes a description of the daily consumption over two consecutive days and a recording of sex, age and body weight of 6250 individuals belonging to 2564 households. Data were collected from April 1997 until April 1998 and were evenly spread over the weeks of the year and the days of the week (Kistemaker et al., 1998). With the Conversion model Primary Agricultural Products (CPAP, Van Dooren et al., 1995) food consumption data were converted into the amount of primary agricultural product, in this case wheat, consumed.

By dividing the provisional TDI (μg DON/kg bw) by the estimated wheat consumption (g wheat/kg bw), the concentration limit of DON in wheat was calculated. It was implicitly assumed that the contribution of other grains (including beer) to the total DON intake was negligible.

Analysis of DON in Food Products

The Dutch Inspectorate for Health Protection monitored DON in various food products from September 1998 onwards. Up to January 2000, data comprised 584 DON analyses in 14 different food categories. Wheat constituted the majority of the samples (n = 219). Sampling was carried out more or less at random in wheat containing food products, resulting in a representative picture of the DON concentrations in Dutch consumer products. DON was analyzed by mixing and blending fifty gram of ground sample with 200 ml water and 10 g polyethylene glycol. After filtration 15 ml of the solution was cleaned through an immunoaffinity column. After evaporation and dissolution 50 μl was injected in a HPLC system consisting of a C18-column and acetonitrile-water mobile phase. DON was detected at 218 nm (Cahill et al., 1999).

Probabilistic Exposure Assessment: Dietary Intake of DON

Human dietary intake of chemicals is usually estimated by combining data on concentrations of chemicals in different food products and the consumption rate of these products. Figure 1 displays the principal flow scheme, which has been employed to analyze human dietary intake of chemicals in the Netherlands (Liem and Theelen, 1997). The flow scheme shows the dependency of different submodels and databases.

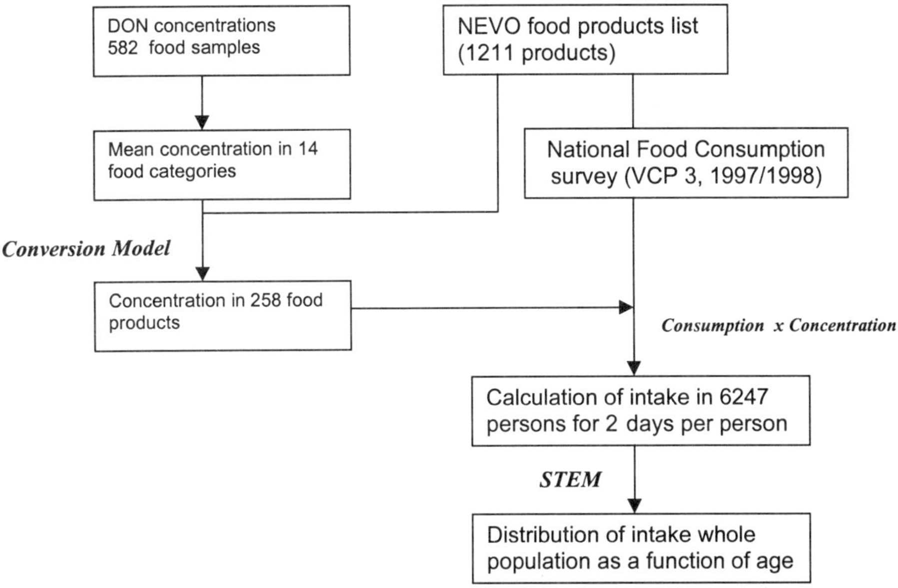

Figure 1. Overview of databases and submodels used in analyzing human exposure to DON.

Since only a limited number of all food products were monitored, we did not have concentration data on all consumed food products as described in detail in the VCP. We therefore calculated for each monitored food product the concentration of DON per kg wheat content, classified the food products into different food categories and calculated for each food category (for example, bread/biscuits/crackers) an average concentration of DON/kg wheat. By multiplying the (estimated) wheat content of a consumed food with the average DON concentration of the food category, DON concentrations of the various food products were calculated. In this way, for each food category, we corrected for possible DON loss due to food processing. For those food categories for which data on DON concentrations were scarce (e.g., composites), we estimated the DON intake from these foods by multiplying the (estimated) wheat content of a consumed food with the average DON concentration in wheat, i.e., 446 μg/kg (n = 219). Data on wheat content were derived from the Conversion model Primary Agricultural Products (CPAP, Van Dooren et al., 1995), or were based on cooking recipes or on the wheat content of similar food products in the same food category.

In the Netherlands, consumer food is coded and described (nutrients, energy content etc.) in the Dutch Nutrient Database (NEVO). The measured or estimated DON concentrations of NEVO food products were combined with the individual food consumption data recorded on two consecutive days, resulting in two daily intakes of all individuals included in the survey. To distinguish the variation between individuals from the daily fluctuations in consumption, we use the statistical exposure model STEM (Slob,

1993; Slob, 1996) which estimates the mean dietary intake as a function of age. It combines regression analysis on age by fitting an appropriate curve to the data with a nested analysis of variance.

Probabilistic Effect Assessment

Toxicity data from animal studies were analyzed by dose response modeling (the benchmark approach). In the benchmark approach (Slob, 1999) a dose response model is fitted to the data. The best fit model was selected by applying likelihood ratio tests to the members of a nested family of models. Subsequently, the selected model can be used for setting human limit values as well as for an actual risk assessment (Slob and Pieters, 1998). For the setting of human limit values, a critical effect size (CES) is selected for each effect parameter and the associated critical effect dose (CED) is derived. With a bootstrapping technique (Monte Carlo sampling, n = 1000) a distribution of the CED is generated. Extrapolation to humans is carried out by combining the distribution of the CED with distributions of appropriate assessment factors in a probabilistic manner (Slob and Pieters, 1998; Vermeire, et al., 1999). To estimate the effects occurring in humans at an actual exposure level, the estimated DON intake is used as input for the model. This results in an estimate of the effect size in the experimental animal under study which is subsequently extrapolated to humans.

RESULTS

Derivation of a provisional TDI

To derive a provisional TDI we evaluated relevant toxicity studies. At low concentrations in the diet DON reduces growth and feed consumption (anorexia) in experimental animals. At higher acute doses it induces vomiting (emesis). Both effects, which are also seen with other trichothecene toxins, are thought to be mediated by affecting the serotoninergic activity in the CNS. DON inhibits protein synthesis at the ribosomal level, and it has been demonstrated to inhibit DNA and RNA synthesis. DON affects the immune system and alters various blood parameters. In addition, it is a potent gastrointestinal irritant (Rotter et al., 1996; Eriksen and Alexander, 1998). There are no indications for carcinogenic and/or mutagenic properties. The provisional TDI can therefore be based on the no observed adverse effect levels (NOAELs) from toxicity studies by applying uncertainty factors. The NOAELs and LOAELs of various toxicity studies vary between 0.04 - 0.75 mg/kg bw/day and are summarized in Table 1. Taking into account the quality of the studies evaluated and the relevance of the toxicological endpoints, only the NOAELs mentioned in Table 2 are considered for the derivation of a provisional TDI.

Extrapolation from experimental animals to humans implies the application of an uncertainty factor (UF). Commonly an UF of 10 is used for extrapolating from rodents to humans, and an additional UF of 10 to cover for (human) interindividual differences. In the case of the swine as the experimental species these UFs are 2 and 10, respectively. For the data above this would result in tolerable daily intakes which are all in the same order of magnitude (i.e., 1.1-5.0 μg/kg bw/day).

The chronic diet study with mice (Iverson et al., 1995) is of good quality and yields the lowest NOAEL. Swine might resemble humans more with regard to its physiology. However, since in these studies the DON preparation was contaminated with other mycotoxins we decided to use the NOAEL of the chronic diet study with mice (0.11 mg/kg bw/day) for the estimation of a provisional TDI. Applying an uncertainty factor of 100,

Table 1. Summary of toxicity studies with DON

Species	Study	Effect	Parameter	Dose *)	Reference
Mouse	Acute	Mortality	LD_{50} oral	46-78	Eriksen and Alexander, 1998
Hen	Acute	Mortality	LD_{50} oral	140	IARC, 1993
Swine	Acute	Vomiting	-	0.05-0.2	Eriksen and Alexander, 1998
Swine	Subacute	reduced food uptake	-	0.03-0.07 (1-2 mg/kg feed)	Eriksen and Alexander, 1998
Mouse	Subacute	reduced food uptake	-	0.6-1.2 (4-8 mg/kg feed)	Rotter et al., 1996
Rat	Subacute	reduced food uptake	-	0.75-1.0 (15-20 mg/kg feed)	Rotter et al., 1996
Mouse	5 weeks	decreased $\alpha 1/\alpha 2$ globulin ratio	NOAEL	0.25	Eriksen and Alexander, 1998
Rat	9 weeks	reduced growth, reduced food uptake	LOAEL	0.25	Arnold et al., 1986
Rat	90 days	reduced growth	LOAEL	1.0	Morrissey et al., 1985
Mouse	2 years	reduced growth, reduced relative liver weights	NOAEL	0.11	Iverson et al., 1995
Mouse	Immunotoxicity	increased susceptibility for infections	NOAEL	0.25	Tryphonas et al., 1986
Mouse	Immunotoxicity	increased susceptibility for infections	LOAEL	0.22	Deijns et al., 1994
Mouse	Teratogenicity	fetal skeleton abnormalities	NOAEL	0.5	Khera et al., 1982
Mouse	Reproduction-toxicity	mortality of pups	NOAEL	0.375	Khera et al., 1984
Rat	Reproduction-toxicity	maternal and/or embryotoxicity	NOAEL	1.0	Khera et al., 1984
Rat	Reproduction-toxicity	reduced fertility	LOAEL	≤ 2.0	Morrissey & Vesonder, 1985
Rabbit	Teratogenicity	reduced fetal weight	NOAEL	0.6	Khera et al., 1986
Swine	Reproduction-toxicity	reduced growth (maternal toxicity)	LOAEL	0.03-0.07 (1-2 mg/kg feed)	Eriksen and Alexander, 1998
Swine	28 days	reduced food uptake, decreased thyroid weight and α-glob., increased T4, serum-albumin and A/G ratio	LOAEL	0.03 (0.75 mg/kg feed)	Rotter et al., 1994
Swine	42 days	reduced growth, reduced food uptake, stomach corrugation	LOAEL	≤ 0.15 (4 mg/kg feed)	Rotter et al., 1995
Swine	90 days	reduced growth, reduced food uptake	NOAEL	0.04 (1 mg/kg feed)	Bergsjø et al., 1992
Swine	95 days	reduced growth, reduced food uptake, increased liver weight, decreased serum albumin	NOAEL	0.06 (1.7 mg/kg feed)	Bergsjø et al., 1993

*) All dosages in mg/kg bw/day, unless indicated otherwise.

Table 2. NOAELs considered for the derivation of a TDI

Study	NOAEL (mg/kg bw/day)	Reference
Mouse, chronic (2 years)	0.11	Iverson et al., 1995
Mouse, immunotoxicity	0.25	Tryphonas et al., 1986
Mouse, teratogenicity	0.5	Khera et al., 1982
Mouse, reproduction toxicity	0.375	Khera et al., 1984
Swine, subchronic (90-95 days)	0.04-0.06	Bergsjø et al., 1992, 1993

a provisional TDI of 1.1 μg/kg bw/day is estimated (Pieters et al., 1999), in line with Ehling et al. (1997), and Eriksen and Alexander (1998). Recently, the EU adopted a temporary TDI of 1 μg/kg bw/day, based on the study of Iverson et al., 1995 (European Commission, SCF 09/12/99).

Concentration Limits

We considered children (1-4 yr) to be the group at risk since they not only have the highest relative wheat consumption, but are also considered to be vulnerable to the critical effect growth reduction. We based the safe concentration limit of (cleaned) wheat on a child with a high wheat consumption, i.e. 8.5 g/kg bw, which is the 95th percentile. Based on a provisional TDI of 1.1 μg DON /kg bw, the concentration limit of DON in wheat was calculated according to:

Concentration limit of DON in (cleaned) wheat =

TDI (μg/kg bw)/ wheat intake (g/kg bw) = 1.1 / 8.5 = 0.129 μg/g wheat
= 129 μg/kg wheat (129 ppb)

Depending on the wheat content of food products, concentration limits for DON can be derived for each food product (Pieters et al., 1999). Since we used the high wheat consumption of children for the calculation of the concentration limit of DON, the derived concentration limit of 129 μg/kg wheat should prevent the major part of the general population from exceeding the provisional TDI.

The derived concentration limit appeared to be low compared to actual levels in wheat and wheat containing food products. Therefore, we carried out a more detailed exposure and effect assessment. Possible human health risks at current exposures are estimated using a probabilistic risk assessment approach.

Probabilistic Exposure Assessment: Dietary Intake of DON

By combining the measured or estimated DON concentrations in NEVO products with the food consumption data, the personal intakes for all individuals included in the survey for two consecutive days were calculated. Figure 2 shows the frequency distribution of intake rates consisting of 12494 values of daily average intakes of DON, 2 consecutive days for 6247 individuals recorded in the Food Consumption Survey database.

The frequency distribution gives insight in the total variation in daily intakes. This variation has two components: a within-subject and between-subject variation. The high tailing value of the distribution (around 7 μg/kg bw/day) should therefore be carefully interpreted, as it represents a one-day event of an individual. Therefore, this distribution is unsuitable for a comparison with the TDI, because the latter is intended for long-term exposure. A distribution of individual long-term exposures would be considerably narrower than the distribution of daily averaged intakes because the latter incorporates within-subject fluctuations.

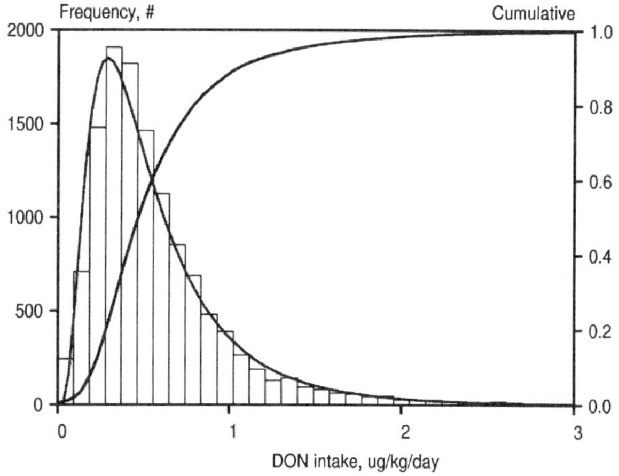

Figure 2. Frequency distribution of DON intakes in 6247 individuals for two consecutive days in the Netherlands.

The long-term exposure distribution, representing interindividual differences, was estimated by using the Statistical Exposure Model STEM (Slob, 1993; Slob, 1996). Analyzing the data displayed in Figure 2 by STEM yields the results as presented in Figure 3, showing that the relative intake decreases with age. The percentiles depicted in the figure represent the variation between individuals after correcting for the within-subject variation between days.

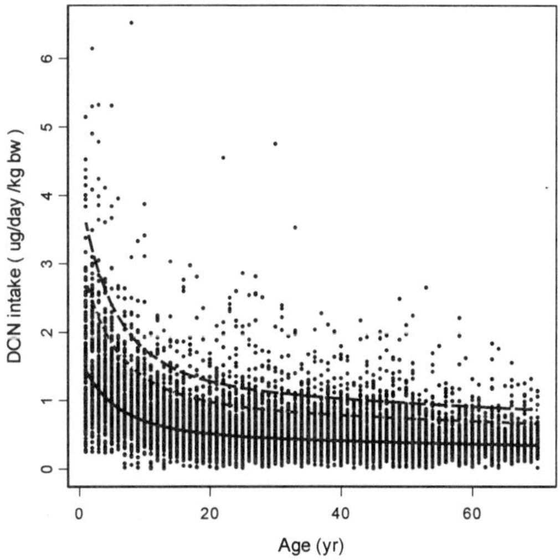

Figure 3. Daily intake of DON as a function of age. Each dot denotes one daily intake of a single individual (6247 individuals, intakes for two consecutive days each). The bold curve represents the estimated geometric mean intake estimated by fitting a regression function. Dashed curves denote the 95th and 99th percentiles, indicating the long-term variation between individuals.

For each age-class the intake distribution can be deduced from the median intake and the estimated between-subject geometric variance. Figure 4 shows a cross-section for ages 2 and 40. Figure 5 shows the percentage of the population associated with a (long-term average) daily DON intake exceeding 1 μg/kg bw/day or 2 μg/kg bw/day for several age classes. It clearly shows that young children are exposed to levels exceeding the provisional TDI. Eighty percent of the one-year-old children are exposed to levels exceeding the provisional TDI and 20% of this age group is exposed to DON levels exceeding twice the provisional TDI. At the age of 4, these percentages are 50 and 5, respectively.

Table 3 shows the relative contribution of different food categories to the total DON intake. Bread/biscuits/crackers are the main source for DON exposure. Porridge made of wheat and other grains is categorized in baby/toddler food and forms only a relevant source of DON for the very young. The DON intake through grains other than wheat (oat, maize, and rye) is negligible since the consumption of these grains is rather low in the Netherlands. Barley in the form of beer forms the only exception. When we assume a DON concentration in beer at the reporting limit (100 μg/kg), the consumption of beer contributes approximately 4% of the total DON intake in adults (>20 yr.).

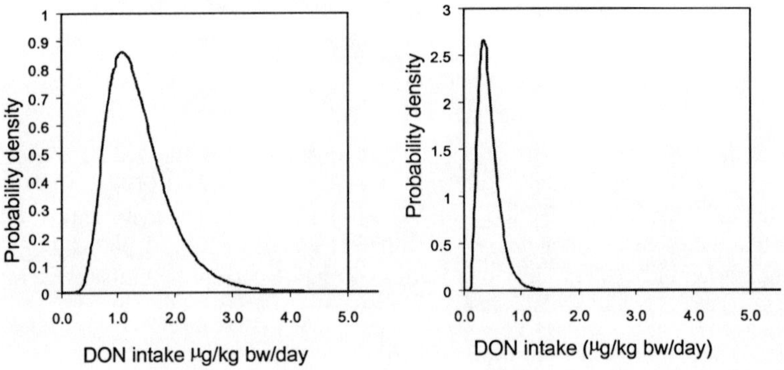

Figure 4. Distribution of DON intake at age 2 (left panel) and age 40 (right panel)

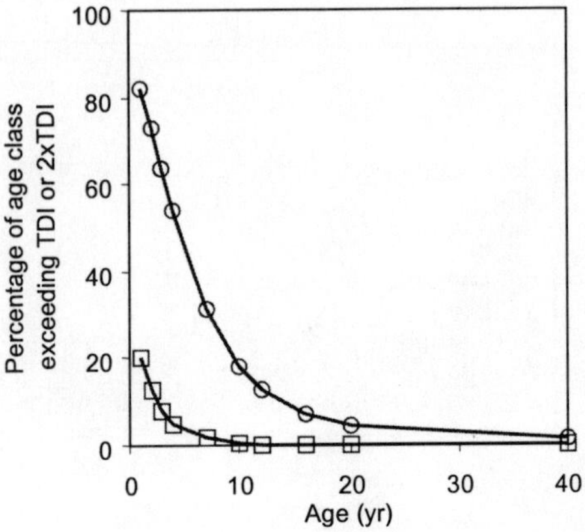

Figure 5. Percentage of age class with a DON intake exceeding the provisional TDI (circles) or twice the provisional TDI (squares).

Table 3. Relative contribution of food categories to the total DON intake

Food Category	Age (yr.)				
	1	2-4	5-10	11-20	>20
Bread/biscuits/crackers	53	64	71	69	65
Baby/toddler food	22	6	0.8	0.6	0.6
Cookies/cakes/pastry	9	11	11	12	11
Paste	3	4	4	5	5
Composites	2	3	3	5	7
Barley (mainly beer)*	0	0	0	2	4
Other	11	12	10	6	7

*) assumed concentration: 100 μg/kg (reporting limit; concentrations below 100 μg/kg were reported as <100 μg/kg)

Probabilistic Effect Assessment

The effects described in the studies mentioned in Table 2 were judged on their suitability for dose response modeling. The data allowed the dose response analysis of the following effect parameters: reduced body weight gain and reduced liver weight (males only) as reported in the chronic mice study of Iverson et al. (1995) and the frequency of resorptions, frequency of anomalies and frequency of affected sternebrae as reported in the reproduction study of Khera et al. (1982). Other effect parameters were not suitable for analysis due to the lack of a clear dose response relationship or to an inadequate report of the data.

Males and females show a similar response to DON on body weight reduction (Figure 6A). The background response (parameter *a*) differs between the two sexes since females have a lower body weight than males. The relative liver weights were decreased in males only and are shown in Figure 6B. Figure 6C-6E shows the observations and fitted models on embryotoxicity. A 5% body weight reduction was chosen as the critical effect size, and similarly a 5% decrease in relative liver weight. For resorptions a 1% additional risk was taken as the benchmark response, while a 5% additional risk level was chosen for the frequency of anomalies and frequency of affected sternebrae.

The distributions of the critical effect doses (CED) associated with the critical effect sizes were calculated. Each of the $CEDs_{animal}$ was extrapolated to a CED_{human}. (Slob and Pieters, 1998) using distributions of assessment factors as described in Vermeire et al. (1999). Table 4 summarizes the results.

We also estimated the effect size in the sensitive human population due to the current exposure of DON. With regard to reduced body weight and reduced liver weights we considered the current intake of children. Children not only show the highest relative DON intake (caused by the high relative wheat intake of this group) but will also be most vulnerable to the effects. Children at the age of 1 yr. have the highest DON intake (95th percentile was 2.90 and 3.24 μg/kg bw for boys and girls, respectively). The dietary DON-intake of a 20-year-old female was 0.88 μg/kg bw (95th percentile), and this value was used as the relevant exposure measure for the embryotoxic effects.

As Table 5 shows, effects on body weight and liver weight are likely to occur. However, the magnitude of these effects is highly uncertain: the confidence interval ranges from 0.2 to approximately 25%. Adults have a lower DON intake per kg bw than children. At the 95th percentile intake level of a 20-year-old female, embryotoxic effects appeared to be small.

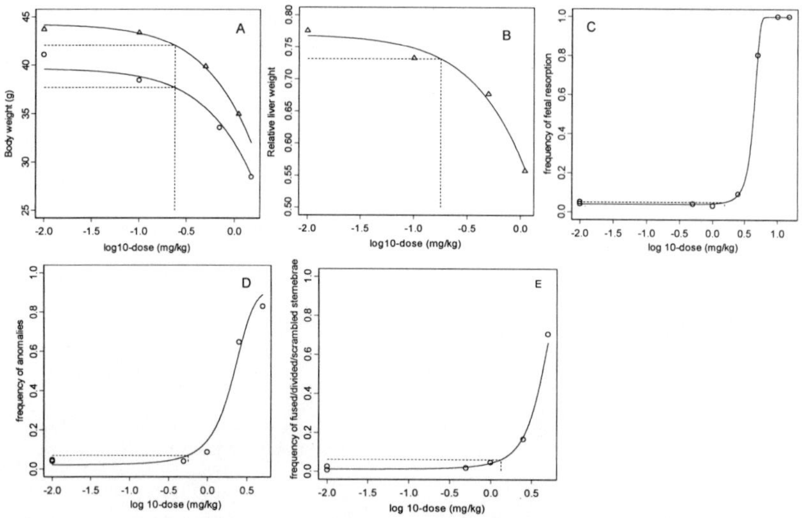

Figure 6. Dose-response data and fitted models for body weight (A), triangles: males, circles: females; relative liver weight (B), triangles: males; frequency of resorptions (C); frequency of anomalies (D); frequencies of fused, divided or scrambled sternebrae (E). Horizontal dashed lines indicate critical effect sizes, i.e., 5% (A, B, D, E) and 1% (C). Vertical dashed lines indicate associated doses.

Table 4. Summary of the probabilistic derivation of a human limit value

Effect parameter	CES	CED_{human} (90% confidence interval[3]), μg/kg bw
body weight reduction[1]	5%	8.6 (**0.6** - 12.2)
decreased relative liver weight[1]	5%	6.0 (**0.5** - 75.5)
frequency of total resorptions[2]	1%	52 (**4.2** - 677)
frequency of anomalies[2]	5%	0.6 (**0.5** - 0.6)
Frequency of fused, divided, or scrambled sternebrae[2]	5%	1.3 (**1.2** - 1.5)

[1] data from Iverson et al. (1995)
[2] data from Khera et al. (1982)
[3] human limit value is indicated in bold

Table 5. Summary of the probabilistic effect assessment

Effect parameter	DON intake, μg/kg bw	Effect Size (%) (90% confidence interval)
body weight reduction[1]	3.2 (girl, 1 yr., P95)	2.2 (0.2 - 24.6)
decreased relative liver weight[1]	2.9 (boy, 1 yr., P95)	2.7 (0.2 - 26.8)
frequency of total resorptions[2]	0.88 (female, 20 yr., P95)	0.0 (0.0 - 0.04)
frequency of anomalies[2]	0.88 (female, 20 yr., P95)	0.1 (0.0 - 2.6)
Frequency of fused, divided, or scrambled sternebrae[2]	0.88 (female, 20 yr., P95)	0.0 (0.0 - 0.6)

[1] data from Iverson et al. (1995)
[2] data from Khera et al. (1982)

DISCUSSION

In 1998 and 1999 DON contamination levels up to 2000 µg/kg and higher in wheat were detected in the Netherlands. In Europe a harmonized regulation of fusarium toxins is absent and guideline values of fusarium toxins in food, including DON, differ markedly between countries. Guideline values for DON in wheat range from 500-2000 µg/kg. On request of the Dutch Ministry of Public Health, Welfare and Sports (VWS), we therefore evaluated the toxicity data on DON. Taking into account the relevant studies and endpoints we derived a provisional TDI of 1.1 µg/kg bw/day. (Pieters et al., 1999). This provisional TDI is in line with Ehling et al. (1997), and Eriksen and Alexander (1998) and has also recently been adopted in the EU (rounded off to 1 µg bw/day).

Based on the provisional TDI and the estimated wheat intake (child, 95th percentile) we calculated concentration limits for DON in (cleaned) wheat (129 µg/kg). We thus assumed that other grains are of minor importance for the DON intake in the Netherlands. We recognized (Pieters et al., 1999) that in view of the high contamination levels of DON in (cleaned) wheat, the suggested concentration limits may be difficult to enforce. Considering growth retardation as a reversible toxic effect, we considered two-fold higher concentration limits temporarily acceptable.

From 1998 onwards the Dutch Inspectorate for Health Protection has been monitoring the DON-concentrations of wheat and food products. The concentration of DON in wheat in the period September 1998 – January 2000 generally exceeded the suggested concentration limit of 129 µg/kg. The average concentration DON in wheat was 446 µg/kg (n = 219). At present, a (temporary) concentration limit of 500 µg DON/kg wheat has been proposed in the EU. The current average DON concentrations in wheat is in line with the EU proposal for a concentration limit, and the performed effect assessment is therefore also representative for a possible future exposure.

The probabilistic effect assessment can be used for the derivation of human limit values (such as provisional TDI) as well as for the estimation of possible health effects due to actual exposures. For the derivation of human limit values a critical effect size has to be chosen for each effect parameter. We chose critical effect sizes of 5% for body weight, relative liver weight, frequency of anomalies and frequency of affected sternebrae and a critical effect size of 1% for frequency of fetal resorptions. By probabilistic combination of the distributions of the critical effect dose in the experimental animal with distributions of the appropriate assessment factors we yield distributions of human CEDs. The 5% lower confidence limits may be used as a human limit value and are 0.6 and 0.5 µg/kg bw for body weight and relative liver weight, respectively. For embryotoxic effects the 5% lower confidence limits were 4.2, 0.5 and 1.2 µg/kg bw for frequency of fetal resorptions, frequency of anomalies and frequency of fused, divided or scrambled sternebrae, respectively. These values are in the same order as the classically derived provisional TDI.

We show that, for people younger than 20 years, the dietary intake of DON in the period studied exceeded the provisional TDI. Young children have the highest relative exposure to DON. These children also form the group at risk since they are most vulnerable to growth retardation. Eighty percent of the one-year-old children are exposed to levels above the provisional TDI and 20% of the one-year-olds are exposed to levels twice the provisional TDI or higher. Another group that may be vulnerable to growth retardation are teenagers. However, the intake of DON rapidly declines with age and DON will thus have less effect on growth in teenagers than in babies and toddlers.

Bread constitutes the major dietary source for DON in the Netherlands, and contributes 60-70% to the total DON intake. For one-year-olds wheat containing porridges also contribute significantly since these children consume less bread. The intake of DON through the consumption of other grains is negligible for the whole population. Beer (barley) may form the only exception. If we assume the DON concentration in beer at the reporting limit (100 µg/kg), the contribution of beer to the total DON intake will be approximately 4% in adults. To gain insight in the real concentrations of DON in beer more sensitive, analytical measurements of DON will be carried out in the near future.

Preliminary results of these analyses indicate that DON levels in beer are below the limits of detection (25 μg/kg). Since wheat (as bread) is a staple food for the Dutch population, the only way to reduce the DON intake is to reduce the concentrations of DON in wheat. Unfortunately, there are no data available on the DON concentrations in wheat and wheat containing foods in the Netherlands in previous years, so we do not have insight in whether the period studied is representative for the exposure to DON in general.

From our probabilistic risk assessment (Table 5) we conclude that the intake of DON in the period September 1998 – January 2000 might have caused health effects in children, with growth retardation and reduction of relative liver weight as the most relevant effects. For this assessment we used data of the chronic mice study of Iverson et al. (1995). In this study the suppressive effect of DON on body weight gain was not associated with reduced food intake. Though we modeled the reported terminal body weights (i.e. the body weights after two years of exposure), the figures reported by Iverson et al. (1995) indicate that the effects on body weight are caused in the first 20% of the total life span. We therefore consider the data appropriate to estimate the effects of DON in children. If DON would only affect body weights, an increased relative liver weight would be expected. Instead, the relative liver weights decreased in males (Iverson et al., 1995). Since this effect may also be explained by inhibition of protein synthesis (as in the case of by growth retardation), we also analyzed this effect parameter in our effect assessment.

Considering the 95th percentile of DON intake of one-year-olds, we show that at this intake level effects on body weight and liver weight are likely to occur. However, the magnitude of these effects is highly uncertain: the confidence interval ranges from less than 1% to approximately 25%. The best estimates of these effects are 2.2% and 2.7%, respectively. The large confidence intervals are caused by the rather broad distributions of the assessment factor used to correct for interspecies variation. We used scaling to caloric demand ($BW^{0.75}$) for the toxicokinetic extrapolation from mouse to human. The remaining uncertainty of the assessment factor was assessed by combining all available experimental data into a distribution (Vermeire et al., 1999, Luttik and van Raaij, 2001). Since this distribution is based on NOAELs the distribution also contains the statistic noise surrounding the NOAELs (Slob and Pieters, 1998). The distribution used will therefore be too broad, however, it is not possible to indicate by how much.

If the classical factor of 100 were used to extrapolate from mice to humans, the estimate of the effect size would have been 7.0% for body weight reduction and 8.4% for decrease of relative liver weight. Adults have a lower DON intake per kg bw than children. At the 95th percentile intake level of a 20-year-old female, embryotoxic effects appeared to be low.

Whether an effect size of approximately 2.5% for reduction of body weight and increased relative liver weight may be (un)acceptable, is a matter of debate and forms part of the risk management process. It should be realized that, similar to the derivation of the provisional TDI, the estimation of the health effects mentioned above are based on extrapolation of animal data. At present, there is no evidence (yet) that these effects will occur in the human population. On the other hand, it should be noted that this effect assessment has been carried out for only one compound (DON). The effects of exposure to other compounds have thus not been taken into account.

CONCLUSIONS

- We derived a provisional TDI of 1.1 μg/kg bw. This provisional TDI is based on a chronic mice study (Iverson et al., 1995). The critical effects were reduction of body weight (both sexes) and relative liver weight (males only). Based on this TDI and a high wheat consumption of children, a concentration limit of 129 μg DON/kg wheat has been proposed (Pieters et al., 1999).

- Monitoring data (September 1998-January 2000) show that the average DON concentration in wheat was 446 μg/kg (n = 219). The major source of DON intake is bread. For one-year-olds porridges also contribute significantly. The DON-intake through other grains is negligible in the Netherlands since the consumption of these grains is low.
- The dietary intake of DON in the Netherlands during the period September 1999-January 2000 exceeds the provisional TDI, especially in children. Eighty percent of the one-year-olds have a DON intake above the provisional TDI and 20% of these children exceed twice the provisional TDI.
- At these exposure levels of DON, health effects might occur in children. The suppressive effects on body weights (growth retardation) and relative liver weight are estimated at 2.2 and 2.7%. However, the large confidence intervals around these estimates indicate that the magnitudes of these effects are uncertain. Whether such estimated effect levels are considered (un)acceptable, is a matter of debate and forms part of the risk management process. It should be noted that the probabilistic effect assessment is based on extrapolation of observations in animal experiments to humans. At present, there is no evidence (yet) that the estimated effects will occur in the human population.

REFERENCES

Arnold, D.L., Karpinski, K.F., McGuire, P.F., Nera, E.A., Zawidzka, Z.Z., Lok, E., Campbell, J.S., Tryphonas, L.and Scott, P.M., 1986, A short-term feeding study with deoxynivalenol (vomitoxin) using rats, *Fundam Appl Toxicol* 6: 691-696.

Bergsjø, B., Matre, T. and Nafstad, I., 1992, Effects of diets with graded levels of deoxynivalenol on performance in growing pigs, *J Vet Med* A39: 752-758.

Bergsjø, B., Langseth, W., Nafstad, I., Høgset Jansen, J. and Larsen, H.J.S., 1993, The effects of naturally deoxynivalenol-contaminated oats on the clinical condition, blood parameters, performance and carcass composition of growing pigs, *Vet Res Commun* 17: 283-294.

Cahill, L.M., Kruger, S.C., McAlice, B.T., Ramsey, C.S., Prioli, R. and Kohn, B., 1999, Quantification of deoxynivalenol in wheat using an immunoaffinity column and liquid chromatography. *J Chromatography* A 859:23-28.

Deijns, A.J., Egmond, H.P. van, Speijers, G.J.A., Loveren, H. van, 1994, Immunotoxicity of natural toxins. A literature overview. (in Dutch), National Institute of Public Health and the Environment. RIVM-report 388802 007, p. 16-17, Bilthoven, the Netherlands.

EHC, 1994, Assessing human health risks of chemicals: Derivation of guidance values for health-based exposure limits. In Environmental Health Criteria, 170. IPCS, WHO, Geneva, Switzerland.

Ehling, G., Cockburn, A., Snowdon, P. and Buchhaus, H., 1997, The significance of the Fusarium toxin deoxynivalenol (DON) for human and animal health, *Cereal Research Commun* 25: 433-447.

Eriksen, G.S., Alexander, J. (eds.), 1998, Fusarium Toxins in Cereals – a Risk Assessment, Copenhagen, Nordic Council of Ministers, TemaNord, 502, pp 45-58, Copenhagen.

IARC, 1993, Monographs on the Evaluation of Carcinogenic Risks to Humans; Vol. 56: Some naturally occurring substances, food items and constituents, heterocyclic aromatic amines and mycotoxins, International Agency for Research on Cancer, World Health Organization, pp 397-444, Lyon.

Iverson, F., Armstrong, C., Nea, E., Truelove, J., Fernie, S., Scott, P.M., Stapley, R., Hayward, S., and Gunner, S., 1995, Chronic feeding study of deoxynivalenol in B6C3F1 male and female mice, *Teratogenesis Carcinogenesis Mutagenesis* 15: 283-306.

Khera, K.S., Whalen, C., Angers, G., Vesonder, R.F. and Kuiper-Goodman, T., 1982, Embryotoxicity of 4-deoxynivalenol vomitoxin) in mice, *Bull Environm Contam Toxicol* 29: 487-491.

Khera, K.S., Arnold, D.L., Whalen, C., Angers, G. and Scott, P.M., 1984, Vomitoxin (4-deoxynivalenol): effects on reproduction of mice and rats, *Toxicol Appl Pharmacol* 74: 345-356.

Kistemaker, C., Bouman, M. and Hulshof, K.F.A.M., 1998. Consumption of separate products by Dutch population groups - Dutch National Food Consumption Survey 1997 – 1998 (in Dutch). Zeist, TNO-Nutrition and Food Research Institute, TNO-report V98.812.

Liem, A.K.D. and Theelen, R.M.C., 1997. Dioxins. Chemical Analysis, Exposure and Risk Assessment. PhD Thesis. Research Institute of Toxicology (RITOX), University of Utrecht, The Netherlands.

Luttik,, R., van Raaij, M.T.M. (eds)., 2001 Factsheets for the (eco)toxicological risk assessment strategy of the National Institute of Public Health and the Environment (RIVM), RIVM report 601516 007, Chapter 5 by T. Vermeire, M.N. Pieters, M. Rennen and P. Bos, RIVM, Bilthoven, The Netherlands

Morrissey, R.E., Norred, W.P. and Vesonder, R.F., 1985, Subchronic toxicity of vomitoxin in Sprague-Dawley rats. *Food Chem Toxicol* 23: 995-999.

Morrissey, R.E. and Vesonder, R.F., 1985, Effect of deoxynivalenol (vomitoxin) on fertility, pregnancy, and postnatal development of Sprague-Dawley rats, *Appl Environm Microbiol* 49: 1062-1066

Pieters, M.N., Fiolet, D.C.M. and Baars, A.J., 1999, Deoxynivalenol. Derivation of concentration limits in wheat and wheat containing food products, National Institute of Public Health and the Environment, RIVM report 388802 018. Bilthoven, the Netherlands.

Rotter, B.A., Prelusky, D.B. and Pestka, J.J., 1996. Toxicology of desoxynivalenol (Vomitoxin). *J Toxicol Environm Health* 48: 1-34.

Rotter BA, Thompson BK, Lessard M, 1995. Effects of desoxynivalenol-contaminated diet on performance and blood parameters in growing swine. *Can J Anim Sci* 75: 297-302.

Rotter BA, Thompson BK, Lessard M, Trenholm HL, Tryphonas H, 1994. Influence of low-level exposure to Fusarium mycotoxins on selected immunological and hematological parameters in young swine. *Fundam Appl Toxicol* 23: 117-124

Slob, W., 1993, Modeling long-term exposure of the whole population to chemicals in food. *Risk Analysis* 13: 525-530

Slob, W. and Pieters, M.N., 1998, A probabilistic approach for deriving acceptable human intake limits and human health risks from toxicological studies: general framework. *Risk Analysis* 18: 787-798.

Slob, W., 1999, Deriving safe exposure levels for chemicals from animal studies using statistical methods: recent developments. In: Statistics for the Environment 4: Pollution Assessment and Control, V. Barnett, A. Stein and K.F. Turkman (eds), John Wiley and Sons, Ltd, p 153 -175, Chichester.

Tryphonas, H., Iverson, F., Ying So, E.A., McGuire, P.F., O'Grady, L., Clayson, D.B. and Scott, P.M., 1986, Effects of deoxynivalenol (vomitoxin) on the humoral and cellular immunity of mice, *Toxicol Lett* 30: 137-150.

VanDooren, M.M.H., Boeijen, I., Van Klaveren, J.D. and Van Donkersgoed, G. (1995). Conversion of consumer food to primairy agricultural products (in Dutch). Wageningen, State Institute for Quality Control of Agricultural Products (RIKILT), report 95.17.

Vermeire, T. , Stevenson, H., Pieters, M.N., Rennen, M., Slob, W. and Hakkert, B.C., 1999, Assessment factors for human health risk assessment: a discussion paper. *Critical Reviews in Toxicology* :29, 439-490

RISK ASSESSMENT OF OCHRATOXIN: CURRENT VIEWS OF THE EUROPEAN SCIENTIFIC COMMITTEE ON FOOD, THE JECFA AND THE CODEX COMMITTEE ON FOOD ADDITIVES AND CONTAMINANTS

Ron Walker

School of Biological Sciences
University of Surrey, Guildford
GU2 5XH, U.K.

ABSTRACT

The chlorinated isocoumarin compound, ochratoxin A (OTA), together with some related derivatives (ochratoxins B, C, α, β) are produced by *Penicillium verrucosum* and by several spp. of Aspergillus, most notably *A. ochraceus*. *P. verrucosum* is the principal source of OTA contamination of stored foods in temperate climates while Aspergillus spp. predominate in warmer countries. The major dietary sources of OTA are cereals but significant levels of contamination may be found in grape juice and red wine, coffee, cocoa, nuts, spices and dried fruits. Because of the chemical stability of OTA and long half-life in mammalian tissues, contamination may also carry over into pork and pig blood products and into beer. OTA is potently nephrotoxic and carcinogenic, the potency varying markedly between species and sexes; it is also teratogenic and immunotoxic. There have been different approaches to the risk assessment of OTA in different jurisdictions, largely arising from whether or not the carcinogenicity of OTA is considered to arise through a thresholded or non-thresholded mechanism. Consequently the tolerable intakes have variously been estimated at 100 ng/kg bw/week (JECFA), 1.5 to 5.7 ng/kg bw/day (Canada) and not more than 5 ng/kg bw/day (European Commission). These differences are also reflected in risk management measures that have been implemented or proposed with different maximum contamination levels being applied to different commodities and to the same commodity in different countries. Prevention of contamination at source is considered to be the most effective public health measure. There is also a need to harmonise the risk assessment and management processes to a greater extent than currently exist if barriers to trade are to be avoided.

INTRODUCTION

Ochratoxin A (OTA) has long been recognised as a mycotoxin contaminant of

Mycotoxins and Food Safety, edited by Trucksess *et al.*
Kluwer Academic/Plenum Publishers, 2002

stored cereals and has been associated with porcine nephropathy following exposure through contaminated feed (Krogh, 1976). A possible association with human endemic nephropathy also has been claimed, based on similarities in the pathology, but to date the evidence remains inconclusive (Pleština, 1996).

OTA has a pentaketide skeleton, and contains a chlorinated isocoumarin moiety linked through a carboxyl group to L-phenylalanine via an amide bond. The chemical structure of OTA is shown in Figure 1.

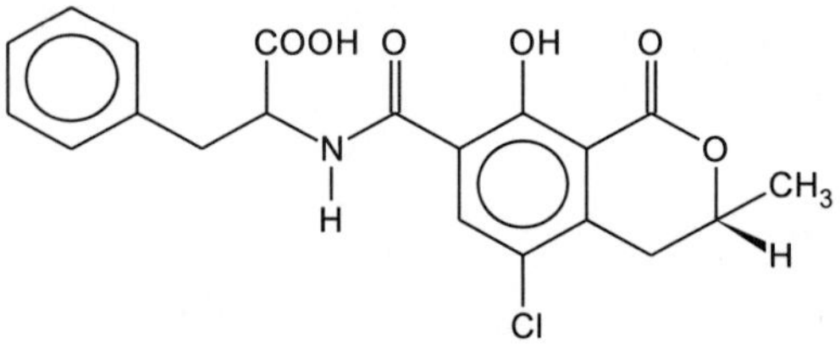

Figure 1. Structure of ochratoxin A

OTA is one of a family of mycotoxins including ochratoxins B and C, which are the dechlorinated analogue and the ethyl ester respectively. The isocoumarin carboxylic acid (ochratoxin α) and its dechlorinated analogue (ochratoxin β) are also detectable in cultures of OTA-producing strains of Aspergillus and Penicillium and 4-hydroxy OTA has also been found in cultures of *Aspergillus ochraceus*.

OCCURRENCE

OTA is produced on stored cereals by *Penicillium verrucosum* in temperate climates and by several species of *Aspergillus* in products of tropical and subtropical climates. The best known species of OTA-producing Aspergillus is *A.ochraceus* but others include *A. sulphureus, A. ostanius* and *A. sclerotiorum*. Of the Aspergilli, only *A. ochraceus* and *A. ostanius* are reported to be significant sources of OTA in cereals (Frisvad, 1995). Although Aspergillus spp. have not been shown to be responsible for contamination with OTA of cereals from warm climates, there are indications that such moulds could be the source of contamination of rice; OTA-producing Aspergillus spp. have also been found in coffee beans and spices contaminated with the mycotoxin and in dried vine fruit (Heenan et al. 1998). Aspergillus spp. probably are the source of OTA production in stored cocoa beans and in grapejuice and wine.

Contamination with ochratoxin is principally found in cereals and some pulses (International Programme on Chemical Safety, 1990; International Agency for Research on Cancer, 1993) but can also occur in coffee, cocoa, nuts and dried vine fruits (Majerus et al. 1993). Significant contamination of grape juice and wine also has been reported (Zimmerli & Dick 1996) and beer may also contain significant residues of OTA. Furthermore, because of the stability of OTA and its long half-life in blood and tissues, pork and food products containing pig blood derived from animals fed on contaminated grain are potential dietary sources.

In 1994, the European Commission established a collaborative scientific co-operation task (SCOOP Task 3.2.2) to provide data on dietary exposure to OTA in the European Union; 13 member states collaborated in this survey. Eight countries estimated intakes based national consumption data and these ranged from 0.7 to 4.6 ng/kg bw/day with a mean of 1.8 ng/kg bw/day. Because of the long half-life in human blood (ca 35 days), plasma levels are a useful biomarker of intake over several days and five countries estimated intakes from these data ranging from 0.2 to 2.4 ng/kg bw/day (mean 0.9 ng/kg bw/day) (European Commission (1997). The similarity in the estimates from the two approaches suggests that the main sources of OTA are known and accounted for in the food surveys.

In updating the position paper on OTA for the Codex Committee on Food Additives and Contaminants (Codex Alimentarius Commission, 1998), intake estimates were made based on the European data for occurrence and diets and by using consumption data from GEMS/Food regional diets, modified in relation to coffee, red wine and beer consumption. The estimated dietary contributions from the major sources are shown in Table 1.

The validity of some of the assumptions made in deriving these mean intakes may be questioned (e.g. that the mean consumption of beer, wine and coffee may added, implying that the mean beer drinker is also the mean wine drinker etc.). In addition, the use of the mean for skewed distributions of contamination is dubious. Nevertheless, the data do identify the main sources of OTA in the diet and highlight a number of points relating to contamination of the diet. Firstly, cereals are clearly the major contributors to dietary exposure while heavy drinkers of red wine may receive a significant intake from that source. Secondly, the heavy contamination of some products, such as raisins and spices, suggests inappropriate conditions may be arising during production and storage with potential for amelioration. A further analysis of regional consumption patterns indicated that, assuming a similar level of contamination, cereals would provide higher intakes of OTA than in Europe in a number of regions, most notably the Middle and Far East regions and Africa.

Table 1. Ochratoxin intake based on European data (adapted from CCFAC Position paper CX/FAC99/YY)

Food	Mean OTA concentration µg/kg	Consumption g/day	Daily OTA Intake ng/kg bw*	% Total Intake
Cereals	0.5	226	1.9	54
Red wine	0.19	171	0.54	15
Coffee	0.9	29	0.43	12
Beer	0.07	234	0.27	7.6
Pork	0.1	76	0.13	3.7
Raisins	2.8	2.3	0.11	3.1
Spices	11	0.5	0.09	2.6
Poultry	0.03	53	0.03	0.9
Pulses	0.1	12	0.02	0.6
Grape juice	1.0	?(50)?**	?(0.8)?	?(19)?
Total			3.5	

* Assumed body weight 60 kg

** No reliable data on consumption; not considered in total intake figure

TOXICOLOGICAL STATUS

The toxicological status of OTA has been reviewed several times and detailed monographs have been published by The International Agency for Research on Cancer (International Agency for Research on Cancer, 1993) and by the Joint FAO/WHO Expert Committee on Food Additives (JECFA) (WHO 1991,1996). From the available evidence it is clear that OTA is potently nephrotoxic and carcinogenic. It is also teratogenic and immunotoxic, affecting both humoral and cell-mediated immunity. The nephrotoxic and carcinogenic properties have been the major focus of the safety evaluation/risk analysis by regulatory authorities. In this regard, the potency varies markedly with species and sex, with the mouse being much less sensitive than the rat and male rats more sensitive than females; with respect to nephropathy, the pig is remarkably more sensitive than the rat. The Lowest-Observed Effect Level (LOEL) and No-Observed Effect Levels (NOEL) for nephrotoxicity and kidney tumours are shown in Table 2. Noteworthy in relation to the safety evaluation/risk assessment is the LOEL or NOEL in the pig compared to the rat since the data from the pig were used to derive the Provisional Tolerable Weekly Intake by the JECFA while some other authorities have based the assessment on the rat data (see below).

Table 2. LOELs and NOELs for nephrotoxicity/carcinogenicity

Species	Effect	Duration of study	LOEL (μg/kg bw/day)	NOEL (μg/kg bw/day)
Mouse (male)	Kidney tumours	2 years	4,400	130
Rat (male)	Karyomegaly of cells of proximal tubule	90 days	15	not demonstrated
	Kidney tumours	2 years	70	21
Pig	Impaired renal function	90 days	8	not demonstrated
	Progressive nephropathy	2 years	40	8

With regard to the genotoxicity of OTA, the situation remains rather unclear. Most short-term assays for gene mutations were negative, and assays for unscheduled DNA synthesis (UDS), sister chromatid exchange (SCE) and chromosomal aberrations in CHO cells were similarly negative or equivocal. However, OTA has been reported to cause single strand breaks in mouse spleen cells *in vitro* and in kidney, liver and spleen cells *in vivo* following i.p. dosing or gavage treatment for 12 weeks at a dose equivalent to 2 mg/kg diet (Dirheimer, 1998).

It has been claimed that treatment of rats and mice with OTA results in adduct formation, detected by ^{32}P-post labelling, but to date the DNA adducts have not been characterised and it is not certain that they derive directly from metabolites of OTA. In some instances, controls had similar post-labelling profiles, although at lower intensity while in others, adduct profiles induced by OTA were dissimilar (Turesky, 1998). Analysis of mutations induced in NIH-3T3 cells in a modified Ames assay revealed predominantly large deletions and these may have arisen from an oxidative mechanism rather than direct DNA-adduct formation.

There are a number of other anomalies in ascribing the carcinogenicity of OTA to a direct genotoxic mechanism. Firstly, the mutagenic metabolite or its breakdown products have not been isolated or identified. Indeed, the major metabolite after oral administration is the relatively non-toxic ochratoxin-α produced by the gut microflora. Secondly, there is little mammalian metabolism of OTA *in vivo* (resulting in the long half-life in tissue and plasma) or in human or rat liver tissue preparations *in vitro*; the major metabolite in rat liver *in vitro* is the 4-hydroxy derivative. Thirdly, the data do not explain the extreme organ and tissue selectivity for toxicity and tumorigenicity.

OTA is a potent competitive inhibitor of phenylalanine-tRNA ligase which results in inhibition of protein synthesis and, secondarily, of RNA and DNA synthesis. Indeed, the acute nephrotoxic effects of OTA are inhibited by co-administration of phenylalanine. The carcinogenic and putative genotoxic effects of OTA may also arise by secondary mechanisms such as impairment of DNA repair, chronic cytotoxicity or induction of oxidative stress and may be thresholded. Differences in the interpretation of these observations has led to differences in the approaches to, and conclusions of, safety evaluation/risk assessment procedures in different jurisdictions.

SAFETY EVALUATION/RISK ASSESSMENT

JECFA

The JECFA evaluated OTA at its 37th meeting (WHO 1991) when a PTWI of 112ng/kg bw was established. This was rounded to 100 ng/kg bw at the 44th meeting (WHO, 1996) when the evaluation was essentially confirmed. The PTWI was derived from the pig data (see table 2) where a NOEL of 8 μg/kg bw/day was reported in the two year study. However, since this level was reported to be a LOEL in the 90-day study based on kidney function, an increased safety factor of 500 was used to derive the PTWI. It has been claimed that the JECFA did not take account of the carcinogenicity of OTA in its evaluation (Kuiper-Goodman, 1996; Codex paper CX/FAC99/YY), but this is inaccurate. The JECFA did review the carcinogenicity of OTA, including the NTP studies, and concluded that renal tumours were only observed at above the Maximum Tolerated Dose (MTD). The data (including the available genotoxicity data) suggested that a secondary mechanism was operative, involving chronic toxicity to cells of the renal tubule, and could be considered thresholded by the chronic toxic NOEL. Consequently, if tumours arise only above the MTD, the most conservative approach was to use the most sensitive index (renal toxicity) in the most sensitive species (pig). In fact this approach leads to an effective safety factor of 1300 if it were based on the NOEL for tumours in the rat. It should also be noted that the PTWI approach was adopted because of the long tissue half-life of OTA. This approach has a more logical basis for relating food intake data to risk than the TDI (based on the human half-life, it would take several weeks to achieve steady state levels at a steady intake level).

Other Risk Assessments

Conversely to the JECFA conclusions, a group convened under the aegis of the Nordic Council of Ministers (Olsen et al., 1991) based their risk assessment on a presumed non-thresholded, genotoxic mechanism and proposed a TDI of 5 ng/kg bw/day. An assessment by the Canadian authorities (Kuiper-Goodman, 1996) used both a quantitative risk assessment and the safety factor approaches. This evaluation made particular note of the aggressive, invasive nature of the renal tumours in rodents, which it was considered most probably reflected a genotoxic mode of action. The risk assessment used in this evaluation was made by a model-independent low dose extrapolation with an incremental risk of 10^{-5} and resulted in a TDI of 1.8 ng/kg bw. The safety factor approach was applied to the NEL, the NOEL, the benchmark dose and the TD_{50} and a safety factor of 5000 was used (50,000 for the TD_{50}) giving rise to TDI estimates of 1.5 to 5.7 ng/kg bw. The reason for using a factor of 5000 was not scientifically derived according to the principles of Environmental Health Criteria 170 (WHO, 1994) but appeared to be arbitrary due to "concerns about the toxic and pharmacokinetic properties". This has no real rationale for a number of reasons: (a) if it is assumed that there is no threshold for the carcinogenic effect, it is inappropriate to use the safety factor approach; (b) conversely, if the mechanism of carcinogenesis ***is*** thresholded, then the nature of the toxicity has no meaning below the threshold, and (c) the relationship between the pharmacokinetics and toxicity is unclear since, if metabolism to a reactive electrophile is supposed, slower metabolism may lead to reduced risk. With regard to the toxicodynamics, it seems more appropriate to use data derived from the most sensitive species, the pig, to calculate a data-derived safety factor.

The Scientific Committee for Food (European Commission, 1998) reviewed its opinion on OTA, cognisant of the earlier evaluations mentioned above. This committee concluded that OTA is a potent nephrotoxic agent, a carcinogen, and that it has genotoxic properties. Hence, it would be prudent to reduce exposure as much as possible, ensuring that exposures are "towards the lower end of the range of tolerable daily intakes of 1.2 to 14 ng/kg bw estimated by other bodies e.g. below 5 ng/kg bw" This conclusion is not transparent since it is not obvious what calculation was used to derive the TDI nor why this figure is optimal within the cited range, or whether it is merely arbitrary.

Health and Trade

From the foregoing, it is clear that different jurisdictions are arriving at different interpretations of the toxicological data and hence at different PTWIs or TDIs. In reviewing the epidemiological evidence, IARC have concluded that OTA is a possible human carcinogen but that there is inadequate evidence of carcinogenicity in humans and, as indicated earlier, the evidence of a causal association with endemic nephropathy remains inconclusive. Thus the potential for conflict in relation to standards for commodities in international trade is obvious. Indeed, nine countries have specific legislation for OTA with limits from 5 – 50 μg/kg applied either to particular cereals, some foods or all foods (van Egmond & Dekker, 1997). There is an obvious need to arrive at a consensus both in relation to the risk assessment and the necessary risk management procedures.

Because of these divergent risk assessments and consequent risk management measures, The Codex Alimentarius Committee on Food Additives and Contaminants has placed OTA on its priority list for re-evaluation and referred this to the JECFA. This review was conducted at a meeting of the JECFA in February 2001.

While the introduction of maximum limits for cereals may be justifiable, since these commodities are the major source of exposure, protection of human health is best assured by preventive measures to minimise the contamination with OTA-producing fungi and the conditions that give rise to production of the toxin. A programme aimed at identifying and controlling the causes of contamination of coffee is currently being conducted by the FAO, and such an approach is likely to be most beneficial, both from the point of view of human health and economically through reduced spoilage and rejection rates.

REFERENCES

Codex Alimentarius Commission, 1998, Position paper on ochratoxin A, CX/FAC 99/YY

Dirheimer, G., 1998, Recent advances in the genotoxicity of mycotoxins, *Revue Méd.Vét.* 149:605-616.

European Commission, 1997, Scientific Committee on Food Opinion on ochratoxin A, CS/CNTM/MYC/14 final, Annex II to Document XXIV/2210/98, 28 September 1998

Frisvad, J., 1995, In: *Stored grain ecosystems*, D. Jayas, N. White and W. Muir, eds., New York: Marcel Dekker, pp. 251-288.

Heenan, C.N., Shaw, K.J. and Pitt, J.I., 1998, Ochratoxin A production by *Aspergillus carbonarius* and *A.niger* isolates and detection using coconut cream agar, *J.Food Mycology,* 1:67-72.

International Agency for Research on Cancer, 1993, IARC Monographs on the Evaluation of Carcinogenic Risks to Humans, Vol. 56: Some naturally occurring substances: food items and constituents, heterocyclic aromatic amines and mycotoxins, Lyon:IARC

International Programme on Chemical Safety, 1990, Environmental Health Criteria 105. Selected Mycotoxins: Ochratoxins, Trichothecenes, Ergot. Geneva: WHO.

Krogh, P., 1976, Epidemiology of mycotoxic porcine nephropathy. *Nordisk Veternaermedicin,* 28:452-458.

Kuiper-Goodman, T., 1996, Risk assessment of ochratoxin A: an update, *Food Additives and Contaminants*, 13, suppl.:53-57

Marjerus, P., Cutka, I., Dreyer, A., El-Dessouki, S., Eyrich, W., Reusch, H, Schurer, B and Waiblinger, H.U., 1993, Zur belastungssituation von Ochratoxin A in Lebensmitteln pflanzlichen Ursprungs, *Deutsche Lebensmittel Rundschau,* 89:112-114.

Olsen, M., Thorup, I., Knudsen, I., Larsen, J.-J., Hald, B. and Olsen, J., 1991, Health evaluation of ochratoxin A in food products. *Nordiske Seminar- og Arbejds-rapporter*. Vol. 545. Copenhagen: Nordic Council of Ministers.

Pleština, R. , 1996, Nephrotoxicity of ochratoxin A, *Food Additives and Contaminants,* 13 (suppl.):49-50.

Scott, P.M. and Kanhere, S.R., 1995, Determination of ochratoxin A in beer, *Food Additives and Contaminants,* 12:591-598.

Turesky, R., 1998, Paper submitted to American Association of Cancer Research meeting, Philadelphia, April, 1999 (personal communication to ILSI Europe)

Van Egmond, H.O. and Dekker, W.H., 1997, Worldwide regulations for mycotoxins in 1995 – a compendium. FAO Food and Nutrition Paper 64. Rome: FAO.

WHO, 1991, Ochratoxin A, Toxicological evaluation of certain food additives and contaminants, WHO Food Additives Series: 28, 365-417. Geneva: WHO

WHO, 1996, Ochratoxin A, Toxicological evaluation of certain food additives and contaminants, WHO Food Additives Series: 35, 363-376. Geneva:WHO

Zimmerli, B. and Dick, R., 1996, Ochratoxin in table wine and grape juice: Occurrence and risk assessment. *Food Additives and Contaminants*, 13:647-654.

WORLDWIDE REGULATIONS FOR MYCOTOXINS

Hans P. van Egmond

National Institute of Public Health and the Environment
Laboratory for Residue Analysis
P.O. Box 1, 3720 BA Bilthoven
The Netherlands

ABSTRACT

Since the discovery of the aflatoxins in the 1960s, regulations have been established in many countries to protect the consumer from the harmful effects of mycotoxins that may contaminate foodstuffs. Various factors play a role in the decision-making process of setting limits for mycotoxins. These include scientific factors such as the availability of toxicological data, survey data, knowledge about the distribution of mycotoxins in commodities, and analytical methodology. Economical and political factors such as commercial interests and sufficiency of food supply have their impact as well.

International enquiry's on existing mycotoxin legislation in foodstuffs and animal feedstuffs have been carried out several times in the 1980s and 1990s and details about tolerances, legal basis, responsible authorities, official protocols of analysis and sampling have been published. Recently a comprehensive update on worldwide regulations was published as FAO Food and Nutrition Paper 64. It appeared that at least 77 countries now have specific regulations for mycotoxins, 13 countries are known to have no specific regulations, whereas no data are available for about 50 countries, many of them in Africa. Over the years, a large diversity in tolerance levels for mycotoxins has remained. Some free trade zones (EU, MERCOSUR) are in the process of harmonizing the limits and regulations for mycotoxins in their respective member states, but it is not likely that worldwide harmonized limits for mycotoxins will soon be within reach.

INTRODUCTION

Food legislation serves to safeguard the health of food consumers and the economic interests of food producers and traders. In the days of early food regulations, protection of food was mostly a local affair and municipal ordinances were promulgated for the purpose. Inspections were relatively simple as there were no auxiliary sciences. Later, when bacteriology, chemistry and microscopy developed, plans for statutory regulations gradually emerged in many countries, leading at the beginning of the twentieth century to the adoption of official legislation and regulations. Today, the food laws not only prohibit

Mycotoxins and Food Safety, edited by Trucksess *et al.*
Kluwer Academic/Plenum Publishers, 2002

the introduction, delivery for introduction, or receipt in commerce of adulterated or misbranded food, but often they include specific regulations that impose limits or tolerances on the concentrations of particular contaminants in foods. Such contaminants may be of industrial or natural origin. Of natural contaminants, mycotoxins are the most recent to be considered because of their implications for public health and their economic consequences, in particular with relation to the international food trade.

At present, food laws often impose limits on the concentrations of specific mycotoxins in human foods and animal feeds. At least this is the case in many countries of the industrialized part of the world, where fully developed market economies exist. In contrast, many developing countries where subsistence farming is significant, have not (yet) established specific regulations for mycotoxins. In this paper, the factors that may influence the establishment of regulations for mycotoxins are discussed and a brief summary will be given of current mycotoxin regulations and developments.

FACTORS INFLUENCING THE SETTING OF MYCOTOXIN REGULATIONS

Several factors, both of scientific and non-scientific nature, may influence the establishment of mycotoxin limits and regulations. These include:

- availability of toxicological data.
- availability of data on the occurrence of mycotoxins in various commodities.
- knowledge of the distribution of mycotoxin concentrations in lots.
- availability of analytical methods.
- legislation in other countries with which trade contacts exist.
- need for sufficient food supply.

Risk Assessment

Measures are primarily taken on the basis of known toxic effects. For some mycotoxins (ochratoxin A, patulin, aflatoxins), the Joint Expert Committee on Food Additives (JECFA) of the World Health Organization and the Food and Agriculture Organization, has partly evaluated the hazard. JECFA is a scientific advisory body. It provides a mechanism for assessing the toxicity of additives, veterinary drug residues and contaminants. Safety evaluation of contaminants incorporates various steps in a formal health risk assessment approach. The qualitative indication that a contaminant can cause adverse effects on health (hazard identification) is usually included in the information presented to JECFA for evaluation. Similarly, qualitative and quantitative evaluation of the nature of the adverse effects (hazard characterization) is embodied in the data sets that are present.

Toxicological evaluation carried out by JECFA normally results in the estimation of a Provisional Tolerable Weekly Intake (PTWI). The use of the term "provisional" expresses the tentative nature of the evaluation, in view of the paucity of reliable data on the consequences of human exposure at levels approaching those with which JECFA is concerned. In principle, the evaluation is based on the determination of a No-Observed-Effect-Level (NOEL) in toxicological studies, and the application of a safety factor. The safety factor means that the lowest NOEL in animal studies is divided by 100, 10 for extrapolation from animals to humans and 10 for variation between individuals, to arrive at a tolerable intake level. In cases where the data are inadequate, JECFA uses a higher safety factor. This hazard assessment approach does not apply for toxins where carcinogenicity is the basis for concern as is the case with the aflatoxins. Assuming that a no-effect concentration limit cannot be established for genotoxic compounds such as the aflatoxins, any small dose will have a proportionally small effect. Imposing the absence of any amount of aflatoxins would then be appropriate, if these were not natural contaminants that can never completely be eliminated without outlawing the contaminated food or feed. In these cases, JECFA does not allocate a PTWI. Instead it recommends that the level of the contaminant in food should be reduced so as to be As Low As Reasonably Achievable (ALARA). The ALARA level, which may be viewed as the irreducible level for a contaminant, is defined as the concentration of a substance that cannot be eliminated from a food without involving the discard of that food altogether or without severely compromising the availability of major food supplies. This covers the case of the JECFA evaluation of aflatoxins made in 1987, 1997 and 1999.

In February 2001, a special JECFA session will be devoted to mycotoxins. The mycotoxins scheduled for evaluation or re-evaluation at this 56th JECFA meeting include fumonisins B_1, B_2, B_3 and B_4, ochratoxin A, deoxynivalenol, T-2 and HT-2 toxins, and aflatoxin M_1. In the further development of tolerable daily intake (TDI) levels to maximum tolerated levels in food and feed for national or international (Codex Alimentarius) purposes, factors other than hazard assessment play a role; these will be discussed below.

In addition to information about toxicity, exposure assessment is another ingredient necessary to make risk assessment possible. Reliable data on the occurrence of mycotoxins in various commodities and food intake data are needed to prepare exposure assessments. They also allow estimating the effects of regulations on the availability of the foods and feeds concerned. Risk assessment is in fact the product of hazard assessment and exposure assessment. Factors it involves, among others, are (Kuiper-Goodman, 1989):

- hazard assessment based on animal toxicity and possibly epidemiological data.
- exposure assessment based on data concerning actual concentrations in commodities.
- extrapolation of hazard assessment from high exposure animal data to low exposure human data.
- comparison of the product of hazard assessment and exposure assessment with accepted risk.

In the European Union, efforts to assess exposure are undertaken within SCOOP projects (SCOOP: Scientific Cooperation on Questions relating to Food), funded by the European Commission. The SCOOP projects are targeted to arrive at the best estimates of intake of several mycotoxins by EU inhabitants. In the 1990s these activities resulted in reports on the exposure assessment of aflatoxins (European Commission, 1997[a]) and ochratoxin A (European Commission, 1997[b]). At the time of this writing an update was in preparation for a SCOOP exposure assessment of ochratoxin A in the EU. Ochratoxin A is a mycotoxin of serious concern to the EU, where specific regulations are in preparation for this toxin in various foodstuffs, including cereals, coffee, wine and beer. For the years 2001 and 2002 similar SCOOP activities in the EU are foreseen for fumonisins, zearalenone and for some of the trichothecenes, including deoxynivalenol. The SCOOP data are being used by the EC's Scientific Committee for Food (SCF) for its evaluation and advisory work on the risks to public health arising from dietary exposure to certain mycotoxins.

Sampling Procedures

The distribution of the concentration of mycotoxins in products is an important factor to be considered in establishing regulatory sampling criteria. The distribution can be very inhomogeneous, as is the case with aflatoxins in peanuts. The number of contaminated peanut kernels in a lot is usually very low, but the contamination level within a kernel can be very high. If insufficient care is taken for representative sampling, the mycotoxin concentration in an inspected lot may therefore easily be wrongly estimated. Also, consumption of peanuts may lead to an accidental high single dose of aflatoxins, instead of a chronic intake at a relatively low level. A similar situation may occur with pistachio nuts and figs. The risk to both consumer and producer must be considered when establishing sampling criteria for products in which mycotoxins are inhomogeneously distributed. The design of sampling procedures has been an international concern for a long time, and discussions are continuously carried out to find a harmonized international approach. The Codex work (Codex Alimentarius Commission, 2000) as well as the FAO Food and Nutrition Paper 55 (Food and Agriculture Organization, 1993) are worth mentioning in this context.

Methods of Analysis

Legislation calls for methods of control, so reliable analytical methods will have to be available. Tolerance levels that do not have a reasonable expectation of being measured are both wasteful in the resources that they utilize, and may well condemn products that are perfectly fit for consumption (Smith et al., 1994). In addition to reliability, simplicity is required, as it will influence the amount of data that will be generated and the practicality of the ultimate measures taken. The reliability of analysis data can be improved if use is made of methods that fulfill certain performance criteria (as can be demonstrated in interlaboratory studies).

AOAC International and the European Standardization Committee (CEN) have a number of standardized methods of analysis available that have been validated in formal collaborative studies, and this number is gradually growing. The latest edition of Official Methods of Analysis of AOAC International (Horwitz, 2000) contains approx. 40 validated methods for mycotoxin determination. CEN has produced a document that provides criteria for mycotoxin methods (Comité Européen de Normalisation, 1999). This document gives information concerning method performance, which can be expected from experienced analytical laboratories.

In addition, the application of Analytical Quality Assurance (AQA) procedures is recommended, including the use of (certified) reference materials, especially when a high degree of comparison and accuracy is required. Further developments in AQA and the use of reference materials are likely to emerge in the future for the control of mycotoxins in foods and feeds. In Table 1 an overview is given of (certified) reference materials for mycotoxins that have been developed or are currently being developed in projects funded by the European Commission's Standards, Measurements and Testing Programme (previously named Bureau Communautaire de Référence [BCR]). They are currently available worldwide.

Table 1. Certified reference materials for mycotoxins, developed in the EC's Standards, Measurements and Testing Programme

BCR reference materials for mycotoxins, August 2000			
Reference material	available	development	considered
aflatoxin M_1 in milk powder	X	X	
aflatoxin M_1 in chloroform	X		
aflatoxins in peanut butter	X		
aflatoxin B_1 in peanut meal	X		
aflatoxin B_1 in animal feedstuff	X		
ochratoxin A in wheat	X		
deoxynivalenol in corn and wheat	X		
zearalenone in corn		X	
trichothecenes other than deoxynivalenol			X
fumonisins			X

Trade Contacts

Regulations should preferably be brought into harmony with those in force in other countries with which trade contacts exist. In fact this approach has been applied both in the EU and MERCOSUR, where now (draft-) harmonized community regulations for mycotoxins exist. Strict regulative actions may lead importing countries to ban or limit the import of commodities, such as certain food grains and animal feedstuffs, which can cause difficulties for exporting countries in finding or maintaining markets for their products. For example, the stringent regulations for aflatoxin B_1 in animal feedstuffs in the European Union (Commission of the European Communities, 1991), led European animal feed manufacturers to switch from groundnut meal to other protein sources to include in feeds; this had an impact on the export of groundnut meal of some developing countries (Bhat, 1999). The distortion of the market caused by regulations in importing countries may lead to export of the less contaminated foods and feeds leaving those inferior foods and feeds, which do not meet the standards for export, for home consumption.

Food Supply

The regulatory philosophy should not jeopardize the availability of some basic commodities at reasonable prices. Especially in the developing countries, where food supplies are already limited, drastic legal measures may lead to lack of food and to excessive prices. It must be remembered that people living in these countries cannot exercise the option of starving to death today in order to live a better life tomorrow.

Conclusion

Weighing the various factors that play a role in the decision-making process of establishing mycotoxin tolerances is not trivial. Common sense is a major factor for reaching a decision. Public health officials are confronted with a complex problem: mycotoxins, and particularly aflatoxins, should be excluded from food as much as possible. Since the substances are present in foods as natural contaminants, however, human exposure cannot be completely prevented, and exposure of the population to some level of mycotoxins has to be tolerated. Despite the dilemmas, mycotoxin regulations have been established in the past decades in many countries, and newer regulations are still being drafted.

CURRENT MYCOTOXIN REGULATIONS AND DEVELOPMENTS

Overviews of limits and regulations for mycotoxins and their rationales have been published several times, e.g. by Krogh (1977), Schuller et al. (1983), Van Egmond (1989), Stoloff et al. (1991), Gilbert (1991), Van Egmond (1991), Van Egmond and Dekker (1995), Boutrif and Canet (1998), Rosner (1998) and Van Egmond (1999). The paper of Van Egmond (1989) was based on a worldwide inquiry made in 1987 on behalf of FAO for the Second Joint FAO/WHO/UNEP International Conference on Mycotoxins, held in Bangkok, Thailand. A comprehensive update, based on a similar inquiry carried out in 1994/1995 was published as FAO Food and Nutrition Paper 64 (Food and Agriculture Organization, 1997).

The International Inquiry 1994-1995

The agricultural attachés or counsellors of Dutch embassies were approached with a request to gather up-to-date information on the situation regarding mycotoxin regulations from the local authorities in as many countries of the world as possible. For this purpose, inquiry forms together with background information in English, French, German and Spanish about the mycotoxin problem were sent out through 34 Dutch embassies covering more than 100 countries. The questions concerned: the existence of mycotoxin regulations; the types of mycotoxins and products for which regulations are in force, together with maximum permissible levels; the authorities responsible for control of mycotoxins; the use of official and published methods of sampling and analysis; and the disposal of consignments containing inadmissible amounts of mycotoxins. The inquiry items and format are shown in Table 2.

Table 2. Inquiry items

Inquiry items
1. Country
2. Commodity
3. Mycotoxins:
(a) Sum of:
(b) Max. conc.:
4. Legal basis:
5. Authority:
6. Sampling method:
(a) Status:
(b) Published in:
7. Analytical method:
(a) Status:
(b) Type:
(c) Published in:
8. Remarks:

Together with the forms, tables were supplied with the data gathered during the previous enquiry in 1987 for the countries covered by each embassy as well as a table with data from the Scandinavian countries. These tables were prepared in order first, to show the data already available or lacking for the countries in each embassy's jurisdiction, and second, to give an idea of the kind of information needed. For the latter purpose, the Scandinavian data were ideal, being neither too elaborate nor too concise. Recipients were asked either to fill in the form or to correct the relevant table. Responses were obtained from 41 countries. Missing data were supplemented with literature data, if available, and with data from the earlier surveys (1987 and 1981).

The activities resulted in a review, with details on the mycotoxin regulations, country by country, as they existed worldwide by 1 October 1996. At that date information was available for 90 countries, which was an increase of 36% with respect to the previous update presented at the FAO/WHO/UNEP Conference in Bangkok in 1987. Most new information came from the South American countries. Seventy-seven countries were known to have (some) regulations on mycotoxins, an increase of 38% with respect to 1987. Thirteen countries were known to have no specific regulations. No data were available for about 50 countries, most of them in Africa. This is illustrated in the world map of Figure 1.

General Observations

Most of the existing mycotoxin regulations concern aflatoxins in food; in fact all countries with mycotoxin regulations at least have regulatory levels for aflatoxin B_1 or the sum of the aflatoxins B_1, B_2, G_1 and G_2 in foods and/or animal feedstuffs. Less frequently, specific regulations also exist for aflatoxin M_1 in milk and milk products. Comparisons of various regulatory levels are only meaningful for aflatoxins. Table 3 compares the 1987 situation with the 1996 update for some aflatoxin/commodity combinations.

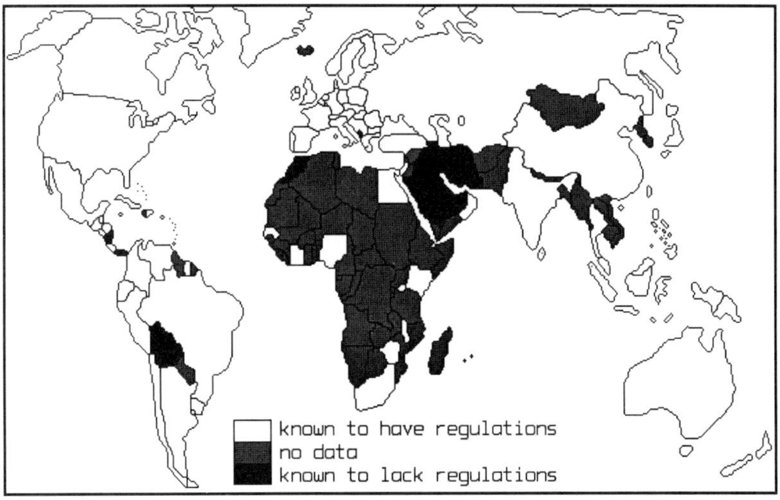

Figure 1. Countries known to have regulations (white), unknown whether or not they have regulations (grey) and known to have no specific regulations (black) for mycotoxins in foodstuffs and feedstuffs.

Table 3. Medians and ranges in 1987 and 1996 of maximum tolerated levels (ng/g) for some (groups of) aflatoxins and numbers of countries that have regulations for these.

	1987			1996		
	median	range	no of countries	median	range	no of countries
B_1 in foodstuffs	4	0-50	29	4	0-30	33
$B_1+B_2+G_1+G_2$ in foodstuffs	7	0-50	30	8	0-50	48
B_1 in foodstuffs for children	0.2	0-5	4	0.3	0-5	5
M_1 in milk	0.05	0-1	13	0.05	0-1	17
B_1 in feedstuffs	30	5-1000	16	20	5-1000	19
$B_1+B_2+G_1+G_2$ in feedstuffs	50	10-1000	8	50	0-1000	21

As compared to the situation in 1987 the maximum tolerated levels for aflatoxins have not changed dramatically. The data from 1996 are also graphically presented in the form of frequency distributions in Figures 2 to 7. In these figures the symbol **n** refers to the number of regulations that exist for the specific aflatoxin/commodity combination. This is not necessarily the same as the number of countries that have these specific regulations and the figures should be interpreted with some caution, since they strongly depend upon the degree of detail in the regulations. For example, some countries have many regulations specifying slightly different tolerated levels for individual foods, whilst others set one tolerated level for all foods. The former case, of course, has much greater impact on the figures than the latter, although it is not possible to state that one regulation is more significant than the other.

As in 1987, also in 1996 many countries regulated the aflatoxins with limits for the sum of the aflatoxins B_1, B_2, G_1, and G_2, sometimes in combination with a specific limit for aflatoxin B_1. It is debatable whether a regulatory level for the sum of the aflatoxins, which requires more analytical work than for aflatoxin B_1 alone, contributes significantly to better protection of public health than a regulatory level for aflatoxin B_1 alone. Aflatoxin B_1 is the most important of the aflatoxins, considered from both the viewpoints of toxicology and occurrence. It is unlikely that commodities will contain aflatoxins B_2, G_1 and G_2 and not aflatoxin B_1, whereas the concentration of the sum of the aflatoxins B_2, G_1 and G_2 is generally less than the concentration of aflatoxin B_1 alone. Typical occurrence rations for aflatoxins B_1 and B_2 (mainly produced by *Aspergillus flavus*) average approx. 4:1. Typical occurrence ratios for aflatoxins B_1 and the sum of the aflatoxins B_2, G_1 and G_2 (mainly produced by *Aspergillus parasiticus*) average approx. 1:0.8, although variations do occur for both ratios. Monitoring agencies in those countries that apply regulatory level for the sum of the aflatoxins should inspect their analytical data to see how frequently the availability of data on the sum of the aflatoxins (above that on aflatoxin B_1) has been indispensable to adequately protect the consumer.

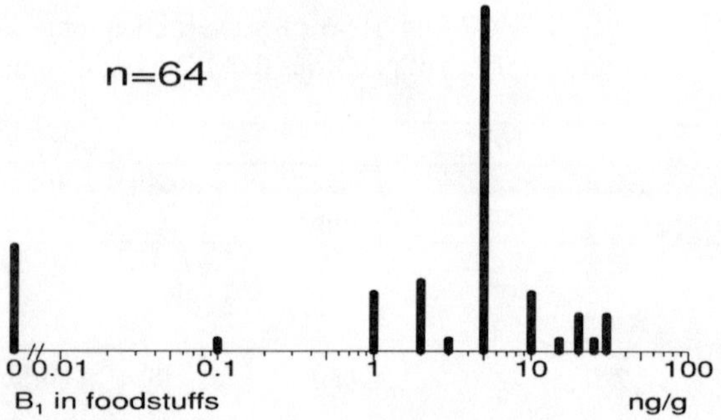

Figure 2. Frequency of specific regulations for aflatoxin B_1 in foodstuffs, as a function of maximum tolerated level.

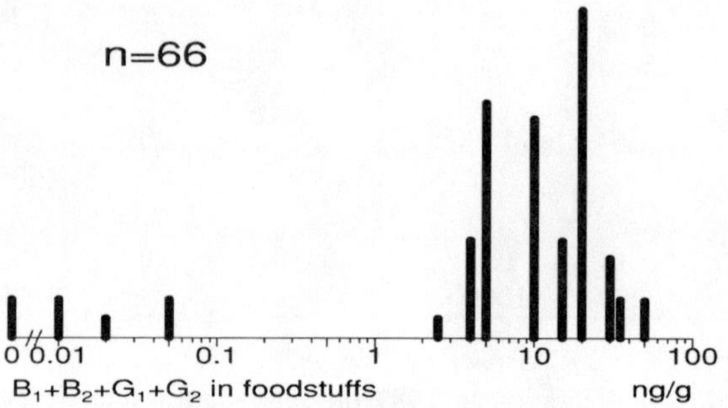

Figure 3. Frequency of specific regulations for the sum of the aflatoxins B_1, B_2, G_1 and G_2 in foodstuffs, as a function of maximum tolerated level.

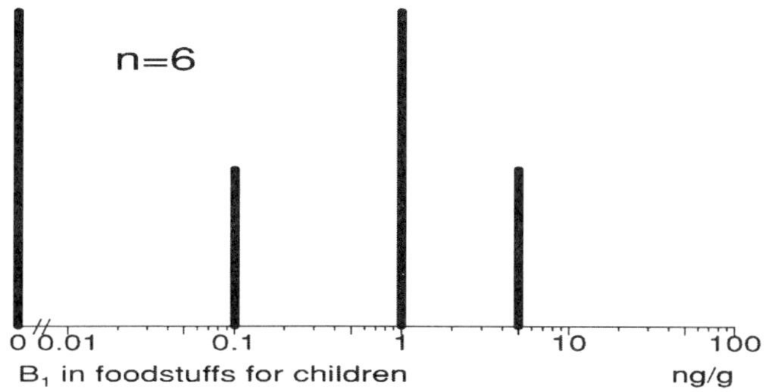

Figure 4. Frequency of specific regulations for aflatoxin B_1 in foodstuffs for children, as a function of maximum tolerated level.

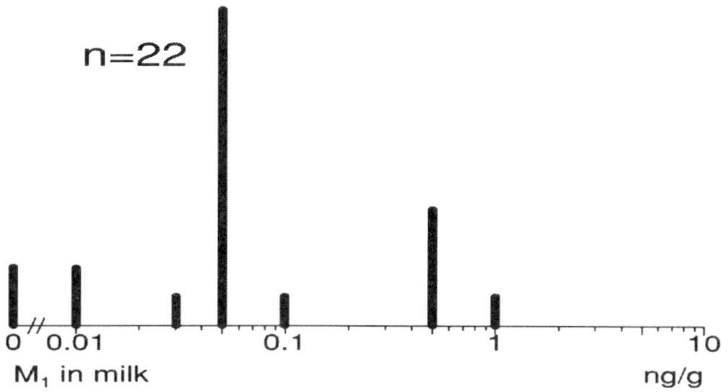

Figure 5. Frequency of specific regulations for aflatoxin M_1 in milk, as a function of maximum tolerated level.

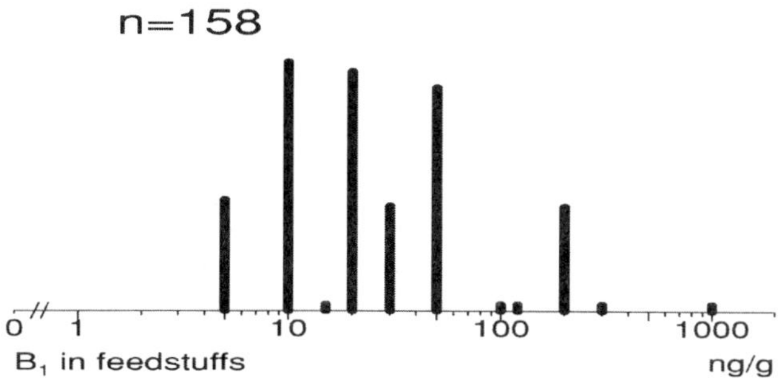

Figure 6. Frequency of specific regulations for aflatoxin B_1 in animal feedstuffs, as a function of maximum tolerated level.

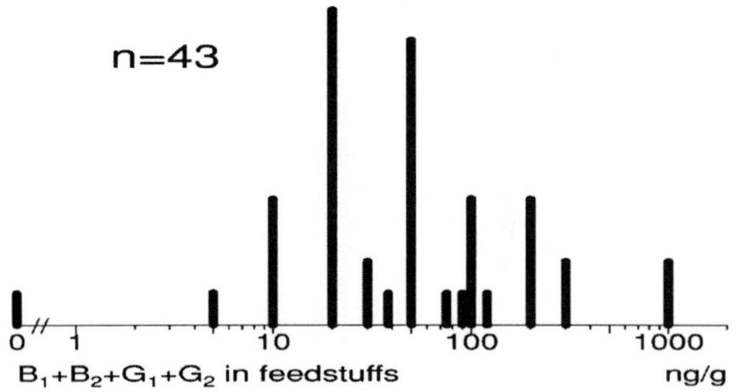

Figure 7. Frequency of specific regulations for the sum of the aflatoxins B_1, B_2, G_1 and G_2 in animal feedstuffs, as a function of maximum tolerated level.

For several other mycotoxins specific regulations exist as well. (i.e. patulin, ochratoxin A, deoxynivalenol, diacetoxyscirpenol, zearalenone, T-2 toxin, chetomin, stachybotryotoxin, phomopsin and fumonisins B_1 and B_2). Although the fumonisins are a major target now in surveillance programmes, regulating authorities seem to be waiting for the outcome of the JECFA 56th meeting (February 2001) where fumonisins will be evaluated, before drafting tolerance levels for these toxins (See also section: **Risk assessment**).

Some free trade zones (EU, MERCOSUR) are in the process of harmonizing the limits and regulations for mycotoxins in their respective member states. In the EU, this harmonization includes a tightening of the limits, in particular for the aflatoxins (European Commission, 1998), which will lead to further debates between the EU, the USA and the developing countries in organizations as the World Trade Organization and Codex Alimentarius. It is not likely that worldwide harmonized limits for mycotoxins will soon be within reach.

Specific Observations

Some specific observations from FAO Food and Nutrition Paper #64 (Food and Agriculture Organization, 1997) relating to zero tolerance levels, and to the relation between tolerance levels and commodities are worth mentioning.

Several countries are applying zero tolerances for one or more mycotoxins in some commodities. From the hazard point of view, it might be desirable to limit the toxin of concern to the lowest possible level. However, a zero tolerance level may be impractical for several reasons. First, mycotoxins are natural contaminants that often cannot be completely excluded from the food chain. Another important consideration is the available analytical methodology, because the enforcement of the regulations requires reliable methods of control (see the section: **Methods of Analysis**). It should be noted that no method of analysis exists for detecting zero tolerance. It is recommended at the international level that tolerance levels be based on a risk assessment approach rather than on analytical limitations (i.e. below which no detection is possible by analytical means).

Several countries reported tolerance levels (or zero tolerance levels) for mycotoxins in commodities in which the presence of such toxins has not been ascertained. For example, tolerance levels for total aflatoxins (B_1, B_2, G_1 and G_2) or some of them (G_1, G_2

for instance) in milk products have been established in addition to tolerance levels for aflatoxin M_1. Aflatoxins G_1 and G_2 are normally not to be expected as carryover toxins in dairy products. In principle they could be formed there as a result of fungal growth for fermentation, or unintentional fungal growth. There are published reports of aflatoxin B_1 and aflatoxin B_2 occurring in cheese, but the occurrence of G_1 and G_2 is most unlikely. According to present knowledge, aflatoxin M_1 is usually the main aflatoxin occurring in milk. It is a metabolite of aflatoxin B_1, occurring in body fluids, organs and tissues, normally at low concentrations. Aflatoxin M_2, a carryover of aflatoxin B_2, is of less significance.

Similarly, no aflatoxin has been reported to date in starch, although one country has been establishing a regulatory level for it. Other products that are also unlikely to become contaminated with naturally occurring aflatoxins include vegetables and milk, but regulatory levels have been established. In addition, although carryover to eggs and meat hardly occurs, specific regulations for aflatoxins in these commodities exist. Furthermore, some regulations set tolerance levels for aflatoxins in butter and animal fat, although aflatoxins are not lipophilic and are therefore hardly to be expected in animal fat.

Some regulations not only cover aflatoxins B_1, B_2, G_1 and G_2 in peanuts and peanut products, but also aflatoxins M_1, M_2 and P_2 and aflatoxicol. According to present knowledge, aflatoxin P_1 and aflatoxicol are not directly produced by *Aspergillus* strains. Although some *Aspergillus parasiticus* and *flavus* strains are able to produce minute amounts of aflatoxin M_1 and aflatoxin M_2 in addition to large amounts of aflatoxins B_1, B_2, G_1 and G_2, occurrence of aflatoxins M_1 and M_2 remains negligible in such products. Last but not least remarkable is a regulation requiring a zero tolerance for all mycotoxins in ice cream.

CONCLUSIONS AND RECOMMENDATIONS

The number of countries with specific regulations for mycotoxins in food and animal feed is increasing over the years. This reflects the general concern that governments have about the potential effects of mycotoxins on the health of humans and animals. Differences are seen, however, in the legal limits that countries have laid down in their regulations. Transparency as to why these limits were chosen, is often lacking.

Harmonization of tolerance levels is taking place in some free trade zones, and harmonization efforts are being undertaken for goods moving in international commerce. This harmonization is a slow process, because of the different views and interests of those involved in the process. Whereas harmonized tolerance levels would be beneficial from the point of view of trade, one might argue this would not necessarily be the case from the point of view of (equal) protection of consumers around the world. Risk assessment is the product of hazard assessment and exposure assessment. The hazard of mycotoxins to individuals is probably more or less the same all over the world (although other factors sometimes play a role as well, e.g. hepatitis B virus infection in relation to the hazard of aflatoxins. Exposure is not the same, because of differences in levels of contamination and dietary habits in various parts of the world.

National governments or regional communities should encourage and fund activities that contribute to reliable exposure assessment of mycotoxins in their regions. Examples of such activities are the SCOOP tasks, undertaken in the EU in support of risk evaluations on some mycotoxins (see section: **Risk Assessment**). The availability of validated (and easily performed) analytical methodology and the application of Analytical Quality Assurance are basic ingredients that can result in obtaining meaningful data on the occurrence of mycotoxins, and must therefore be stimulated.

Efforts involving hazard assessments should preferably be coordinated and funded at the international level. Chronic toxicity studies carried out under Good Laboratory

Practice conditions are time consuming, very expensive, and not necessarily bound to certain regions. These studies should be carried out in internationally recognized centers of excellence and their results evaluated by international groups of experts, e.g. JECFA.

The enacted regulations and those under development should be the result of sound cooperation between interested parties, drawn from the scientific sector, from the ranks of the consumers, from industry and from official circles. Only then can realistic mycotoxin legislation be achieved.

ACKNOWLEDGEMENTS

The author wishes to thank the Food and Agriculture Organization of the United Nations for permission to reproduce parts of the material published in FAO Food and Nutrition Paper 64: *Worldwide regulations for mycotoxins 1995.* A compendium and FAO Working Document *MYC-CONF/99/8a*, prepared for the Third Joint FAO/WHO/UNEP International Conference on Mycotoxins, Tunis, Tunisia, 3-6 March 1999. The views and opinions expressed are those of the author, and do not necessarily represent the decisions or the stated policy of the National Institute of Public Health and the Environment.

REFERENCES

Bhat, R., 1999, Mycotoxin contamination of foods and feeds. Working Document Third Joint FAO/WHO/UNEP International Conference on Mycotoxins. *MYC-CONF/99/4a.* Tunis, Tunisia, 3-6 March 1999.

Boutrif, E. and Canet, C., 1998, Mycotoxin prevention and control: FAO programmes. *Revue de Médicine Vétérinaire* 149:681-694.

Codex Alimentarius Commission , 2000, Joint FAO/WHO Food Standards Programme. *Report of the 32nd Session of the Codex Committee of Food Additives and Contaminants.* Beijing, People's Republic of China, 20-24 March 2000. Rome, Italy.

Comité Européen de Normalisation, 1999, Food Analysis-Biotoxins-criteria of analytical methods of mycotoxins. CEN report CR 13505.

Commission of the European Communities, 1991, Commission Directive of 13 February 1991 amending the Annexes to Council Directive 74/63 EEC on undesirable substances and products in animal nutrition (91/126/EEC). *Official Journal of the European Communities* L60:16-17.

European Commission, 1997[a], Reports on tasks for scientific cooperation. Report of experts participating in Task 3.2.1. Risk assessment of aflatoxins. *Report EUR 17526 EN,* Directorate-General for Industry, Office for Official Publications of the European Communities, Luxembourg.

European Commission, 1997[b], Reports on tasks for scientific cooperation. Report of experts participating in Task 3.2.2. Revised edition – November 1997. Assessment of dietary intake of ochratoxin A by the population of EU Member States. *Report EUR 17523 EN,* Directorate-General for Industry, Office for Official Publications of the European Communities, Luxembourg.

European Commission, 1998, Commission Regulation (EC) No 1525/98 of 16 July 1998, amending Regulation (EC) No 194/97 of 31 January 1997 setting maximum levels for certain contaminants in foodstuffs. *Official Journal of the European Communities* L201:43-46.

Food and Agriculture Organization, 1993, Sampling plans for aflatoxin analysis in peanuts and corn. *FAO Food and Nutrition Paper* 55, Rome, Italy.

Food and Agriculture Organization, 1997, Worldwide Regulations for Mycotoxins 1995. A compendium. *FAO Food and Nutrition Paper* 64, Rome, Italy.

Gilbert, J., 1991, Regulatory aspects of mycotoxins in the European Community and USA, in: *Fungi and Mycotoxins in Stored Products*: proceedings of an international conference, Bangkok, Thailand, 23-26 April 1991. B.R. Champ, E. Highley, A.D. Hocking and J.J. Pitt, eds. ACIAR Proceedings no. 36, 194-197.

Horwitz, W., 2000, *Official Methods of Analysis of AOAC International.* Seventeenth Edition 2000, Chapter 49 "Natural Toxins". AOAC International, Gaithersburg, USA.

Krogh, P., 1977, Mycotoxin tolerances if foodstuffs. *Pure and Applied Chemistry* 49:1719-1721.Kuiper-Goodman, T., 1989, Risk assessment of mycotoxins, in: *Mycotoxins and phycotoxins.* S. Natori, K. Hashimoto and Y. Ueno, eds. Amsterdam: Elsevier pp. 257-264.

Rosner, H., 1998, Mycotoxin regulations: an update. *Revue de Médicine Vétérinaire* 149: 679-680.

Schuller, P.L., Van Egmond, H.P. and Stoloff, L., 1983, Limits and regulations on mycotoxins, in: *Proceedings International Symposium on Mycotoxins*, K.Naguib, M.M. Naguib, D.L. Park and A.E. Pohland, eds. Cairo, Egypt, 6-8 September 1981. pp. 111-129.

Smith, J.W., Lewis C.W., Anderson J.G. and Solomons G.L., 1994, Mycotoxins in Human and Animal Health. *Technical report, European Commission, Directorate XII: Science, Research and Development*, Agro-Industrial Research Division, EUR 16048 EN, Brussels, Belgium.

Stoloff, L., Van Egmond, H.P. and Park, D.L., 1991, Rationales for the establishment of limits and regulations for mycotoxins. *Food Additives and Contaminants* 8:213-222.

Van Egmond, H.P., 1989, Current situation on regulations for mycotoxins. Overview of tolerances and status of standard methods of sampling and analysis. *Food Additives and Contaminants* 6:139-188.

Van Egmond, H.P., 1991, Regulatory aspects of mycotoxins in Asia and Africa. In: *Fungi and Mycotoxins in Stored Products*: proceedings of an international conference, Bangkok, Thailand, 23-26 April 1991. B.R. Champ, E. Highley, A.D. Hocking and J.J. Pitt, eds. ACIAR Proceedings no. 36, 198-204.

Van Egmond, H.P. and Dekker W.H., 1995, Worldwide Regulations for Mycotoxins in 1994. *Natural Toxins* 3:332-336.

Van Egmond, H.P., 1999, Worldwide Regulations for Mycotoxins. Working Document Third Joint FAO/WHO/UNEP International Conference on Mycotoxins. *MYC-CONF/99/8a.* Tunis, Tunisia, 3-6 March 1999.

ECONOMIC CHANGES IMPOSED BY MYCOTOXINS IN FOOD GRAINS: CASE STUDY OF DEOXYNIVALENOL IN WINTER WHEAT

Arthur W. Schaafsma
Ridgetown College
University of Guelph
Ridgetown Ontario N0P 2C0

ABSTRACT

The *Fusarium* epidemic of 1996 in Ontario winter wheat resulted in direct losses of well over $100 million Canadian dollars (CDN). More importantly, wheat marketing in Ontario has changed. The market focus remains primarily food grade. Thus, the awareness of deoxynivalenol (DON) entering the food chain has influenced marketing and trade. New market tolerances for DON have been set. The Chicago Board of Trade will only accept up to 5 ppm (DON), while a new tolerance of 0.5 ppm DON has been set in the breakfast cereal markets. Advance contracts are avoided and there is a new reluctance to service export customers, all because of the liability associated with DON. Furthermore, there are no markets for process by-products which contain concentrated levels of DON. Before 1996, DON problems were handled largely by blending grain across the province. A reluctance to blend DON contaminated wheat and site specific sourcing of grain is growing.

INTRODUCTION

In 1996, the winter wheat industry in Ontario experienced the worst *Fusarium* epidemic on record. This epidemic increased the awareness of the entire industry in the province (producers, grain handlers, buyers, and end-users) about the mycotoxin, deoxynivalenol. This event changed, probably permanently, how winter wheat is viewed and marketed in Ontario. The scope of this paper is relatively narrow, in that it applies to one mycotoxin in one commodity framed within a small geographic region. However, the purpose of this discussion is to use a detailed assessment of the economic changes that occurred due to the 1996 epidemic, to illustrate the broader economic impact that mycotoxins have on food grains.

FUSARIUM AND DEOXYNIVALENOL IN WINTER WHEAT

Miller and Trenholm (1994) provide an excellent review on mycotoxins in grain.

Mycotoxins and Food Safety, edited by Trucksess *et al.*
Kluwer Academic/Plenum Publishers, 2002

Fusarium graminearum Schwabe is the most important and more common species of *Fusarium* found in wheat and in corn in Ontario. *F. graminearum* epidemics are favored by warm and moist conditions at and after anthesis in wheat, with the timing of precipitation being the most crucial factor. In winter wheat, *F. graminearum* produces the trichothecene, deoxynivalenol (DON), whereas, in corn, zearalenone may also be produced.

Swine are the most sensitive species to DON, where diets containing 1-2 ppm DON have resulted in reduced feed consumption (Prelusky et al., 1994). The guideline for dairy cattle is also 1 ppm, but there is evidence that lactating cattle can tolerate levels higher than 8 ppm. (Allen and Dupchak, 1999). Similarly the guideline for beef cattle and sheep stands at 5 ppm, but these species can handle levels greater than 10 ppm. Poultry are the most tolerant species. In humans, DON is a potential teratogen and embryotoxin. Furthermore DON may affect food intake and body weight, plus suppress humoral and cellular immune function (Kuiper-Goodman, 1994). Canada's guideline for DON in uncleaned soft wheat (including bran) for non-staple foods is 2.0 ppm, whereas for wheat destined for baby foods the guideline is 1.0 ppm (Kuiper-Goodman, 1994). Most of the winter wheat produced in Ontario is destined for human consumption, however, when quality misses the mark for milling, swine consumption is generally the next major destination.

Before 1996, DON-contaminated wheat was avoided by grading suspicious samples visually, where fusarium damaged kernels (FDK) were expressed as a percentage of the sample by weight. At that time, the allowable level was set by the Canadian Grain Commission to be 1% FDK (Allen and Dupchak, 1999). The assumption with this method is that the relationship between FDK and DON is linear, and that 1% FDK will ensure the grain will contain no more than 2 ppm DON. Potential problems in Ontario were prevented by grading and blending large grain pools. Samples were rarely tested at the receiving point.

WINTER WHEAT PRODUCTION

Winter wheat production has ranged from 300,000 to 740,000 acres in Ontario since 1981 (Figure 1). Corn and soybeans are the most important field crops, with about 2 million acres planted for each crop. From 1981 to 2000, soybean production increased four-fold (Figure 1). Wheat and corn have followed parallel production trends running opposite to soybeans in the last 10 years.

In Ontario, net per acre returns for winter wheat have been about $50 CDN lower than for either corn or soybeans, hence the lower acreage planted (Figure 1). However, the consistent acreage that has been planted in the last ten years reflects the role it plays in crop rotation patterns. The crop rotation most widely recommended in Ontario for field crops is a three year cycle including corn, followed by soybeans, followed by winter wheat. In Ontario, winter wheat has been considered more for its rotational value than as a cash crop to allow for better production of corn and soybeans.

There are three main economic benefits to including winter wheat in the rotation. The first and most difficult to quantify is soil conservation. Winter wheat provides excellent cover over winter on soybean stubble reducing soil erosion. The second benefit of wheat in the rotation is disease management, primarily for soybeans. This economic benefit is also difficult to measure. The soybean cyst nematode arrived as a major pest in Ontario in 1987 (Wallace and Favrin, 1999) resulting in losses of up to 40%. Rotation to non-host crops is a pivotal strategy to manage this pest. Phytophthora root rot has also been a production-limiting disease in clay-textured soils. Phytophthora root rot is best managed through crop rotation to improve soil structure and drainage via the organic matter that comes from a wheat crop and through a reduction of inoculum potential by growing a non-host crop. The third economic benefit of growing winter wheat in the rotation is an overall improvement of soil tilth, structure and organic matter. In studies conducted in Ontario from 1997 to 1999, compared with continuous soybean production, Young, (personal communication, University of Guelph) reports an average yield increase in soybeans of 7% when grown after corn and 10% after wheat in a two year rotation, and 14% when grown after corn and wheat in a three year rotation. Compared with a continuous corn rotation, Young reports a 5% yield increase with a two year corn/soybean rotation and a 10% yield increase for three year rotation of corn/soybeans/wheat. The economic benefits of winter wheat production on farm are, therefore, realized over a longer period than three years, and difficult to measure. Any catastrophe in the wheat crop will potentially upset the balance established by the three-year rotation. There are no other viable crops available to replace winter wheat as a rotational crop in Ontario.

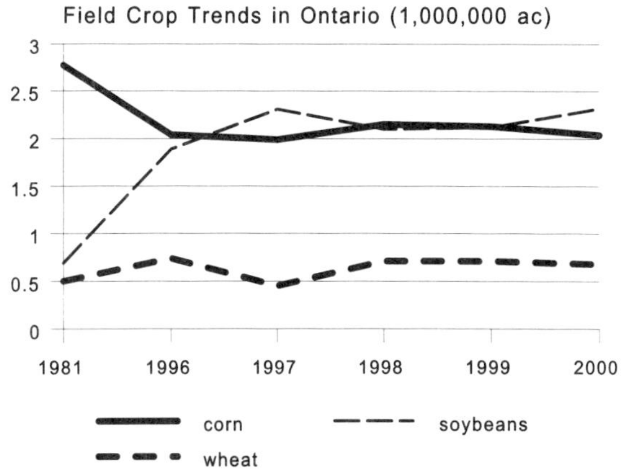

Figure 1. Production trends for the three main crops of Ontario in the past 20 years. From Anon (2000).

In the last 20 years, Ontario experienced four *Fusarium* epidemics (Figure 2) and in each year following the epidemic there was a marked decline in acreage planted with winter wheat. This decline reflected the lack of confidence in winter wheat as a cash crop. The drop in acreage only lasted for one season, and growers resumed planting similar acreages to those planted before the epidemic. Other forces negatively affected acreages in 1990 and 1992. Growers demonstrated their hope that, after 1995, winter wheat was finally going to yield net returns similar to soybeans and corn as a cash crop, by planting another aggressive acreage for 1996.

DIRECT COST OF FUSARIUM EPIDEMICS

The *Fusarium* epidemic of 1996 in Ontario winter wheat was a financial disaster. The Ontario Wheat Producers' Marketing Board (OWPMB, Jim Whitelaw personal communication) estimates a 30% yield loss equating to a 8,441,000 bu loss in volume. At \$4.08/bu CDN this resulted in a direct loss to producers of \$34.5 million CDN. Deoxynivalenol (DON) contamination led to a quality penalty on about 27,525,000 bu of on average \$1.22/bu CDN, or another \$33.75 million CDN loss to the industry. The crop was extremely difficult to market resulting in an additional \$0.41/bu CDN marketing cost totaling about \$11.25 million CDN. Another \$20-30 million CDN was lost by the milling industry to buy replacement wheat from the Pacific Northwest. The total loss to the *Fusarium* epidemic in

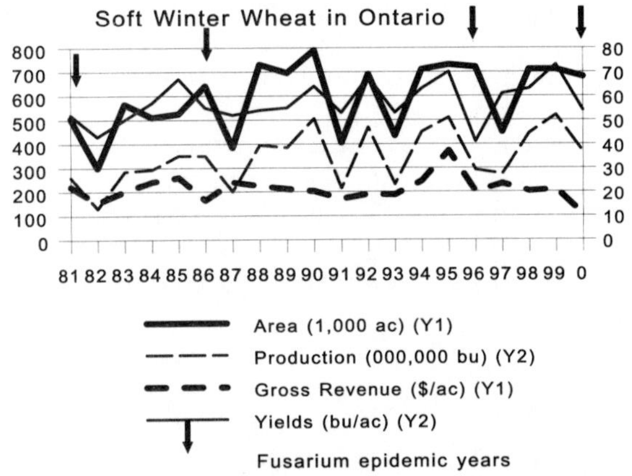

Figure 2. Production history of winter wheat in Ontario in the past 20 years relative to *Fusarium* epidemics. From Anon (2000).

1996, was well over $100 million CDN. All but the latter $20-30 million CDN were losses sustained by producers at the farm gate.

In 2000, Ontario was hit by another epidemic. This time yield was unaffected because the weather favorable for DON accumulation did not arrive until the crop was mature. Wheat can reabsorb moisture when mature and the inclement weather during delayed harvest allowed DON to accumulate to levels above tolerance in many cases. According to Jim Whitelaw (personal communication) the 605,000 tonnes of winter wheat were produced within food grade guidelines resulting in about $51 million CDN in revenue to the industry. A further 520,000 tonne was down graded to feed wheat mainly as a result of samples containing in excess of 1% FDK. Over 50% of these samples were graded between 1.0 - 2.5 % FDK; the remaining were over this level. This down-grading resulted in a net loss of about $22 million CDN at the farm gate. About 12,000 tonnes were graded down to sample because of FDK levels exceeding 10.0 %; another $1 million CDN loss at the farm gate.

These estimates are reliable because the OWPMB controls the marketing of all the wheat produced in Ontario and records the entire acreage. It is difficult to obtain similar partitioned estimates from adjacent U.S. jurisdictions because of the proprietary nature of production and marketing in those regions.

Windels (2000) reported direct yield losses in U.S. dollars of about $2.5 billion in the 1990's and about $300 million from 1993 to 1998 in the Canadian province of Manitoba. Windels (2000) did not partition the losses further in her account, but did suggest that *Fusarium* head blight epidemics crippled the wheat and barley industry in the Red River Valley, driving producers to financial ruin and human hardship, which resulted in the dwindling of rural communities.

MARKETING CHANGES

The 1996 epidemic has changed the wheat market in Ontario. The market focus remains primarily food grade. As such, the awareness of DON getting into the food chain has increased greatly and has become a key component of marketing and trade. If there is a potential fusarium epidemic, the fear of DON contamination puts wheat for food in peril.

To address these fears, several new market restrictions for DON have been set. The Chicago Board of Trade will only accept up to 5 ppm DON (Spring 1998), while a new limit of 0.5 ppm DON has been exercised in some breakfast cereal markets (Jim Whitelaw OWPMB personal communication). The cereal markets no longer make advanced purchase contracts, but the markets have changed to post-harvest sales.

The marketers of Ontario wheat have a new reluctance to service export customers, because of the liability associated with DON. Furthermore, DON is a special problem because there are no markets for process by-products which contain concentrated levels of DON.

Before 1996, DON problems were handled largely by blending across the province. A reluctance to blend DON contaminated wheat is growing, and if there is any suspicion of DON contamination, markets will move to checking individual loads at the source. Traditionally all wheat in Ontario has been marketed through the OWPMB. There has been sudden pressure in the last three years for growers to market directly to end users through identity preservation of the crop. Contracts used in these cash sale agreements often contain a DON clause (Hayhoe, 2000).

RISK MANAGEMENT

The main strategy that Ontario has adopted to reduce the incidence of *Fusarium* epidemics and DON contamination is to develop *Fusarium* resistant crops. While noble and important, this strategy is not likely to bear fruit in the near future. In the meantime, other tools are recommended in an integrated approach. Ontario is recommending good crop rotation, selecting more resistant varieties (note that there are no resistant commercial cultivars available), forecasting epidemics, treating with fungicides, modified harvesting and cleaning strategies, and selective marketing (Schaafsma et al., 2000).

Most of the data from around the world and recently from North America suggest that use of the fungicide, tebuconazole (FOLICUR), consistently reduces DON levels in grain by about 50%. If this fungicide had been applied during the epidemic of 1996, the 750,000 tonnes that averaged from 8-10 ppm DON could have been marketed through the Chicago Board of Trade, falling below the 5-ppm tolerance. Consequently, losses incurred during the epidemic could have been reduced by $37.5 million.

In 2000, when Ontario wheat producers experienced another *Fusarium* epidemic, the fungicide FOLICUR was registered under emergency use for the management of *Fusarium*. Less than 1% of the total acreage was treated; the fungicide was tested by a few growers on a small part of their acreage. Several growers sprayed their entire acreage and these were able to keep most of their wheat within DON guidelines, while neighbors produced grain above guidelines for DON. The use of FOLICUR has been monitored, and, while the data are still being analyzed, early indications show that the use of the fungicide is a viable method risk management tool.

Risk management aside, the problem that continues to frustrate producers is the subjective nature of grading grain for FDK at the point of delivery. We tested this subjectivity by having a single grain sample graded at 11 different receiving points covering three countries and five grain handlers. About 0.5 kg of grain was used as the test sample. The sample scores ranged from 0.4% to 1.8% FDK with a mean of 1.1% FDK. With the industry guideline at 1.0% FDK, growers could have easily been graded above or below which meant a difference between an initial payment of $2.00/bu and $1.00/bu for wheat below and above the tolerance level, respectively.

Use of analytical methods, such as immunoassays, to monitor DON levels in grain is not without concerns. In the 2000 harvest, a cargo load of grain was tested using a commercial enzyme-linked immunosorbent assay (ELISA) kit and found to be well above the 2.0 ppm DON guideline. The same load was re-tested by the Canada Grain Commission with the result being well below the guideline of 2 ppm. This example indicates that the methods used to sample and analyze grain for DON are not without error. The sources of error include sampling, sub-sampling, grading, grinding, extraction, and quantitation (analysis) of the toxin. When grain is near the tolerance levels, such as during the 2000 harvest, subjectivity in grading needs to be reduced through the use of standardized sampling and testing. Producers are less likely to adopt risk management tools, especially those that come at a price, if the benefits are unclear, because of clouded grading systems.

CONCLUSIONS

The epidemics of 1996 and 2000 have certainly increased the awareness of *Fusarium* in the market place. Standards for DON and FDK, whether official guidelines or commercial tolerances, limit the trade of grain more stringently now than before, with the producer absorbing most of the risk. If wheat loses its appeal as a cash crop, several hidden costs will be incurred. Insect and disease problems will rise as producers move to more soybeans and corn in the rotation. Soil productivity will slow because of increased risk of soil erosion, decline in soil structure, and reduced organic matter.

To maintain confidence in wheat as a cash crop, growers need to be rewarded for risk management through better pricing via a better grading system and crop segregation. End users also need to share more of the risk through tighter contracts and better quality incentives.

REFERENCES

Anonymous, 2000, Field crops statistics. Ontario Ministry of Agriculture Food and Rural Affairs. http://www.gov.on.ca/omafra/english/stats/crops/index.html

Allen, D, and Dupchak, K., 1999, Grading tolerances for fusarium-damaged grain and DON feeding guidelines. Canada Grain Commission. http://www.cga.ca/Pubs/fusarium/backgrounder/don-e.htm

Hayhoe, M. 2000, Hayhoe Mills Ltd. cash wheat prices. http://www.hayhoe.com/cash%20wheat%20prices.htm

Kuiper-Goodman, T., 1994, Prevention of human mycotoxicosis through risk assessment and risk management. In: *Mycotoxins in Grain: Compounds other than aflatoxin*. Miller, J.D. and Trenhom, H.L., Eds. , Eagen Press, MN, pp. 439-469

Miller, J.D. and Trenhom, H.L., 1994, In: *Mycotoxins in Grain: Compounds other than Aflatoxin,* . Miller, J.D. and Trenhom, H.L., Eds. Eagen Press MN.

Prelusky, D.B, Rotter, B.A., and Rotter R.G., 1994, Toxicology of mycotoxins. In: *Mycotoxins in Grain: Compounds other than aflatoxin*. Miller, J.D. and Trenhom, H.L., Eds. , Eagen Press MN

Schaafsma, A., L, Tamburic-Ilincic, J., Winter, D., Hooker, A., Tenuta, and Spieser, H., 2000, Management of Fusarium Head Blight in Wheat Ridgetown College, University of Guelph, Ridgetown Ontario N0P 2C0. http://www.ridgetownc.uoguelph.ca/fusarium/

Spring, K. , 1998, Chicago Board of Trade quality requirements for wheat contract approved by CFTC. Chicago Board of Trade. http://www.cbot.com/points_of_interest/pressbox/pressreleases/p9811241.htm.

Wallace, S. and Favrin, R., 1999, *Heterodera glycines* Ichinohe, Soybean Cyst Nematode. Plant Health Survey Unit, Canadian Food Inspection Agency, Agriculture and Agrifood Canada http://www.cfia-acia.agr.ca/english/ppc/science/pps/datasheets/hetglye.shtml

Windel, C.E., 2000, Economic and social impacts of Fusarium head blight: changing farms and rural communities in the Northern Great Plains. *Phytopathol.* 90, 17-21.

U.S. PERSPECTIVE ON MYCOTOXIN REGULATORY ISSUES

Douglas L. Park and Terry C. Troxell

Office of Plant and Dairy Foods and Beverages
Center for Food Safety and Applied Nutrition
Food and Drug Administration
Washington, DC 20204

ABSTRACT

Control programs set up by the Food and Drug Administration (FDA) for aflatoxin, an unavoidable natural contaminant produced by specific molds that invade a number of feedstuffs and basic foods, provide an example of forces that affect risk assessment and management strategies by a regulatory agency. More recently, on an international scale, efforts to establish international food standards for fumonisin, deoxynivalenol, ochratoxin A, zearalenone, and patulin, as well as for aflatoxin, demonstrate the complexity of developing regulations and/or standards designed to protect consumer health and ensure fair trade practices on a global scale. Current FDA regulations for aflatoxins address public health concerns for potential contamination in basic foods, residues in milk, and animal feeds for numerous commodities and applications. Regulatory limits, sampling and analytical procedures, decontamination and/or diversion to less risk uses for contaminated product are components of mycotoxin control programs. Current efforts by FDA to establish regulatory controls for deoxynivalenol, fumonisin, and patulin add further insight on the role that safety and risk assessment procedures play in the development of action levels and advisories for mycotoxins.

INTRODUCTION

Mycotoxins can contaminate human foods and animal feeds through the growth of molds before and during harvesting, or because of improper storage after harvest. Post-harvest contamination can be minimized when proper storage conditions are maintained. Pre-harvest contamination most often occurs when environmental conditions consistent with mold growth and mycotoxin formation occur. For the most part, these conditions, are beyond the control of man.

Mycotoxins and Food Safety, edited by Trucksess *et al.*
Kluwer Academic/Plenum Publishers, 2002

The need to limit mycotoxins in food and feed is based on two major concerns: (1) the adverse effects of mycotoxin-contaminated commodities on human and animal health and (2) the presence of mycotoxins in human foods derived from animals fed mycotoxin-contaminated feed. The manner that the Food and Drug Administration (FDA) managed the risk posed by mycotoxin contamination can be best illustrated by viewing procedures taken initially to minimize hazards associated with aflatoxin contamination. These procedures involved the risk assessment process, including hazard evaluation, exposure determination, risk determination, and management of risk (Park and Stoloff, 1989). Although the risk assessment process for aflatoxin started in the early 1960's, the process has been modified as more information became available and other mycotoxins were discovered. Factors such as the nature of the hazard, availability of analytical and sampling procedures and the effect that regulations have on the availability of the food/feed supply, were all considered in the establishment of mycotoxin regulations.

The food safety concept, which includes efforts to address hazards associated with mycotoxin contamination in foods, is focused on providing a safe, wholesome food supply and providing nutritional needs as well as disease prevention measures; it is the driving force for reducing risks associated with mycotoxin contamination in human foods and animal feeds. Since mycotoxins are naturally occurring toxic secondary mold metabolites, and their occurrence in foods is unavoidable and unpredictable, food safety control efforts must focus on minimizing their presence to the greatest extent feasible. These efforts include prevention control programs, regulatory limits and monitoring programs, processing and/or decontamination procedures, as well as programs for informing the public of what is feasible and practical. Although numerous mycotoxins have been identified, the aflatoxins, fumonisins, patulin, and selected trichothecenes such as deoxynivalenol (DON) have been identified in the U.S. as having a significant public health concern for which regulatory programs have been established.

FOOD SAFETY MANAGEMENT PROGRAMS

Theoretically, prevention of mycotoxin formation is believed to be the best means of managing hazards associated with mycotoxin contamination. Pre-harvest mycotoxin formation can be partially controlled through good agricultural practices (GAPs) which include insect management, good irrigation and mineral nutrition, crop rotation, and the use of crop varieties which resist mold contamination (Lopez-Garcia and Park, 1998). Prevention procedures for post-harvest operations include good storage conditions such as minimizing exposure to moisture, preventing insect infestation, and cleaning and disinfecting storage and transportation equipment to prevent mold cross-contamination (Boutrif and Canet, 1998). Should mycotoxin contamination occur as a result of ineffective preventive measures, a safety management program must be established (FAO, 1979). Factors crucial to the establishment of an effective food safety management program are presented in Table 1. The collective use of these integrated mycotoxin management components can significantly reduce human and animal exposure to mycotoxins.

SETTING REGULATORY LIMITS

In 1961, when information was first disseminated that a mold-produced toxin (aflatoxin) was isolated from peanut meal, both FDA and the industry followed the research developments closely, since early indications suggested that the toxin could be a hepatocarcinogen (Fischbach, 1984; Stoloff, 1976; Stoloff, 1977). By 1963, the evidence was clear that the toxins, now identified as aflatoxins B_1, B_2, G_1, and G_2, were potent hepatocarcinogens in rats and were found in U.S.- produced peanuts and peanut meal. The peanut industry took the initiative in setting up a cooperative industry/Department of Agriculture (USDA)/FDA effort to develop practical analytical methods and to obtain information on the extent of contamination in peanut products (Bauer and Parker, 1984). This effort resulted in the creation of a voluntary agreement between peanut growers, sellers, and the USDA which allowed an inspection of in-shell peanuts for aflatoxin contamination and certification of raw in-shell peanut lots as either positive or negative for aflatoxin.

Table 1. Components of an effective food safety management program

Component
Prevention
Setting regulatory limits
Establishment of monitoring programs
Control through good agricultural practices
Control through processing
Decontamination through specific treatments
Consumer/producer education

From: Lopez-Garcia et al. (1998)

Numerous factors must be considered in efforts to set regulatory limits for naturally occurring toxicants in foods and animal feeds. Although the primary goal of a regulatory program is the provision of a safe, wholesome food supply, the limits established must maintain an appropriate availability of food and feed.

Although no direct evidence linked aflatoxins to liver cancer in humans in the U.S., the Commissioner of FDA concluded that the observation of severe carcinogenic effects in experimental animals, and reports of positive correlation between dietary aflatoxins and primary liver cancer in humans in other parts of the world, was sufficient justification to take action to control human exposure to aflatoxins. In 1965, the FDA established an informal action level of 30 μg/kg (ppb) for total aflatoxins in peanut products. Due to the lack of information available on the relative carcinogenicity of the individual aflatoxins (B_1, B_2, G_1, and G_2), the sum of all four aflatoxins was used for the action level. By 1969, based on research and experience gained with analytical methodology, the FDA reduced the action level to 20 μg/kg and applied it to all commodities susceptible to aflatoxin contamination.

The justification for seizures of commodities due to aflatoxin contamination was based on Section 402 (a)(1) of the Food, Drug and Cosmetic Act This section considers a food to be adulterated if it contains an added poisonous or deleterious substance, which may render it injurious to health. Under this section, the FDA needs to demonstrate a reasonable possibility of harm from exposure to the poisonous substance; there is no need to prove the substance will cause harm at the level stated. Thus, it is evident that FDA risk assessment and management actions during this early period were based on prudence and pragmatism.

In 1974, FDA considered a proposal to control unavoidable poisonous and deleterious substances through Section 406 of the Food, Drug and Cosmetic Act. This section allows the Commissioner to set a tolerance for unavoidable contaminants at a level necessary for the protection of the public health. Based in part on the lack of sufficient toxicological data to set a safe level on which to base a tolerance, the FDA did not proceed with the establishment of a tolerance through Section 406. Instead, the Commissioner sought to bring four factors into balance: the need to minimize human exposure to aflatoxins; the capabilities of sampling procedures and analytical methods to detect, measure and confirm aflatoxins; the capability of agriculture and manufacturing technology to prevent and remove contaminated peanuts; and the need for continued availability of a low-cost protein source (i.e., peanuts). Prompted by comments received on the tolerance proposal, FDA conducted an assessment of the risk associated with aflatoxins in consumer peanut products and other food commodities (Food and Drug Administration, 1978). The results of that assessment supported the action level of 20 µg/kg (established in 1969) as being adequate to protect the public health of consumers. Under 402(a)(1), this action level served the Agency as a bench mark for enforcement action and allowed FDA the flexibility to manage the risk despite of the lack of sufficient toxicological data.

Table 2 summarizes the factors that should be considered in the establishment of regulatory limits for mycotoxins. Human exposure to mycotoxins of public health concern must be held to "safe" levels (levels resulting in negligible risk) or, if this is not possible, to the lowest practical level while at the same time providing for an adequate food supply. Regulatory limits set for animal feed consider the health of the animal and potential residues in foods derived from animals fed mycotoxin-contaminated feed. Using aflatoxin as an example, Table 3 shows the relative aflatoxin levels in feeds to resulting residues levels in edible tissues. The only food commodity from animals that shows a high probability of aflatoxin residues in human foods at levels greater than 0.1 µg/kg levels is milk. Therefore, in the United States, permitted levels for aflatoxins in animal feeds vary according to the animal species and the food product obtained from that animal. The current aflatoxin action levels established by the Food and Drug Administration are presented in Table 4.

Table 2. Factors affecting the setting of regulatory limits

Human exposure to mycotoxins
Source of the contaminant
Toxicological characteristics of the contaminant
Analytical capability
Animal feed to edible tissue ratio
Effect of the control on the availability of food or feed
Practicability and effectiveness of various regulatory enforcement strategies

Table 3. Relative aflatoxin levels in feed to aflatoxin residue levels in edible tissue.

Animal	Tissue	Aflatoxin	Feed/Tissue Ratio
Beef	Liver	B_1	14,000
Dairy Cattle	Milk	M_1	75
Swine	Liver	B_1	800
Poultry, Layers	Eggs	B_1	2,200
Poultry, Broilers	Liver	B_1	1,200

From Park and Pohland (1986)

Guidelines have been issued in the US for other mycotoxins of public health concern including DON, fumonisins and patulin. DON advisories to protect human and animal health are 1mg/kg (ppm) for finished wheat products for human use; 10 ppm for ruminating beef and feedlot cattle older than 4 months and for chickens (not to exceed 50% of the diets); 5 ppm for swine at <20% of the diet; and 5 ppm for all other animals at <40% of the diet. FDA has issued no regulations or action levels for fumonisins thus far. However, a draft document to provide guidance to industry for levels adequate to protect human and animal health has been published in the Federal Register, June 6, 2000 and the comments are under review (Food and Drug Administration, 2000a). The proposed guidance levels for fumonisins (total FB_1, FB_2 and FB_3) for human foods and animal feeds are presented in Table 5. For patulin in apple juice and apple juice ingredients, FDA has proposed an action level of 50 μg/kg (ppb) on June 16, 2000 (Food and Drug Administration, 2000b); the comments received are under review. For ochratoxin A, FDA supports the use of good agricultural practices and good manufacturing practices to keep the levels of this toxin to the lowest level feasible. FDA has issued no regulations or guidelines for ochratoxin A at this time.

Table 4. Current aflatoxin action levels (total established by the Food and Drug Administration)

Food and Feed	***Action Levels (μg/kg, ppb)***
Human foods (except milk)	20
Milk	0.5
Animal feeds (except as listed below)	20
Cottonseed meal (as a feed ingredient)	300
Corn and peanut products for breeding beef cattle, swine and mature poultry	100
Corn and peanut meal for finishing swine	200
Corn and peanut meal for feedlot beef cattle	300
Corn for immature animals and dairy cattle	20

Food and Drug Administration (1996)

In addition to efforts to control mycotoxin contamination levels in human foods and animal feeds in the U.S. through action levels and advisories, specific regulations have been developed in many countries (FAO, 1996). In 1995, seventy-seven countries with more than 5 million inhabitants had regulations for aflatoxins. All countries with mycotoxin regulations have regulatory limits for aflatoxins in food and/or feed. Regulatory limits also exist for other mycotoxins (i.e., patulin, ochratoxin A, DON, diacetoxyscirpenol, zearalenone, T-2 toxin, chetomin, stachybotryotoxin, phomospin, and fumonisins).

ESTABLISHMENT OF MONITORING PROGRAMS

In the establishment of mycotoxin monitoring programs, it is necessary to identify which mycotoxin(s), product(s) and/or commodities should be included. This is governed by prior knowledge concerning which commodities are susceptible to specific mycotoxin contamination, which agricultural crops are grown in the area, and whether the potential for pre-harvest contamination is apparent. This can be determined by a review of published literature, surveys of crops for toxigenic molds and/or mycotoxins, and the nature of regional weather conditions.

Crucial components of the monitoring program include the establishment of an appropriate sampling plan (FAO, 1993). The sampling plan components include the collection of a representative sample, the preparation of a sub-sample for analysis and an analytical method. It is crucial that the sample collected be a true representation of the entire lot. Homogeneity must be maintained throughout the grinding, mixing and sub-sampling process to obtain the analytical test portion. Numerous sampling plans have been established for selected mycotoxins and various commodities. In order to maintain the highest degree of credibility the analytical method should be validated through an interlaboratory validation program such as the system offered by the AOAC International. The final component of the monitoring program is to have a predetermined plan for the use of the products that are found to be out of compliance, i.e., alternative lower-risk uses.

Table 5. Proposed fumonisin guidance levels

Product	Total Fumonisins ($FB_1 + FB_2 + FB_3$)
Human Foods	
Degermed dry milled corn products	2 ppm
Whole/partially degermed dry milled corn products	4 ppm
Dry milled corn bran	4 ppm
Cleaned corn intended for masa production	4 ppm
Cleaned corn intended for popcorn	3 ppm
Animal Feeds	
Corn and corn by-products for:	
Equids and rabbits	5 ppm, <20% diet
Swine and catfish	20 ppm, <50% diet
Breeding ruminants, poultry and mink; dairy cattle and laying hens	30 ppm, <50% diet
Ruminants >3 mos before slaughter and mink for pelt production	60 ppm, <50% diet
Poultry being raised for slaughter	100 ppm, <50% diet
All other species or classes of livestock and pet animals	10 ppm, <50% diet

From: Federal Register, June 6, 2000

DECONTAMINATION STRATEGIES

Procedures currently available for reducing the presence of mycotoxin contamination in foods include: 1) strict adherence to moisture control, 2) implementation of GMPs, 3) good quality assurance efforts, and 4) following HACCP (hazards analysis critical control point) principles (Park et al., 1999; Lopez-Garcia and Park, 1998).

The HACCP concept (FAO, 1995; Park et al., 1999) is built on several principles and actions including conducting a hazards analysis and identifying preventive measures (critical control points), establishing critical limits, monitoring each critical control point during the processing operation, and, where necessary, implementing corrective actions. It is also necessary to maintain accurate records and to verify the effectiveness of the process.

When evaluating the effectiveness of various food processing techniques for mycotoxin reduction, it is necessary to consider the chemical stability of the mycotoxin, the nature of the process, the potential interaction between the mycotoxin and the food matrix, and the possible interaction between different mycotoxins present. Although numerous procedures or processing operations are available to reduce mycotoxin levels, the addition of chemical agents to human foods specifically for reducing mycotoxin levels is not permitted in the United States. Traditional procedures, such as cleaning and the separation of immature, damaged, or mold-infected kernels/grains/nuts, wet and dry milling operations, and thermal treatment can result in significant reduction of mycotoxin levels.

Effective strategies for decontamination fall into several categories including physical separation, physical decontamination, biological decontamination (e.g., fermentation), chemical inactivation and/or removal, and chemoadsorption (Phillips et al., 1994; Park, 1993). The special considerations for decontamination/detoxification procedures are presented in Table 6. The most successful decontamination strategies include physical separation/decontamination by the cleaning and separation of immature, damaged, or mold-infected kernels/grains/nuts, thermal and chemical inactivation, and removal by extraction. Chemical inactivation of aflatoxin by ammonia has been widely used in several locations worldwide including several states in the United States, France, Senegal, Sudan, Brazil and South Africa. Chemoadsorption clays, that tightly bind toxins thus making them not biologically available, show strong promise. Indiscriminate use of clays can pose a risk because essential nutrients may also bind to selected clays.

Table 6. Special considerations for decontamination/detoxification procedures

- Inactivate, destroy, or remove the toxin
- Not produce or leave other toxic residues
- Retain nutritive value/acceptability of the product
- Not alter significantly the technological properties of the product
- If possible, destroy fungal spores

From: Park et al. (1988)

PRODUCER AND CONSUMER EDUCATION

It is important for public health officials to work closely with food producers and processing operations in the dissemination of information regarding procedures that prevent the formation of mycotoxins before and after harvest. This educational effort should increase awareness of safety concerns, identify and assist the implementation of preventive measures, and help in the establishment of controls and regulatory limits. The sharing of important information dealing with GAPs and GMPs can play an important role in reducing the exposure of mycotoxin hazards to the consumer.

RISK ANALYSIS

U.S. regulatory strategies for mycotoxins are based on risk analyses, which have components encompassing risk assessment, risk management and risk communication. Risk assessment is a data gathering effort; the information obtained is used to determine relative risks associated with mycotoxin contamination. Hazard identification is under this phase of the analysis; toxicological data for the mycotoxin of interest, data on target organ(s) and toxic events, and information on the relevance of these data to humans are evaluated. A dose-response assessment is then conducted. The next important block of information is the exposure assessment. In this phase, commodities susceptible to mycotoxin contamination and levels of contamination are identified; target populations and susceptibilities such as the young, elderly, medically or nutritionally compromised are identified. Using this information, a risk characterization is determined where quantitative estimates of exposure are made and the probability of a disease event estimated. For risk management, mycotoxin control programs are established and alternative uses for the contaminated product are identified. The final phase, risk communication, involves the interaction and exchange of information by risk managers concerning foodborne risks, regulatory options, industry efforts and consumer actions. This must be an open communication process.

INTERNATIONAL MYCOTOXIN CONTROL EFFORTS

The development of internationally recognized regulations and control measures for mycotoxins that protect public health and promote fair trade at the international level must be vigorously pursued. Several agreements, the World Trade Organization (WTO), Agreements on Sanitary and Phytosanitary Measures (SPS) and Technical Barriers to Trade (TBT) promote greater harmonization and transparency in the establishment of food regulations that protect the consumer and facilitate trade. Codex standards, guidelines and other recommendations are considered acceptable requirements under the WTO Agreement for consumer protection for commodities in international trade. Countries may set more stringent levels if necessary, based on unique population considerations or the mandate of the country's societal level of public health protection. The Joint FAO/WHO Expert Committee on Food Additives (JECFA) provides evaluations of contaminants based on sound scientific and risk assessment principles. The Codex Committee on Food Additives and Contaminants (CCFAC) uses JECFA risk assessments to develop relevant risk management standards. It is anticipated, on the international scale, that maximum levels for mycotoxins in foods will be set to protect public health and facilitate trade. This will ultimately facilitate the development of an integrated Federal/State/International food safety system.

CONCLUSIONS

Selected mycotoxins pose public health concerns. These include aflatoxins, fumonisins, deoxynivalenol, patulin and ochratoxin A. The primary agricultural commodities affected with mycotoxin contamination include cereal grains, oilseeds, dairy products, animal feeds and animal products. Prevention, through pre-harvest procedures and post-harvest management strategies, is the best method for controlling mycotoxin contamination. Mycotoxin control programs, including the establishment of regulatory limits, monitoring programs and decontamination procedures, can minimize hazards to the consumer. Risk assessment, management and communication procedures provide a sound science base for the establishment of mycotoxin regulatory programs.

REFERENCES

Bauer, F. J. and Parker, W. A., 1984, The aflatoxin problem: Industry-FDA-USDA Cooperation. *J. Assoc. Off. Anal. Chem.*, 67:1.

Boutrif E. and Canet, C., 1998, Mycotoxin prevention and control: FAO programs, *Revue Med. Vet.* 149:681.

Fischbach, H., 1984, Coping with the aflatoxin problem in the early years. *J. Assoc. Off. Anal. Chem.*, 67:1.

FAO, 1979, Recommended practices for the prevention of mycotoxins in food, feed and their products. FAO Food and Nutrition Paper No. 10, Rome

FAO, 1993, Sampling plans for aflatoxin analysis in peanuts and corn. A technical consultation. FAO Food and Nutrition Paper No. 55, Rome

FAO, 1995, The use of hazard analysis critical control point (HACCP) principles in food control. FAO Food and Nutrition Paper No. 58, Rome

FAO, 1996, Worldwide regulations for mycotoxins, A compendium FAO Food and Nutrition Paper No. 64, Rome

Food and Drug Administration, 1978, Aflatoxins in shelled peanuts and peanut products used as human foods, proposed tolerance: Reopening of comment period. *Fed. Regist.* 43:8808.

Food and Drug Administration, 1996, Action levels for aflatoxin in animal feeds, Compliance Policy Guides CPG 7126:33, 384 (Section 683.100), CPG 7106-10 (sec 527.400), CPG 7120.26 (sec 555.400)

Food and Drug Administration, 2000a, Draft guidance for industry: fumonism levels in human foods and animal feeds; availability. *Fed. Regist.* 35945.

Food and Drug Administration , 2000b, Foods: Apple juice, apple juice concentrates, and apple juice products-adulteration with patulin; draft compliance policy guide; availability and patulin in apple juice, apple juice concentrates, and apple juice products; draft supporting document availability. *Fed. Regist.* 37791.

Lopez-Garcia, R. and Park, D. L., 1998, Effectiveness of post-harvest procedures in management of mycotoxin hazards. In *Mycotoxins in Agriculture and Food Safety*, D. Bhatnagar and S. Sinha, eds. New York, Marcel Dekker, pp. 407-433.

Park, D. L., 1993, Controlling aflatoxin in food and feed. *Food Technology,* 47:92

Park, D. L. Njapau, H. and Bontrif, E., 1999, Minimizing risks posed by mycotoxins utilizing the HACCP concept. *Food, Nutrition and Agriculture,* 23:49.

Park, D. L. and Pohland, A. E., 1986, A rationale for the control of aflatoxin in animal feeds. In *Mycotoxins and Phycotoxins*, P.S. Steyn and R. Vleggar, eds., Elsevier, Amsterdam pp. 473-482.

Park, D. L., Lee, L. S., Price, R. L. and Pohland, A. E., 1988, Review of decontamination of aflatoxin by ammoniation: Current status and regulation. *J. AOAC*, 71:685.

Park, D. L. and Stoloff, L., 1989, Aflatoxin control: How a regulatory agency managed risk for an unavoidable natural toxicant in food and feed. *Regul. Toxicol. Pharmacol.* 9:109.

Phillips, T. D., Clements, B. A. and Park, D. L., 1994, Approaches to reduction of aflatoxins in foods and feeds. In *The Toxicology of Aflatoxins-Human Health Veterinary and Agricultural Significance*, D. L. Eaton and J. D. Groopman, eds, San Diego, California, Academic Press, p. 383.

Stoloff, L., 1976, Incidence, distribution and disposition of products containing aflatoxins. *Proc. Ameri. Phytopathol. Soc.* 3:156.

Stoloff, L., 1977, What FDA is doing about the mycotoxin problem. *Agri Fieldman/Farm Technol.* 28:60a.

INDEX